Teaching and Learning in Ecosocial Work

Catherine Forde • Satu Ranta-Tyrkkö
Pieter Lievens • Komalsingh Rambaree
Helena Belchior-Rocha
Editors

Teaching and Learning in Ecosocial Work

Concepts, Methods and Practice

Springer

Editors
Catherine Forde
School of Applied Social Studies
University College Cork
Cork, Ireland

Pieter Lievens
KdG University of Applied Sciences
and Arts
Antwerp, Belgium

Helena Belchior-Rocha
Political Science and Social Policies
ISCTE - University Institute of Lisboa
Lisbon, Portugal

Satu Ranta-Tyrkkö
Department of Social Sciences
and Philosophy
University of Jyväskylä
Jyväskylä, Finland

Komalsingh Rambaree
Department of Social Work,
Criminology and Public Health Sciences
University of Gävle
Gävle, Sweden

ISBN 978-3-031-58707-8 ISBN 978-3-031-58708-5 (eBook)
https://doi.org/10.1007/978-3-031-58708-5

This Springer imprint is published by the registered company Springer Nature Switzerland AG
The registered company address is: Gewerbestrasse 11, 6330 Cham, Switzerland

Foreword

When social work started to emerge in Western countries at the turn to the twentieth century, it was connected to industrialization, urbanization, growth of the capitalist world market, militarization, and expansion of colonial structures of exploiting natural resources. To an extent social work aimed to respond to the social, economic, and environmental consequences of these developments. In diverse local contexts, social work pioneers were involved in facilitating access to fresh nourishment, clean water, healthy environment, and housing, including cohesive local communities and fair systems of work and economy for people. Although this kind of orientation never became the mainstream of professional and disciplinary thinking, the current era invites us to recall such examples of holistic and community-oriented ecological thinking from the early days of social work's development as well as from indigenous social work. They encourage our capacity to imagine social work futures which are based on understanding the interconnectivity between human wellbeing and natural environment.

In the Western world, social work scholars have started to conceptualize the interconnectivity between social work and environmental issues more explicitly since the 1980s. As a reaction to the deepening environmental crisis and encouraged by the global and local ecological movements, a critical role of social work has been identified. Today we can observe that the environmental paradigm in social work is growing, which is also demonstrated by the richness of published international and national literature. Varying between the different locations on the globe, it has been called environmental, ecological, green, or ecosocial social work. Increasingly also, social work research and doctoral training have become involved in the ecosocial paradigm.

However, we can critically reflect on the reality that most of the ecosocial discussion in social work has been theoretical. Not very much has changed toward sustainability in the social work profession, not to talk about the lives of service-users and local communities suffering from environmental injustice. The environmentally oriented practical implementations of social work remain marginal experiments. The question therefore arises as to how to make sustainability transitions and ecological thinking really work across the discipline, practice, and policy making of social work.

Regarding the relatively poor development of ecosocial practice, we can critically evaluate social work education on the basis that it could have much more substantial role in an ecosocial transition of the profession. Particularly in Europe, a shortage of transformative pedagogical orientations, contents, and methods in social work education is apparent. How can the most urgent major questions of the Planet be reflected in social work courses? How can the education of future social workers itself undergo sustainability transitions beyond the human-centered worldview and economic growth? New types of learning environments, teaching inputs, collaborations in field placements, and the scientific foundation of knowledge are surely required.

Consequently, 17 social work teachers and other scholars networked in the biannual conference of the European Association of Social Work (EASSW) in Madrid 2019 and took the initiative to apply for a status of Special Interest Group (SIG) for *Ecosocial Work* of the EASSW. Once the SIG Ecosocial Work got established, it also applied and received funding from the EASSW for a small project called "Teaching and Learning Ecosocial Work in Europe."

One of the SIG's main aims was to co-publish a handbook for social work educators and practitioners across Europe and internationally. This edited book is now presenting the results of the engagement of the SIG Ecosocial work and draws together an extremely useful and exciting collection of shared information, experiences, and reflections. The novelty of this volume is that it puts social work education in the center, compared with the already existing international publications on the environmental paradigm in social work.

Thus, the next urgent task in social work education is to contextualize all learning in a sustainability transitions perspective. Not only individual course contents or mainstream-type of institutional settings of social work education are challenged, but a comprehensive re-conceptualization of social work education is required. Such perspectives may re-connect to the very original and unique key idea of social work back in history, but also mitigate this with the new discourses of transformative post-academic and post-hierarchal-institutional orientations. This means that social work is no longer to be understood in a human-centered but in a holistic eco-centered manner. It underlines that nature and environment are not something outside of social work to be taken into account. Ecosocial work considers that all human life, including the practical, educational, and scientific activities of social work, is inherently based on nature, embedded in and dependent on the bio-physical sphere.

It is affirming to see this book getting published. Only one (self)critical question becomes apparent and remains open: why did European social work education need to wait this long to produce such a needed book? This volume completes the missing cornerstone of the environmental discourse in social work in Europe and historically. It is highest time for this book and for making it easily accessible on the screens of social work educators, practitioners, students, and collaborators.

We would like to congratulate the editors and the authors of this valuable volume.

Jyväskylä, Finland
November 2023

Aila-Leena Matthies

Acknowledgments

The co-editors would like to acknowledge the support of the European Association of Schools of Social Work (EASSW), which funded the preparation of this book. EASSW is home to the Special Interest Group (SIG) on Ecosocial Work for Social Work Education. During the COVID-19 period, online SIG on Ecosocial Work meetings provided the impetus for this book. The co-editors would like to express their gratitude to all members of the SIG on Ecosocial Work.

Thanks to our book contributors for working with us to produce this ecosocial work teaching and learning resource. It was an enriching experience to share knowledge with ecosocial work students, practitioners, and academics from across the globe.

We acknowledge with thanks the help and support of our Editor, Zachary Romano, and Production Editor, Tina Silva, at Springer Nature.

The times that have passed while producing this book have been in many ways turbulent and occasionally also personally challenging for many of us. A hearty thanks to family members, friends, and colleagues who have supported the becoming of this book and its chapters with their wit, skills, connections, love, humor, care, patience, criticism, and just being there.

Contents

About the Contributors

Olivia Ahlin is a Swedish master's student in International Performance Studies at Stockholm University. In 2022 she completed her bachelor's degree in International Social Work at the University of Gävle. In her degree she focused on expressive art as a tool in social work.

Elsa Alves is a dynamic nutritionist whose expertise extends across various domains, epitomizing a career characterized by an unwavering quest for diverse knowledge to tackle communal and individual challenges. Her professional odyssey commenced in a Portuguese municipality, pioneering a role that contributed to the inception of innovative functions in an emerging professional sphere. Transitioning into academia, Elsa served as a research assistant on esteemed European projects at the University of Porto, notably contributing to the European public health nutrition report and Pro Children initiatives. Her journey transcended borders as she assumed roles as an auditor and lecturer at Instituto Colombiano de Credíto Educativo y Estudios Técnicos en el Exterior (ICETEX) and Fundación Universitario Sanmartin in Colombia. Presently, Elsa's altruistic spirit shines through her volunteerism in Portugal's sustainable campus network while also lecturing at Instituto Politécnico de Bragança. Not confined to academic realms, Elsa's diverse roles encompass technical directorships in social solidarity institutions, educator, consultant, and project manager, focusing on community nutrition, local development, sustainability, and food safety. As an entrepreneur, Elsa spearheaded pioneering ventures like healthy cooking workshops and introduced the innovative "Personal Diet Portugal" concept, dedicated to fostering sustainable and healthy dietary practices among citizens.

Janet Carter Anand is an international Social Work academic, educator, and practitioner and has worked in leading higher education facilities in Australia, the UK and the Republic of Ireland. She is currently a Professor in International Social Work, University of Eastern Finland, and her current research topics include developing models of global exchange in higher education, the application of sustainable social wellbeing in professional curriculum, and the measurement social outcomes intervention in health and practice skills including crisis prevention and intervention. Janet is currently a PI on the Finnish government-funded projects, Sostyö-TerVa and Lampe, projects promoting practice-based research networks and cultural expertise in professional practice. Janet had had experience in leading Erasmus programs and research projects with India, Bangladesh, Malaysia, Jordan, Namibia, and multiple European countries and has been active in international professional organizations and networks and editorial roles. Currently, Janet coordinates social work courses on global perspectives in social work, working for diversity, global migration, and social sustainability.

Helena Belchior-Rocha PhD in Social Work, is an Assistant Professor at ISCTE-University Institute of Lisbon in the Department of Political Science and Public Policies; deputy director of the Transversal Skills Laboratory; and integrated researcher at CIES, Centre for Research and Studies in Sociology, linked to national and international research projects, namely, three Marie Curie Actions. She is the author of papers and communications at national and international congresses in the areas of social work theory and methodology, environment, sustainability, community intervention, ethics, human rights, social policies and well-being, education, and soft skills. She is a member of the editorial boards of several national/international journals. Helena is also a member of the Inclusive Policy Lab of UNESCO and a member and Scientific Science Communication Coordinator of the EU Cost Action "Digital Human Rights."

Afonso Borga is a social worker in Lisbon and currently working in the Social Responsibility Department of Auchan. Afonso completed his master's degree at ISCTE University Institute in Lisbon, where his research was focused around eco-neighborhoods in Portugal. With a strong foundation in social responsibility, his professional journey has been enriched by internships at JRS Portugal and at the Paula Elizabete Center in Brazil. Additionally, Afonso has contributed his expertise to GRACE—Responsible Companies—further expanding his impact in the field of social responsibility.

Georgianna L. Dolan-Reilly LMSW, is a dynamic systems thinker with an interdisciplinary background in research, community organizing, education, policy, and editing. Focusing on disease prevention, health promotion, and social and environmental justice, they are dedicated to dismantling systems of oppression, and empowering individuals and communities. As an educator and researcher Georgianna has experience in quantitative and qualitative methods and supports others in enhancing and applying their knowledge and assets. In addition to prior publications Georgianna recently contributed to the textbook *Ecosocial Work: Environmental Practice and Advocacy* and is developing publications on community organizing as group work, and photovoice for environmental justice. Georgianna serves on the Council on Social Work Education's Committee on Environmental Justice (US). Other roles include Advocacy and Outreach Manager at the Institute for Social Work and Ecological Justice, Adjunct at several graduate schools of social work, Community Development Specialist at the Long Island Prevention Resource Center, and Board Member of the International Association of Social Work with Groups. Georgianna is currently pursuing a PhD in Social Work, exploring the intersection of environmental social work and anti-racism. Through the above work, Georgianna combines their academic pursuits and personal and professional identities to advance environmental and social justice.

Lisa Dörfler works as an environment and climate protection advisor for social institutions at Paritätischer Gesamtverband, Berlin, Germany, since 2022. She holds a master's degree in social work from the Alice Salomon University Berlin on social work as a human rights profession. She additionally has a lectureship on climate crisis and social work at Nuremberg institute of technology, since 2021. Her publications focus on the intersection of ecological crisis, human rights, and social work. She is part of the organizing team of the expert group on climate justice and socio-environmental transformation in social work within the German association of social work. Before her current work in environment and climate protection for social institutions she was head of a youth service for refugee youth for four years.

Catherine Forde is a senior lecturer in youth and community work at the School of Applied Social Studies, University College Cork (UCC). She delivers twin modules on environment and sustainability on qualifying social work and youth and community work degree programs at the School of Applied Social Studies. Catherine's main research/publication interests are in the areas of community development; ecosocial work and environmental education; state-civil society relations; and children and young people's participation. She has published widely on these themes and is co-author of the book *Social Work and Community Development: A Critical Practice Perspective* (Palgrave Macmillan, 2015). She is co-editor of the themed section Community Development in Social Work Education: Themes for a Changing World (*Community Development Journal*, October 2021). She is a Co-I on the Horizon Europe project *Intersectional Spaces of Participation: Inclusive, Resilient, Embedded (INSPIRE)* (2024–2027) and co-PI of a baseline research project (2023–2024) for the *Feminist Communities for Climate Change* initiative of the National Women's Council of Ireland/Community Work Ireland. She serves on the editorial board of the journal *Ethics and Social Welfare*.

Daniela Freitas graduated in Social Work in 2011 and completed her master's degree in Sociology in 2015 at the University of Coimbra. In 2012, she moved to Lisbon where she currently lives. She worked for 10 years as the Technical Director of a Private Social Solidarity Institution providing support for children (nursery and pre-school) and support for the elderly (Day Center, Social Center and Home Support Service). She began her PhD in Social Work at ISCTE-IUL in 2020, with a thesis entitled "Social Work and Sustainable Development: Eco-Social Intervention" with research into eco-social practice in community intervention in Portugal. Since 2022, she has been teaching courses such as Critical Thinking, Stress Management, Conflict Management, Technical Report Writing, Bibliographic Research and Information Analysis, Introduction to Sustainability, and Introduction to Design Thinking, at ISCTE-IUL's Soft Skills Laboratory in Lisbon.

Joana Gomes is a social worker on Terceira Island in Azores, currently employed at Associaçao Planeamento Familiar (APF)-AÇORES. Joana completed her master's degree at Iscte in Lisbon, where she did an internship at Animalife—Association for Social Awareness and Animal Support. She is the author of *Textile Circular Economy Initiatives: A Framework for Environmental Social Work* (2020) and co-author of the book chapter "The Contemporary Challenges of Environmental Social Work: Contributions to an Epistemological Reflection on the Potential of Animal-Assisted Interventions."

Luna Haine is a young woman hailing from Antwerp, Belgium. Growing up in a loving family with one sister, mom, and dad, she embarked on a new adventure by cohousing with her two best friends in the heart of Antwerp. During her high school years at the ES Steiner School in Berchem, she actively participated in climate protests and even planted 3600 trees with her family. She pursued a degree in Social Cultural Work and has a vibrant social life, surrounded by intriguing individuals. She finds solace in the woods, where she immerses herself in the wonders of nature and enjoys working with her hands, whether it's crafting with clay, stones, or other materials. She has a deep passion for the fine arts, including photography, sculpture, and painting. She has a penchant for smaller, intricate pieces that pay homage to the beauty of details. Writing is another passion of her, though she does not indulge in it as often as she would like. That's precisely why she eagerly embraced this project, as it offers her an opportunity to share her thoughts and creativity.

Mary Hurley Master of Social Science by research and Diploma in Social Work Studies, is a part-time lecturer in the School of Applied Social Studies, University College Cork. She teaches on the Master in Social Work (MSW) program. She is a registered social worker. She has been an environmental activist for over 20 years in her local community which has challenged the building of an incinerator in Cork Harbour. She was also involved in setting up a local zero waste group in her neighborhood. Her research interests include ecosocial work, community and international social work, and environmental justice. With her colleagues Fiachra Ó Suilleabháin and Catherine Forde, she has developed the first interdisciplinary module for the social professions in Ireland, "Sustainability, Environmental and Social Justice Issues in Professional Practice." In recent collaborations she continues to work with the Irish Association of Social Workers (IASW) to help develop practitioner awareness and knowledge of the connections between social work, the climate emergency, and social justice for service users in Ireland.

Manish K. Jha is Professor of Community Organisation & Development Practice and former Dean of the School of Social Work, Tata Institute of Social Sciences (TISS), Mumbai. His research interests include Migration, Climate Change, Humanitarian & Disaster response, Middle Classes, Poverty, and Social Justice. He teaches courses on Community Organization, Social Policy, Social Action, Advocacy and Movements, and Migration and Politics. He has been a visiting Fellow at School of Oriental and African Studies (SOAS), University College Dublin (UCD), University of Durham, UK, Gothenburg University, Sweden, and other places. He has extensively contributed to international journals and edited volumes on migration, disaster, community organization, politics, populism, and social justice. His recent books include *Beyond Consumption: India's New Middle Class in the Neo-Liberal Times* published by Routledge (2022) and *Data Centres as Infrastructure: Frontiers of Digital Governance in Contemporary India*, published by Orient BlackSwan (2022).

Bipin Jojo PhD, has been teaching at Tata Institute of Social Sciences, Mumbai, India (TISS), Mumbai, since 1997. Currently he is the Professor and Dean at the School of Social Work at TISS. His interest areas are Biodiversity of Indigenous/Tribal peoples and their well-being; Social Work with Tribal Perspective; Internal Migration; and affected communities in extractive industries. He has been teaching Working with Groups, Social Welfare Administration, International Social work and Indigenous Peoples, Participatory Project Planning and Development. His recent publications are *Children and Scars of COVID-19 Pandemic in India: Issues and Challenges* (ed.) (Forthcoming, Routledge, Taylor & Francis Group, UK); '"What Makes Pro-poor Public Policy Dysfunctional?" Evidence from the Select Public Policies in Odisha, India', in *The International Journal of Community and Social Development*, 5(2) 168–186, 2023; and 'Land Alienation and Subversion Strategies on Land Rights: Perspectives from Attappady' in the *Journal of Land and Rural Studies*, 11(1) 7–18, 2023.

Anish K. R. is an Assistant Professor and former head of the department at Rajagiri College of Social Sciences, India. He has served as Chairman of BOS of Social Work at Mahatma Gandhi University, Kannur University, and Rajagiri College of Social Sciences (Autonomous), India. He is a Registered Mental Health Professional (Under the Mental Health Care Act, 2017). His research interests are in social work education, migration and intercultural competence, health and mental health interventions in social work, psychiatric rehabilitation, and mental health promotion through life skills. He has received research grants and fellowships from organizations including the University Grants Commission, the Australian-India Council, the Finnish National Agency for Education, the German Academic Exchange Program (DAAD), and the Indian Council for Social Science Research. He has published eight books and several articles in Scopus-indexed journals. He has served as a consultant to the National AIDS Control Organisation, the National Institute of Social Defence (Ministry of Social Justice and Empowerment), and the National Institute of Rural Development and Panchayati Raj (Ministry of Rural Development).

Emmanuelle Larocque is a registered social worker, yoga therapist (C-IAYT), and professor at the Université du Québec en Outaouais (UQO), in Canada. Her research focuses on the development of ecological identity through nature and outdoor therapy and place-based teaching and learning. With a strong background in narrative therapy and mind-body modalities, Emmanuelle is also interested in the application of collaborative and embodied methodologies to better understand lived experiences of place-attachment, ecological grief, and collective resilience.

Pieter Lievens is a lecturer in the Bachelor of Social Work of KdG University of Applied Sciences and Arts, Belgium, and he is connected to the Thomas More University of Applied Sciences in Flanders since 2003. He was educated as a sociologist and specialized in European Social Policy Analysis. He has done research in the fields of women in vulnerable labor positions, participation of lower classes in local organizations, satisfaction among users of labor counselling services, and older people's care. He was a consultant for municipalities on local social policy planning. He has expertise in the design of e-learning environments. He has coordinated Intensive Programmes in Social Work and is involved in many international projects. He has been teaching international perspectives in social work and ecosocial work since 2005. Global challenges are his main expertise. Between 2010 and 2013 Pieter was chairman of the learning network "Orientation of Social Work towards sustainable development." Pieter has been teaching sustainability, ecology, ecosocial work, and international perspectives and global change for more than 15 years. He is currently engaged in futures thinking, deep history of planet earth, collective governance of commons and other activities related to ecosocial work. More information: www.pieterlievens.be.

Mona B. Livholts is Professor of Social Work in the Department of Social Sciences, University of Helsinki, Finland; Executive Board Member in The European Association of Social Work, EASSW; Founder of the Network for Reflexive Academic Writing Methodologies, RAW (2008–2017). Livholts works with glocal-, post-anthropocentric, feminist-, and postcolonial power analysis for social and environmental justice in the intersection of social work, creative writing, and art-based research. She has invented new forms for creative life writing, such as the thinkingwriting subject, post/academic writing, and the untimely academic novella by uses of literary fiction, memory work, diaries, letters, poetry, and photography. Research themes include media narratives on rape, sexual harassment, gender, space and memory, monuments and narrative inequality, environmental exhaustion, and the body politics of glocal social work. Livholts has published monographs, co-edited and edited volumes in Swedish and English, including *Social Work in a Glocalised World* (with Bryant 2017), *Situated Writing as Theory and Method. The Untimely Academic Novella* (2019), and *The Body Politics of Glocal Social Work: Essays in the Post-Anthropocentric Condition* (2022).

Rita Luís is a social worker in Lisbon, currently working for the European Anti-Poverty Network (EAPN) Portugal. Rita completed her master's degree at ISCTE in Lisbon and her research was centered on the relationship between Social Work and Sustainable Development. She represents Portugal on the IFSW Europe project, "A Social Europe is Possible! The NEW Generation of Social Workers Are the Spearhead of Change." In addition, Rita has developed some work related to the evaluation of social projects and higher education study cycles as a research assistant and student evaluator, respectively.

Deborah Lynch is a Senior Lecturer in Social Work at The University of Queensland in Australia. She is a Senior Fellow of the Higher Education Academy (HEA), an accreditation of the UK Professional Standards Framework. Deborah has published widely on systems, policies, and practices in health, education, and justice that impact on children, young people, families, and communities. Through her international collaborations in social work, Deborah contributes to interdisciplinary scholarship in collaborative projects and journal publications including the *Community Development Journal*, *Australian Social Work*, *European Journal of Social Work*, *The British Journal of Social Work*, and *Ethics and Social Welfare*. She co-authored *Social Work and Community Development: A Critical Practice Perspective* (2015; Palgrave McMillan) with Dr Catherine Forde at University College Cork in Ireland. Her areas of interest are social work and community development, specifically focusing on the ecosocial transition and sustainable development, health equity, and disrupting patterns of education and social disadvantage. She grounds her teaching and learning in social justice and human rights, critical thinking and working collectively to nurture dynamic, innovative forms of practice that can respond to contemporary community, environmental, and societal issues.

Eduardo Marques is a Social Work Professor at University of Azores in Portugal where he is the director of the Social Work Degree. As researcher he is affiliated at CLISSIS—Lusíada Center for Social Research and Social Intervention (Lusíada University) and at CICS-UAc—Interdisciplinary Centre of Social Sciences (University of the Azores). He has a European Doctorate (PhD) in Social Work at the Complutense University of Madrid (Spain) and in the last 30 years he has been involved in projects with a focus on environmental areas—Nature-Based Solutions; Psychosocial Intervention in Disasters; Community Work; Entrepreneurship; Social Innovation, Youth Work; Participatory Arts; Nature Therapy, etc. He is a member of the International Editorial Committee of the Magazine *Espacios Transnacionales—Revista Latinoamericana-Europea de Pensamiento y Acción Social*. He belongs to the Editorial Committee of the Revista Científica UISRAEL and to Comité Científico de la Revista Internacional de Diversidad en Organizaciones, Comunidades y Naciones. Eduardo Marques was Ambassador of the European Climate Pact within the scope of the European Union's climate initiative (DG CLIMA).

Nicole Mattocks LMSW, is an Assistant Professor in the MSW program at Colorado Mesa University. She is engaged in teaching, research, community-academic partnerships, and service focused on: the mental health impacts of climate change and environmental injustice on marginalized communities, integrating nature into social work practice, and promoting ecosocial work education. She is also a Psychotherapist in a private practice and connects her clients with the healing benefits of nature. In her former faculty position she collaborated with a student to co-develop the first ecococial work student group for the campus, and served as faculty advisor, connecting students with networks of climate justice leaders and opportunities. Dr Mattocks has presented her research at numerous conferences and has co-developed and co-taught two International Environmental Social Work courses designed to engage students in the global implications of environmental injustice and climate change. She is currently serving as a guest co-editor of Ethics and Social Welfare for a special issue on environmental justice and partnering with a non-profit to measure the effectiveness of nature-assisted therapeutic interventions for supporting addiction recovery. In 2024, she begins a 3-year community-engaged research project to investigate the impacts of climate change on fisher folk in Zambia.

José María Morán Carrillo is a Full Professor at the Pablo de Olavide University in Seville (Spain). He teaches undergraduate and graduate courses in Epistemology of Social Work, Social Work with Individuals and Families, and Psychosocial Intervention in Crisis and Social Emergencies. He is Coordinator of the Emergency Commission of the Andalusian Council of Professional Associations of Social Work since 2021. He has published numerous articles in specialized journals and is the author of the books *Bases of Social Work: Social Work and Epistemology* (Tirant lo Blanch, 2006) and *Professionalization of Social Work in Spain: Genealogy and Indicators of Organized Social Action* (University of Granada, 2021).

WiseWater was established by **Elaine Moynihan and Katie McShane**, both passionate water advocates living on the south coast of Ireland. With a deep-rooted love for nature and a profound respect for the interconnectedness of ecosystems, Katie has made it her life's mission to educate and raise awareness about the importance of water conservation and protection. Her research interests include culture and heritage, the power of sound as a healing modality and water bodies as therapeutic spaces for community connection. Elaine has a great love and respect for all water environments which has deepened her understanding of the interconnection between people and nature. As a year-round swimmer and surfer, a currach builder and rower she has a profound interest in the research of sea swimming, blue spaces, and their impact on woman's wellbeing. Inspired by their work with young people and communities, Katie and Elaine pursued degrees in youth and community work, which led to the creation of the unique WiseWater wellness program. Elaine and Katie recognize the value of connecting and collaborating with fellow water and environmental workers. If you share this vision please reach out, and let's create ripples of positive change together.

Fiachra Ó Súilleabháin, MSW, DSocSc, is a Social Work academic and researcher in the School of Applied Social Studies, University College Cork, Ireland. His research focuses on sexualities and genders, violence against women and children, the scholarship of teaching and learning, digital social work/technology-assisted practice, ecosocial work, and contemporary social work practice issues. In addition, Fiachra serves on numerous academic committees and actively contributes to civic society by providing voluntary consultancy and governance/board direction to local, regional, and national charities. He currently sits on the board of the Irish Association of Social Workers.

Meredith C. F. Powers (she/her) is an Associate Professor in the Department of Social Work at UNC Greensboro. She teaches, facilitates community-engaged, action-research, and has co-authored and co-edited a growing body of open-access work on topics of climate justice, climate migration, ecosocial worldviews, eco-therapeutic practices for healing, ecosocial work, and justice, equity, diversity, and inclusion. She presents her research locally, nationally, and internationally at professional conferences, including being invited as a keynote speaker at the United Nations for World Social Work Day (2018). Dr Powers is the founding Director of the Climate Justice Program of the International Federation of Social Workers. With this role, she also produces and co-hosts a new podcast project, JAM Sessions. She established and co-administers the global Green/Environmental Social Work Collaborative Network. Her local, community-engaged, action research projects include: "Parks for All People," for the extensive enhancement of outdoor recreation spaces; "Climates of Inequality: Stories of Environmental Justice," focused on global and local climate migration and resettlement; and, most recently, "All That We Share," an inclusive, community public art project that serves to enhance the community as we celebrate diversity and strengthen our bonds to each other and our environment.

Elisa Rainerma MSocSc, MSc (Environmental Technology), MTh, is a university teacher and a doctoral researcher in the faculty of Social Sciences and Philosophy at the University of Jyväskylä, Finland. She has previously worked with refugees and immigrants and as a social worker in disability services. Her doctoral research elaborates the foundations of ecosocial work and how to mainstream ecosocial work in social work with adults.

Komalsingh Rambaree (PhD) is an Associate Professor of Social Work at the University of Gävle, Sweden. He started his career in the year 1990, as a social worker working with youth and adolescents in Mauritius. He has also worked on various sustainable development projects for international organizations such as the European Union, World Bank, and the United Nations. He graduated with a PhD in Social Work and Social Policy from the University of Manchester, England, in the year 2006. He is currently engaged in teaching, learning, and researching at the University of Gävle, Sweden, ecosocial/green social work issues, international social work, adolescent and youth development, and computer-assisted qualitative data analysis with ATLAS-ti. Details on Komalsingh's publications can be found at http://www.hig.se/kolsie.

Satu Ranta-Tyrkkö PhD Title of Docent, is a university lecturer in the faculty of Social Sciences and Philosophy, University of Jyväskylä, Finland. Her teaching and research focus on the interfaces and confluences of social work and environmental issues, including her postdoctoral research (2014–2017) on the consequences of the mining industry for disadvantaged groups in Northern Finland and Eastern India, and the Climate Handbook for Social Work (in Finnish) together the Talentia Union of Professional Social Workers. Her interests also include the diversity of social work as both locally embedded, and transnationally and globally manifest profession, discipline, and social movement, and interfaces of social work with arts (especially theatre) and arts-based practice. Satu has also published on ethical and postcolonial issues in social work. Overall, she has a macrolevel ecosocial and community-based orientation to social work.

Melissa Indera Singh EdD, LCSW, MS-HSA, is a University of Southern California, Suzanne Dworak-Peck School of Social Work Associate Professor in Practicum. She is an NASW Advanced Certified Hospice and Palliative Care Social Worker, a Certified Pharmacy Technician, and an experienced clinician in Eye Movement Desensitization and Reprocessing (EMDR) and Field Traumatology. In 2016, a fellowship that she co-led won the Award for Innovative Teaching in Social Work Education from SAGE/Council on Social Work Education (CSWE) for Educating Social Work Students for Macro Practice. In 2017, Dr Singh was awarded the Hutto Patterson Distinguished Faculty Award for excellence in teaching; service to the university, the school, and the community; and accessibility to students. Dr Singh is known for extensive research and dedication to improving social work education. Her research interests encompass a wide range of topics, reflecting her commitment to addressing societal injustices. Dr Singh is a trailblazer in interdisciplinary research, as demonstrated in her articles on tackling the climate crisis and reproductive justice. Her commitment to inclusivity is underscored by her contributions to publications like "Anti-Racism and Equity-Mindedness in Social Work Field Education." Her recent presentations on topics like online education during COVID-19 and the impact of natural disasters on student learning highlight her dedication to advancing social work education.

Alex Skinner is a Sessional Lecturer and Field Educator in Social Work at The University of Queensland and Griffith University in Australia. He is a community development practitioner, social worker, and permaculturalist. His practice has a strong focus on the intersections of the social world with the environment, and he sees value in interdisciplinary, holistic approaches to teaching, learning, and practice. Alex has experience in the community sector working in several settings such neighborhood centers, in disability, and with young people experiencing homelessness, as well as with First Nations peoples. He teaches social work and community development both in the classroom and as an external fieldwork placement supervisor. He has presented on themes of ecosocial work and community development at conferences in Australia.

Kelly Smith DSW, is the founder and director of the Institute for Social Work and Ecological Justice (ISWEJ), developing programs that build social workers' capacity to confront environmental and social justice challenges related to climate change. Kelly teaches for Columbia School of Social Work and Adelphi University School of Social Work and is a member of the Grand Challenge for Creating Social Responses to a Changing Environment Advisory Council. Her research explores the inclusion of environmental and ecological justice content in social work curricula in relation to the disparate impacts of climate change and environmental degradation on communities with concentrated systemic disadvantages. Kelly employs a multisolving approach to climate and social justice issues. Her work in continuing education bridges gaps across social work practice and supports innovations that address pervasive inequities and systems of oppression. Kelly earned her doctorate in social work from the University of Southern California, where she was honored with The Order of Arête. She also holds a master's degree in Gender and Social Policy from the London School of Economics. Kelly lives with her family on Long Island, serving on her community's Environmental Advisory Board.

Lauryn Smith is a Clinical Research Assistant at the University of Maryland School of Social Work. She has participated in research related to the impact of affinity groups on students of color at predominately white institutions. Lauryn has a Bachelor's degree from Swarthmore College, where she studied Education and Psychology. In her free time, she can usually be found reading fiction books or sketching in the National Arboretum in Washington, DC.

Ingo Stamm is a postdoctoral researcher and lecturer at the University of Jyväskylä, Kokkola University Consortium Chydenius, Finland, since 2015. He finished his PhD in sociology on social security and human rights in Finland and Germany at the University of Siegen, Germany, in 2014. He further holds a master's degree in social work from the Alice Salomon University Berlin on social work as a human rights profession. His research and teaching center on the intersection between social work, social policy and sustainability. Additionally, Ingo Stamm has 10 years of practical experience as a social worker in the fields of family care, child protection, and youth work.

Joe Whelan is an Assistant Professor in the School of Social Work and Social Policy at Trinity College Dublin. Joe's main areas of research include exploring the intersections of poverty, work and welfare, exploring and understanding lived experiences in the context of welfare recipiency with a focus on the processes and effects of welfare conditionality, critical social theory, and sustainable social policy. Joe is a member of the Editorial Collective for the distinguished scholarly journal, *Critical Social Policy*, and is the chairperson of the Sociological Association of Ireland's Work, Welfare and Markets study group. Joe's most recent book, *Hidden Voices: Lived Experiences in the Irish Welfare Space* (2022) is published by Policy Press.

Introduction: Addressing the Role of the Social Professions in the Sustainability Transition

Satu Ranta-Tyrkkö, Catherine Forde, and Pieter Lievens

What Does This Book Offer?

Focusing on teaching and learning, this book aims to introduce and popularise ecosocial work among currently practising and future social professionals, scholars, and educators. The book has been written with educational institutions offering social work, community work, or related programmes in mind. On the one hand, this is the context in which most of the book's authors are rooted; on the other hand, ecosocial work could, and in our view, should, be far more widely taught, known, and practised than it currently is. Overall, the social professions have not yet fully realised that their contribution is much needed and potentially significant in the crucial sector-crossing sustainability transition underway (Ranta-Tyrkkö & Närhi, 2021).

The term 'ecosocial work' has been in use for some time, at least since the turn of the millennium and the publication of the seminal book *The Eco-social Approach in Social Work* by Aila-Leena Matthies et al. (2001). Of late, the term has made a breakthrough, possibly due to its capacity to bring together 'economy, ecology and social issues' (Wendt, 2021, p. 8), in a somewhat similar way to other compound words such as 'socioecological' or 'naturecultures'.

In this book, ecosocial work stands as an umbrella term for a holistic endeavour to connect different dimensions of sustainability—ecological, social, economic, and cultural—into the work with different service user groups, individuals, and

S. Ranta-Tyrkkö (✉)
Department of Social Sciences and Philosophy, University of Jyväskylä, Jyväskylä, Finland
e-mail: satu.ranta-tyrkko@jyu.fi

C. Forde
School of Applied Social Studies, University College Cork, Cork, Ireland

P. Lievens
KdG University of Applied Sciences and Arts, Antwerp, Belgium

C. Forde et al. (eds.), *Teaching and Learning in Ecosocial Work*,
https://doi.org/10.1007/978-3-031-58708-5_1

communities in ways that promote their wellbeing and participation (Boetto, 2016; Närhi & Matthies, 2018; Boetto et al., 2020; Ranta-Tyrkkö & Närhi, 2021; Närhi, 2022; Talentia, 2023). Moreover, although social and community work mostly operate on the microlevel, in different local contexts, the ecosocial perception about the tasks and roles of social work and the field of social welfare extends from micro to macro level and from personal to societal (Närhi 2022, pp. 9–10).

Being in essence a simultaneously ecological, structural, and community-based approach, ecosocial work aspires towards a systemic understanding of the issues at stake at various scales from the local to transnational and global levels, attempting to comprehend the 'glocal' nature and complexity of the issues. This holistic view includes trying to grasp the root causes of and connections between the phenomena that affect individuals, families, communities, or specific service user groups and how these issues connect with broader patterns of sustainability or unsustainability. While upper-level political and economic decision-making to a great extent shapes societal realities, the idea is that change, even global change, can also be effected and put into practice locally and across localities.

In social work scholarship, 'ecosocial' stands as one of the common denominators of different social work discourses and traditions that combine social and ecological perspectives, such as ecological or environmental social work, which are premised on the understanding that ecological sustainability is foundational to social and other forms of sustainability. Many authors use the terms 'ecological', 'environmental', 'ecosocial', and 'green' interchangeably. To briefly trace the history of these traditions, environmental issues, especially those related to the quality of the living environment or access to crucial resources, have been on social work and community work agendas from early on (e.g. Närhi & Matthies, 2016). Amongst other matters, enhancing the health and wellbeing of people has required improving their living conditions, such as sanitation and housing, and acknowledging the prevailing social and environmental circumstances as starting points of the work. Within and beyond social and community work, this kind of work has continued across the globe ever since in diverse forms, addressing, among other things, the connections between environmental and resource politics and marginalisation of various people living on the land (Gadgil & Guha, 1995).

In the north American and west European contexts, Kati Närhi and Aila-Leena Matthies (2016) identify two distinct theoretical and practical strands of ecological or environmental orientation that conjoin in ecosocial work. These are the systems theoretical approach, influential in social work since the 1970s and often referred to as the person-in-environment approach, and the eco-critical perspective, which has its origins in the increasing environmental awareness and resultant environmental movements in the Western world since the 1970s and 1980s. In social work, the core of the systems-theory approach has been that to help their clients, workers benefit from a holistic understanding of the multilayered environments formative to the clients' situations and scopes of action. However, in practice, the person-in-environment framework has not been mindful of the natural environment or keen to explore human-nature relationships as political questions. As for the eco-critical approach, it has been critical of the modern, industrial, capitalist society and professional social work while being at home with alternative projects integrating environmentalism, community orientation, and self-help (ibid.). It is worth noting that outside the western world, Mohandas Gandhi, one of the key leaders of the Indian independence movement, promoted somewhat similar ideas decades earlier as part of his politics and his village-level rural reconstruction programme. Moreover, Gandhi's method of nonviolent resistance has had a significant influence on later environmental and peace movements (Ranta-Tyrkkö, 2009).

In the academic social work literature, the systems theoretical and the ecocritical approaches remained distinct strands throughout the 1990s (Närhi & Matthies, 2016). Since the turn of the millennium, the environmental/ecological discussion has diversified. Ecological social work (e.g. McKinnon & Alston, 2016) has incorporated elements from both earlier approaches and new and alternative conceptualisations, such as the deep-ecological social work (Besthorn, 2003, 2012), the social ecological or ecological social approach in social work (Peeters, 2012a, b), green social work (Dominelli, 2012), environmental social work (Gray et al., 2013; Ramsay & Boddy, 2017; Krings et al., 2020), post-anthropocentric social work (Bozalek & Pease, 2021), or eco-spiritual social work (e.g. Chigangaidze, 2023) (see Rambaree in Chap. 1 for discussion of several of these perspectives). Despite different emphases, all these discussions seek an ecological paradigm change in social work and an overarching sustainability transition across societies, and share the premise that humans are part of and dependent on nature (Närhi & Matthies, 2016).

Why Is This Book Important?

In social and community work and related fields, ecosocial work emerges from a dire need to address the planetary-scale problems that threaten the mission, continuity, and achievements of these fields. While humans, like all other beings, are dependent on the continuity and carrying capacity of the multitude of the Earth's ecosystems, studies on climate (Smith et al., 2009), biodiversity (Ceballos et al., 2020), resource depletion, water scarcity, and other global challenges show that the Earth's resources have been utilised far beyond what the local, and even global, natural capital can sustainably provide (e.g. Brand & Wissen, 2021; Fanning et al., 2022). Outcomes of the extractivist relationship to Earth (Kröger, 2020; Nygren et al., 2022) include the planetary-scale climate and biodiversity crisis, resource scarcity, and problems of waste (Dufva, 2020). Consequently, these crises have both deepened existing and caused new forms of social, environmental, and ecological inequality (Armstrong McKay et al., 2022) that also concern future generations and other than human beings.

To take the climate crisis as an example, global CO_2 emissions have increased by two-thirds over the past 30 years, despite increased awareness, technological advances, policy plans, and corporate promises. September 2023, for example, was the hottest month in the hottest year ever recorded (Wong, 2023), while registering new record temperatures one year after the other has become nearly ordinary.

Globally, the top 10% of income earners are responsible for nearly half of total emissions. To mitigate climate emissions, the richest one billion inhabitants of the planet, almost the entire EU and US, would thus need to reduce their climate emissions by much more than the global average by 2030. This amounts to a 15% reduction *per year* in 'advanced economies', from which the current yearly reduction rate of 1% for the EU and 0.5% for the US between 2015 and 2022 (European Commission, Joint Research Centre, Crippa et al., 2023) is a far cry. The Intergovernmental Panel on Climate Change asserts that 'without a strengthening of policies, global warming of 3.2 °C is projected by 2100' (IPCC, 2023, p. 12), which greatly exceeds the limit of 1.5 °C set by the Paris Agreement. Global warming has already set in motion processes that will continue centuries into the future, such as the global sea level rise (NOAA, 2022).

Besides impairing the conditions and continuity of human and other forms of life on Earth in the future, the ecological and climate crises challenge the meaningfulness and ethical justification of human existence already in the present (Pihlström, 2020, pp. 297–298). Moreover, the crises cut the ground from under social work's mission and operating conditions to reduce inequality and support and protect those vulnerable and marginalized. While the climate crisis alone acts as an epochal threat multiplier of the other crises (McKinnon, 2012), the social, ecological, and economic impacts of the multiple interrelated crises fall highly unevenly and unfairly between individuals and communities. This unevenness is rooted in inequalities in wealth, education, race, ethnicity, religion, gender, age, class, caste, disability, and health status (Cardona et al., 2012; Mathias et al., 2023). Consequently, in an interdependent world, social work's mission leaves no ethical grounds to dismiss the distress of the people who bear the brunt of global environmental problems, even if that takes place far away from one's own location. Furthermore, given that some communities are far more deeply affected by environmental hazards than others, social work should be sensitive to, and address, different scales and faces of vulnerability and inequality. In proportion, social work should seek appropriate ways of working across diverse circumstances and geographical and temporal scales (Mathias et al., 2023).

Prioritising and Working for the Sustainability Transition

As the consequences of the evolving ecological crises are known to be socially, economically, and ecologically catastrophic, adjusting to planetary boundaries (Steffen et al., 2015) should be of utmost priority in social work (Dominelli, 2012, p. 82; Talentia, 2023, p. 7). In other words, there is an urgent need to perform effective and large-scale systemic changes to avoid the worst effects of global warming and other planetary-scale environmental crises (Dominelli, 2012.; BIOS, 2019). In practice, this means that societies must go through a rapid ecosocial transformation, also known as the ecological/ green/ post fossil/ sustainability transition, or the ecological reconstruction of societies (BIOS, 2019). Harking back to the post–World War II reconstruction period, which in wartorn societies meant not only rapid infrastructural but also social and political reconstruction and renewal, we are reminded that rapid and thorough renewal of societies is possible with widespread political will and determination. The transition is essentially about finding ways to provide societal well-being within the carrying capacity of the Earth (Dufva, 2020). As we are not yet quite there, by addressing teaching and learning in ecosocial work, this book highlights the role of social and community workers and educators in the promotion of the sustainability transition in their own fields and beyond.

The transition is a huge, sector-crossing project in which social work can have a crucial role to play. The transition requires in-depth social and cultural changes, which social work and the other social sciences and social workers, together with other social professionals, should be well equipped to facilitate (see, for example, the sector-crossing discussions in Stamm and Dörfler and Whelan this volume). However, to be able to do so, social work needs to reconfigure its own relationship to the natural environment and critique its own modernist premises and the reality that professional social work stems largely from the same anthropocentric and modernist world view as the current environmental problems (Boetto, 2017). In the reconfiguring of social work's relationship to other than human beings and the planetary limits of existence and to foster an ecosocial paradigm change, education plays a pivotal role. This book clarifies the need to include ecosocial sustainability in social work teaching and learning.

Social Work, the Social Professions and Ecosocial Work

We have produced this book from the desire to face current ecological and social challenges in social and community work education. These challenges require research-based analysis and novel perspectives as well as creativity (see for example Livholts and Ahlin this volume), genuine and serious conversation, soul-searching, and the ability to take action.

This book calls us to *Act Now* for *system change, not climate change,* and considers these slogans as deeply rational and founded on scientific evidence. Promoting and organising the ecosocial transition is one of the many roles available to social and community workers and other social professionals. By engaging in these roles, they can articulate a vision of an ecologically sustainable economy and society based on solidarity, equality, and justice.

Ecosocial work requires social workers to engage with people in a vulnerable position while restoring and enhancing ecosystems and ecosystem services. Various existing structures in society lead to social and ecological injustice. By challenging these structures and by proposing and showcasing alternatives, ecosocial work contributes to a process of political change (Peeters, 2012a; Boetto, 2017). Like sustainable development, ecosocial work is a political concept that takes a stand for the direction of society. This requires a willingness to join social movements to reach political ends and to engage in a profound process of 'social learning' (Peeters, 2012a).

The Need for Teaching and Learning in Ecosocial Work

While this book discusses teaching, learning, and practice in ecosocial work and the role of social workers in developing ecosocial sustainability, it also discusses how social work education can move in the direction of ecosocial transformation. Accordingly, this book aims to demonstrate that key aspects of the teaching and learning endeavour are to connect social work education with already emerging ecosocial practices and to provide theoretical, experiential, and practical tools for internalising, assimilation, and integration of ecosocial concepts into thinking, policies, and practice. Education must thus find a balance between the need to educate professionals into the existing system and its frameworks, while also justifying, envisioning, and implementing change.

Social work becoming ecosocial necessitates a change in teaching and learning (see Ranta-Tyrkkö, Chap. 2 and Lievens, Chap. 3, this volume) and, we believe, eventually more community based and collective modes of organising the actual work. As it does not make sense to undertake ecosocial work in an individual manner, schools of social work could try to emphasise community-based modes of working throughout their curricula and thereby influence both mainstream social work and promote the ecosocial transition. Yet, even in community practices, where people find satisfaction and meaning when they are connected to others through ecosocial activities, social work may risk conforming to the status quo, only focusing at generating (otherwise valuable) positive experiences, support and participation without really questioning the ongoing 'politics of unsustainability' (Blühdorn, 2014). As a whole, this book therefore endeavours to demonstrate that education has a key obligation to ecosocial work to ensure that its practices are really transformative and challenge practitioners to see the bigger picture, beyond the 'here and now' and towards 'elsewhere and later'.

Even in a context of acceptance of the seriousness of the state of the world, teaching and practices can give space to joy and fulfilment. Although, the starting point of the need for an ecosocial transition is dire with many dangers and unresolved challenges in proximity, acting on its behalf can be joyful and fulfilling. Hope is born from action, and one can act for a more hopeful and better world even when not feeling particularly optimistic (Virtanen, 2022). It is therefore crucial that higher education institutions engage in forward thinking and commit to transformative practices as well as working at the policy level. Ecosocial work calls for the capacity not only to reshape existing power structures and systems but also to challenge dominant narratives and foster alternative and hopeful stories that support the vision of a transformed society.

Changing teaching and learning in social and community work education is urgent, but 'teaching the future' is not without difficulty. We need to cultivate new kinds of human potential based on (planetary) sensitivity and responsibility, wisdom, intelligence, and solidarity. Many current students were born at the beginning of this century, and most of them can expect to live until 2080 and beyond. This book aims to give these young people a more positive and optimistic perspective on what ecosocial work can be (see, for example, Haine and Marques et al. in this volume). Among other things, this involves leaving intellectual conservativism behind and engaging with communities and the global society towards a better future by drawing on social cooperation to preserve and perpetuate life and together build an ecologically sustainable society based on solidarity, equality, and justice.

How Can This Book Contribute to Change?

This book has been produced in a historical context in which social work in general and ecosocial and related forms of work in particular, along with many other disciplines and professions, are in the process of reconfiguring their relationship to other than human beings and the planetary boundaries of existence. Such reconfiguration, when and if thoroughly carried out, is bound to have implications for teaching, learning, and practice (see, for example, Luís, Gomes and Borga in this volume). While there is clearly a need for a comprehensive systemic renewal, ecosocial work has heretofore proceeded mostly from within the system, developing niches of fairer and more sustainable everyday practices, income-earning possibilities, relationships, and wellbeing.

On one hand, ecosocial practices recognise and utilise the healing power of the natural environment (see Moynihan and McShane, Livholts, and Larocque in this volume), including plant and animal companions, such as in various forms of nature and/or animal assisted care. At the same time, many activities are organised around recycling and upcycling practices so that they both provide sustainable income-earning opportunities and promote and enable resource sparing ways of life (for example, Haine and Freitas in this volume). Sustainable local food production, for example, provides multiple opportunities to work on both social and environmental issues, such as through permaculture (see Skinner and Lynch this volume), community gardening or community assisted forms of agriculture (discussed by Marques, Alves and Morán Carrillo this volume).

On the other hand, notwithstanding that ecosocial work has made some sort of breakthrough and much is thus being done or underway, in relation to the ultimate transition goal, the work is still in its early stage. In the global North, where the transition demands are most pertinent and necessary due to the heavy past and present ecological footprints of northern societies, ecosocial work has not yet broken into the mainstream in educational institutions or in professional practice. While the ground is prepared for takeoff, an actual paradigm shift is still awaiting, as is a broader leap from occasional micro- and meso-level practices to a wide-ranging ecosocial change. Meanwhile, ecosocial work is greatly at odds with prevailing neoliberal, managerialist, relatively short-termist, and narrowed-down expectations that are commonly imposed upon social workers and other social professionals and their work.

Ecological and social—and thus ecosocial—issues are political issues with economic ramifications, and pursuing ecosocial work and transition is necessarily a political project. For social work, the situation is by no means new, for being political is at the core of social work, amongst other reasons due to the pre-political nature of social work ethics and core professional values (Kokkonen, 2024). One of the messages of this book is that ecosocial work and transition can proceed and be promoted in multiple ways and at various layers, depending on what is feasible in a given situation. While influencing macro-level political and economic decision-making is important and sometimes repays the effort, so is smaller scale enabling and empowering of communities, clients, and groups with fewer resources to lead sustainable lives.

When and if the ecosocial project appears overwhelming, solace may be found in the fact that even global change is made in a multitude of localities—where social work and the social professions are typically present. What is more, in the mighty social and cultural challenge to adapt human needs and wants to the planetary boundaries, social work knowledge on promoting change in the lives of individuals and communities is a good starting point. Actual change work may include supporting the emotional and other processes that people need to go through when changing their lifestyles, as well as organising meaningful sustainable living, care, support, and recreational systems locally, and working for equality, diversity, just transition, and societal peace in society (Ranta-Tyrkkö, 2017). Ways to achieve this goal are as important as the goal itself. Whatever we do individually or collectively, it is important to learn and embed more respectful and collaborative ways to relate to, and be part of, the web of life around us across our micro, meso, and macro levels of practice and life. The very minimum requirement is not to hamper but to facilitate those willing and able to act. Encouraging action and providing some signposts, examples, and food for thought is the project of this book.

Chapter Outline

The book begins by considering some of the key concepts that help to inform and shape ecosocial work pedagogy (Part I). It proceeds to explore the possibilities, challenges, and exemplars of teaching and learning ecosocial work in higher education (Part II). Part III offers unique and innovative exemplars of ecosocial work pedagogy and practice from a range of different contexts and practice settings.

Part I of the book introduces and discusses a range of concepts and perspectives on ecology and social work and their implications for the design and use of ecosocial work teaching and learning activities. The aim is to lay a foundation for teaching and learning ecosocial work that transcends the scope of the local and national. This part debates the prerequisites for and implications of developing a global perspective and a planet-oriented worldview in courses and curricula through a focus on critical thinking and practice. The importance of a global perspective is demonstrated by the International Federation of Social Workers (IFSW), which in July 2022 developed and adopted *The People's Charter for a New Eco-Social World*, which states that 'An eco-social understanding of prosperity must critically examine societal and economic structures, practices, and ways of life'. The first part of this book elaborates on these premises and explores their implications for teaching and learning.

The opening chapter by **Komalsingh Rambaree** presents and discusses different perspectives on key concepts, including sustainability, sustainable development, justice, resilience, and eco-social work. The chapter dissects and analyses the diversities and complexities inherent in these concepts, which enable us to consider the mutually interdependent relationship between humans and the bio-physical environment within social work education. The chapter concludes with the implications of the discussion for designing and implementing transformative social work education.

Satu Ranta-Tyrkkö's chapter focuses on materiality and posits the unsustainability of contemporary consumer societies as one of the keys to the global ecological crises at stake. Roughly a quarter of the Earth's population consumes about three quarters of the resources and produces three quarters of waste and emissions. The challenge is to significantly reduce the material imprint of humanity, especially of its over-consuming sections, and prioritise the use of materials over consumables for essential activities. The chapter addresses and discusses these issues and introduces exercises that facilitate awareness of the material and consumerist underpinnings of our lives.

Pieter Lievens argues that acceptance of the magnitude of our global challenges impacts our existing ways of education and thinking. Teaching ecosocial work and being part of transformative education is overshadowed by unsustainable material consumption, increased risks, and profound changes within societies. Fully understanding this has a disruptive effect on mental wellbeing and may lead to eco-anxiety and cognitive dissonance in young students who hold a progress-oriented worldview. Eco-anxiety and cognitive dissonance require understanding of and support for ecosocial work at institutional level. The chapter draws on the author's experience of teaching the global challenges to discuss the need, possibility, and conditions of an eco-centred worldview in general social work education.

Part II of the book offers rich and detailed examples of ecosocial teaching and learning from across the globe, including Asia, North America, Europe, and Oceania. The aim of this part is to elucidate the effectiveness of introducing and integrating an ecosocial work curriculum to social work and cognate human services training programmes, using a diversity of concepts, ideas, and approaches as well as interdisciplinary perspectives. *This part of the book demonstrates that practices in ecosocial work teaching and learning are already engaging across micro, meso, and macro levels to create change and promote eco-social practice by employing* innovative learning ideas and frames that can be used to achieve effective student engagement, learning, and connection with practice.

In their chapter, **Nicole Mattocks, Melissa Indera Singh, Lauryn Smith, Georgianna L. Dolan-Reilly, Meredith C. F. Power, and Kelly Smith** highlight key findings from a recent study conducted to examine current MSW course offerings in the United States of America. The chapter considers the implications of the findings for educators and makes a clear call to action to develop ecosocial work courses that emphasise the important role that social workers should and do play in addressing environmental injustices. Additionally, the chapter offers concrete suggestions and resources for educators who wish to develop such a course in their own programmes and universities.

As educators in the allied social professions of social work and youth and community work, **Mary Hurley, Catherine Forde, and Fiachra Ó Suilleabháin** discuss the development of an interdisciplinary environmental module for students on a professional postgraduate Masters in Social Work (MSW) programme in University College Cork. This elective module is now in its third year, and a sister module subsequently rolled out to students of the qualifying BSocSc (Youth and Community Work) degree. The longer-term goal is to teach these modules as an interdisciplinary offering to students of the social and helping professions across the university in order to foster strong relationships between students and encourage knowledge exchange and collaborative multidisciplinary practice. This chapter explores the development and content of the sister modules and discusses some of the key challenges and opportunities of introducing ecosocial work to the curriculum of the social professions in the university.

The chapter by **Ingo Stamm and Lisa Dörfler** introduces a new teaching concept focused on the link between human rights-based social work and the ecosocial paradigm. In Germany, the discussion on ecosocial work has just recently started, and teaching human rights and ecosocial approaches is a new field in social work education. The chapter reviews and evaluates how this teaching concept was put into practice in 2022 in a seminar of the master's programme 'Social Work as a Human Rights Profession' at the Alice Salomon University Berlin. The seminar focused on the link between human rights, sustainability, and the natural environment, providing a theoretical foundation for the key concepts and discussing how to combine a human right and an ecosocial perspective in social work practice.

Alex Skinner and Deborah Lynch argue that grounding ecosocial practice concepts in social work curricula remains an ongoing challenge for educators. Their chapter explores practical ways social work educators can enable students to critically engage with local and global socio-environmental issues and enact ecosocial practices in a diverse range of social work settings as future practitioners. Conceptually, the chapter draws on the system-based holistic lens of permaculture to offer ecosocial practice narratives for teaching and learning to enrich and deepen social work skills and practices within the ecosocial framework.

In their chapter, **Rita Luís, Joana Gomes and Afonso Borga,** social work students and graduates of ISCTE University Institute in Lisbon, address the relationship between social work and environmental sustainability using a pragmatic approach: the construction of a toolkit as a practical guide for the application of sustainable practices and projects in social work. Their chapter describes the construction of the toolkit, which was developed using a literature review, presentations from experts in the field of sustainability, and six webinars that inform the six chapters of the toolkit: Environmental Social Work in the Community, Circular Economy, Environmental Education, Food Security Sustainability and Social Work, The Consideration of the Animal as an Element in Social Work, and Social Work in

Disaster. This practical guide for the application of sustainability in the social sector is intended as a reference tool for students and professionals who want to develop academic work or social projects with an environmental sustainability component.

In his chapter, **Joe Whelan** explores social work education in the context of climate change. He suggests that a turn towards sustainable social policies that align with the mission of social work might be one useful way to begin to think, talk, and teach about climate justice. In this respect and couched within the overarching framework of a postgrowth aligned capability approach, Universal Basic Services (UBS) and Participation Income (PI) are explored. Ultimately, it is suggested that focusing on ecosocial policies in the context of social work education can give future practitioners something to aspire towards, to engage in activism for and to help shape practice.

The ecological crises increase human inequality and vulnerability, and amongst other factors, force people to migrate. Social workers are usually among the frontline workers trying to meet the needs of the affected people and enhance resilience at the grassroots. Yet, existing social work degree programmes provide hardly any specific training for social workers on crisis preparedness, and research on crisis preparedness is limited. The chapter by **Janet Carter Anand, Anish K. R., Manish Jha, Bipin Jojo, and Satu Ranta-Tyrkkö** reflects on the teaching and learning experiences from the intensive course *Climate Change, Migration and Crisis Preparedness*, taught in Finland in June 2022 and again in India in January 2023. The chapter elucidates the conceptualization, framing, and organisation of preparatory sessions, identification of resource materials, and delivery of the course, engaging critically with the active participation of students, their feedback, and field insights from the global north and south.

The **third and final part** of the book presents innovative, contemporary, and provocative examples of ecosocial work practices from Finland, Canada, and Portugal. Each example illustrates ecosocial work practice and its connections with situated teaching and learning opportunities for social work students and practitioners. The premise of this part of the book is that teaching and learning are only meaningful if there is a strong interaction with practice, especially in ecosocial work, where there is considerable potential for incorporating a global perspective or a planet-oriented worldview into practice as well as into teaching and learning. The contributions underline the reality that what social work students learn in the classroom and on placement can be replicated and reinforced in their practice as qualified social workers and practitioners, thereby enabling sustainable social work teaching and practice.

In her chapter, **Mona B. Livholts** invites social work students and practitioners to make use of environmental writing as a critical and creative methodological practice to include more-than-human perspectives into knowledge and learning in social work. Inspired by feminist, Indigenous, and postcolonial writing, she uses life writing genres such as diary notes, photography, and poetry, to re-think and promote renewed embodied and situated practices for the futures of writing with social work. The chapter includes illustrations and scenes thematically related to childhood memories from a farm, environmental exhaustion as an alternative to social problems in times when both human and non-human life is threatened, and water as a pathway for social justice. Each section includes brief guidelines for how to make use of one or several of the genres to conduct environmental writing. The implications of the chapter are to promote environmental writing in eco-social and post-anthropocentric social work.

Emmanuelle Larocque's chapter offers an overview of the steps undertaken during an intervention that aimed to co-create, with various community actors, an ecosocial sport and nature-based therapy programme for young people. She describes her collaborative and embodied methodology and focuses especially on how environmental and ecological dimensions were integrated into the program. The chapter provides an outline of the intervention protocol, detailing each of the programme's components and specifying how the programme was continually adapted in accordance with participants' trajectories and normative understandings of Nature.

In the closing chapter, **Eduardo Marques, Elsa Alves, and José María Morán Carrillo** assess the impact of mushroom production and consumption on health, the economy, and potential role in social entrepreneurship led by social workers. The chapter offers a reflexive analysis, with a holistic approach, of the contribution of interdisciplinary approaches to the achievement of goals that promote social inclusion and sustainable development in harmony with nature. Departing from a social work case study (Mushroom House) located in Azores (Portugal), the chapter demonstrates the possibilities of an ecosocial work journey through the paths intersecting the social sciences, nutrition sciences, and social economy organisations striving for community well-being through a circular economy.

The book also provides a number of short case studies of ecosocial work practice from social work and community work students, graduates, and practitioners. The case studies are drawn from Finland (Elisa Rainerma), Ireland (Katie McShane and Elaine Moynihan), Belgium (Luna Haine), Portugal (Daniela Freitas), and Sweden (Olivia Alhlin). These reflective pieces illustrate some of the advantages, opportunities, and challenges of using an ecosocial work approach in a range of practice settings including formal social work contexts, communities and community organisations, schools and universities, and illuminate the opportunities for creative and inclusive ecosocial practices that effect real change.

Bibliography

Armstrong McKay, D., Staal, A., Abrams, J., Lenton, T., et al. (2022). Exceeding 1.5°C global warming could trigger multiple climate tipping points. *Science, 377*(6611). https://doi.org/10.1126/science.abn7950/

Besthorn, F. (2003). Radical ecologisms: Insights for educating social workers in ecological activism and social justice. *Critical Social Work, 3*(1), 66–106.

Besthorn, F. (2012). Deep ecology's contributions to social work: A ten-year retrospective. *International Journal of Social Welfare, 21*(3), 248–259.

BIOS. (2019). Ecological reconstruction » BIOS: Ecological reconstruction. Retrieved 11 December 2023.

Blühdorn, I. (2014). Post-Ecologist Governmentality: Post-Democracy, Post-Politics and the Politics of Unsustainability. In E. Swyngedouw & J. Wilson (eds.): *"The Post-Political and Its Discontents Spaces of Depoliticization Spectres of Radical Politics Post-Ecologist Governmentality"* (pp. 146-166). Edinburgh University Press.

Boetto, H. (2016). Developing ecological social work for micro level practice. In J. McKinnon & M. Alston (Eds.), *Ecological social work: Towards sustainability* (pp. 59–77). Palgrave Macmillan.

Boetto, H. (2017). A transformative eco-social model: Challenging modernist assumptions in social work. *The British Journal of Social Work, 47*(1), 48–67.

Boetto, H., Bowles, W., Närhi, K., & Powers, M. (2020). Raising awareness of transformative eco-social work: Participatory action research with Australian practitioners. *International Journal of Social Welfare, 29*, 300–309.

Bozalek, V., & Pease, B. (Eds.). (2021). *Post-anthropocentric social work. Critical posthuman and new materialist perspectives*. Routledge.

Brand, U., & Wissen, M. (2021). *The imperial mode of living: Everyday life and the ecological crisis of capitalism*. Verso. ISBN 9781788739122.

Cardona, O. D., Van Aalst, M. K., Birkmann, J., Fordham, M., Mc Gregor, G., Rosa, P., Pulwarty, R. S., Schipper, E. L. F., Sinh, B. T., & Décamps, H. (2012). Determinants of risk: exposure and vulnerability. In *"Managing the risks of extreme events and disasters to advance climate change adaptation": Special report of the Intergovernmental Panel on Climate Change* (pp. 65–108). Cambridge University Press.

Ceballos, G., Ehrlich, P., & Raven, P. (2020). Vertebrates on the brink as indicators of biological annihilation and the sixth mass extinction. *PNAS, 117*(24), 13596–13602. https://doi.org/10.1073/pnas.1922686117

Chigangaidze, R. (2023). The environment has rights: Eco-spiritual social work through Ubuntu philosophy and Pachamama: A commentary. *International Social Work, 66*(4), 1059–1063.

Dominelli, L. (2012). *Green social work: From environmental crises to environmental justice*. Polity Press.

Dufva, M. (2020). *Megatrend 1: Ecological reconstruction is a matter of urgency*. The Finnish Innovation Fund Sitra, March 6th. Retrieved 6th November 2022, https://www.sitra.fi/en/articles/megatrend-1-ecological-reconstruction-is-a-matter-of-urgency/

European Commission, Joint Research Centre, Crippa, M., Guizzardi, D., Schaaf, E. (2023). GHG emissions of all world countries: 2023, Publications Office of the European Union. https://data.europa.eu/doi/10.2760/953322

Fanning, A. L., O'Neill, D. W., Hickel, J., & Roux, N. (2022). The social shortfall and ecological overshoot of nations. *Nature Sustainability, 5*, 26–36. https://doi.org/10.1038/s41893-021-00799-z

Gadgil, M., & Guha, R. (1995). *Ecology and equity: The use and abuse of nature in contemporary India*. Routledge.

Gray, M., Coates, J., & Hetheringon, T. (Eds.). (2013). *Environmental social work*. New York.

IPCC. (2023). *Climate change 2023: Synthesis report*. A report of the Intergovernmental Panel on Climate Change. Contribution of Working Groups I, II and III to the sixth assessment report of the Intergovernmental Panel on Climate Change [Core Writing Team, H. Lee and J. Romero (eds.)]. IPCC (in press).

Kokkonen, T. (2024). *Polical and social effectiveness of social work research*. Keynote in the Preconference 'Political Nature and Societal Influence of Social Work Research' of the Finnish Social Work Research Conference, 14th February 2024, Tampere.

Krings, A., Victor, B., Matthias, J., & Perron, B. (2020). Environmental social work in the disciplinary literature, 1991–2015. *International Social Work, 63*(3), 275–290.

Kröger, M. (2020). Politics of extraction: Theories and new concepts for critical analysis. *Oxford Bibliographies – International Relations*, https://doi.org/10.1093/OBO/9780199743292-0292;

retrieved 6th November 2022, https://www.researchgate.net/publication/343924759_Politics_of_Extraction_Theories_and_New_Concepts_for_Critical_Analysis

Mathias, J., Krings, A., & Texeira, S. (2023). Which environmental social work? Environmentalisms, social justice, and the dilemmas ahead. *Social Service Review, 97*(3). https://doi.org/10.1086/724522

Matthies, A., Närhi, K., & Ward, D. (2001). *The eco-social approach in social work*. SoPhi.

McKinnon, C. (2012). *Climate change and future justice. Precaution, compensation and triage*. Routledge.

McKinnon, J., & Alston, M. (2016). *Ecological social work. Towards sustainability*. Palgrave.

Närhi, K. (2022). Ekososiaalisen työn reunaehtoja. *Tutkiva sosiaalityö* 22, 8–18. Talentia-lehti & Sosiaalityön tutkimuksen seura.

Närhi, K., & Matthies, A. (2016). Conceptual and historical analysis of ecological social work. In J. McKinnon & M. Alston (Eds.), *Ecological social work: toward sustainability* (pp. 21–38). Palgrave Macmillan.

Närhi, K., & Matthies, A. (2018). The ecosocial approach in social work as a framework for structural social work. *International Social Work, 61*(4), 490–502.

NOAA. (2022). Climate Change: Global Sea Level. Climate Change: Global Sea Level | NOAA Climate.gov, retrieved 17th December 2023.

Nygren, A., Kröger, M., & Gills, B. (2022). Global extractivisms and transformative alternatives. *The Journal of Peasant Studies, 49*(4), 734–759. https://doi.org/10.1080/03066150.2022.2069495

Peeters, J. (2012a). The place of social work in sustainable development: Towards ecosocial practice. *International Journal of Social Welfare, 21*(3), 287–298. https://doi.org/10.1111/j.1468-2397.2011.00856.x

Peeters, J. (2012b). Social work and sustainable development: Towards a social-ecological practice model. *Journal of Social Intervention, 21*(3), 5–26.

Pihlström, S. (2020). Ilmastonmuutos eksistentiaalisena uhkana. In S. Kyllönen & M. Oksanen (Eds.), *Ilmastonmuutos ja filosofia* (pp. 297–316). Gaudeamus.

Ramsay, S., & Boddy, J. (2017). Environmental social work: A concept analysis. *The British Journal of Social Work, 47*(1), 68–86.

Ranta-Tyrkkö, S. (2009). Vähän on kylliksi – sosiaalityöstä Mahatma Gandhin malliin. In M. Mäntysaari, A. Pohjola, & T. Pösö (Eds.), *Sosiaalityö ja teoria. Sosiaalityön tutkimuksen vuosikirja* (pp. 17–46). PS-kustannus.

Ranta-Tyrkkö, S. (2017). Sosiaalityön tulevaisuuden etiikka epävarmuuden ja ympäristöriskien maailmassa. In R. Enroos, M. Mäntysaari, & S. Ranta-Tyrkkö (Eds.), *Mielekäs tutkimus. Näkökulmia sosiaalityön tutkimuksen missioihin* (pp. 113–138). Tampere University Press. http://urn.fi/URN:ISBN:978-952-03-0606-9

Ranta-Tyrkkö, S., & Närhi, K. (2021). Striving to strengthen the ecosocial framework in social work in Finland. *Community Development Journal, 56*(4), 608–625. https://doi.org/10.1093/cdj/bsab030

Smith, J., Schneider, S., Oppenheimer, M., & van Ypersele, J. P. (2009). Assessing dangerous climate change through an update of the Intergovernmental Panel on Climate Change (IPCC) "reasons for concern". *PNAS, 106*(11), 4133–4137. https://doi.org/10.1073/pnas.0812355106

Steffen, W., Richardson, K., Rockstrom, K., Cornell, S. E., Feltzer, I., Bennett, E. M., Biggs, R., Carpenter, S. R., et al. (2015). Planetary boundaries: Guiding human development on a changing planet. *Science, 347*(6223). https://doi.org/10.1126/science.1259855

Talentia. (2023). *Sosiaalialan ilmastokäsikirja*. Talentia. https://www.talentia.fi/wp-content/uploads/2023/08/talentia_ilmasto-opas_2023_final.pdf

Virtanen, H. (2022). *Trauma ja luonto. Eli kuinka selviytyä ekoahdistuksesta*. SKS.

Wendt, G. (2021). What to think of in the ecosocial concept. In *Ecology of common care*. Springer Briefs in Social Work. https://doi.org/10.1007/978-3-030-65699-7_2

Wong, C. (2023). Earth just had its hottest year on record – climate change is to blame. *Nature, 623*, 674–675. https://doi-org.ezproxy.jyu.fi/10.1038/d41586-023-03523-3

Part I
Concepts in Ecosocial Work Teaching and Learning

Chapter 1: Ecosocial Work and Transformational Teaching and Learning: Navigating Through Complex and Contested Concepts

Komalsingh Rambaree

Introduction

A concept is a mental construct defined as a collection of ideas and interpretations represented through word/s based on abstraction and generalization. Concepts play important cognitive and explanatory roles in scientific discourses and academic writing. A concept is often considered a structural explanatory unit that contributes to a better understanding of a phenomenon/phenomena through meaning-making based on descriptive facts/evidence and personal experience. It is, therefore, a central aspect of teaching, learning, and researching social issues and problems.

Taylor and Vosgerau (2021) argue that concepts constitute not single but multiple representational kinds of explanation determined by subjectivity and context. Similarly, Meadowcroft and Fiorino (2017) opine that concepts can be regarded as "thought categories" influencing people's apprehension and constraining reasonings and debates. Concepts are dynamic, as well as relative and subjective. Thus, a concept can have various conceptions—based on different ways of thinking, observing, reasoning, apprehending, and explaining—under different contexts and perspectives (Lalumera, 2014). Concepts must be understood and discussed as a term embedded in complex instrumentalities, social relationships, and learning patterns within their formative activity systems (Engeström, 2020). In a similar manner, Solonchak & Pesina (2015) state:

> a concept may be refracted in the human consciousness due to understanding of diverse events and phenomena of the reality, taking into account the individual and social experience, the specific space, time, geographical, historical and cultural environment, as well as depending on the way of thinking (common or scientific) that prevails in any specific situation. (p. 352)

K. Rambaree (✉)
Department of Social Work, Criminology, and Public Health Sciences, University of Gävle, Gävle, Sweden
e-mail: komal.singh.rambaree@hig.se

C. Forde et al. (eds.), *Teaching and Learning in Ecosocial Work*,
https://doi.org/10.1007/978-3-031-58708-5_2

According to Gerring (1999), aspects that determine concept formation are (a) the events or phenomena to be defined through the concept, (b) the properties or attributes that define the concept, and (c) a label covering both (a) and (b). Philosophers such as Plato and Aristotle posit that humans have a unique cognitive capacity to form concepts based on perception and intellect (as cited in Ebbesen & Gregoric, 2022). Therefore, the diversities within human rationality and capabilities require humans to operate with and navigate complex and often contested concepts. In his thesis, Nylén (2023) provides an overview of different typologies of concepts. Among these he refers to "contested concepts" as those that "inevitably involve endless disputes about their proper uses on the part of their users" (Gallie, as referred in Nylén, 2023, p.39) and to "umbrella concept," a broad concept or "idea used loosely to encompass and account for a set of diverse phenomena" (Hirsch & Levin, as quoted in Nylén, 2023, p.39). Within academic discourses, there are often never-ending debates—both within and across disciplines—about different types and forms of concepts. Such discussions are essential in exploring the added value of a concept in theoretical understanding and practical applications.

This chapter aims to present and critically reflect on the conceptualization of ecosocial work to generate debates and critical reflections that could enhance a better understanding, conceptualization, and application of ecosocial work approaches that promote transformative teaching and learning within social work. After this brief introduction on the basis and process of conceptualization, the chapter outlines the context within which contemporary conceptualization and utilization of ecosocial work is taking place. It describes aspects such as values, goals, and practices of ecosocial work. This part of the chapter intends to help readers navigate various meanings and interpretations of ecosocial work and other leading concepts, such as socio-ecological justice, sustainability, and sustainable development, that influence the construction, understanding, and practice of ecosocial work. This part of the chapter also argues about the "transformative" characteristic and expectation attached to ecosocial work in various discourses. Then, the chapter discusses ecosocial work teaching and learning from a transformative perspective. Finally, the conclusion is based on the discussions on the added value of ecosocial work in promoting sustainable development.

Ecosocial Work: Navigating Complex and Contested Concepts

In an article, Rambaree et al. (2019) argue that ecosocial work is not a specialty within social work and, instead, all social work can and should be ecosocial work. Such reasoning emerges from the founding beliefs and values of social work itself, which is based on a premise of the "person in environment" perspective in conceptualizing social work. Similarly, several authors posit that ecosocial work compels us to consider that the biophysical environment significantly impacts individuals, families, and local and global communities (Boetto, 2016; Boetto et al., 2020; Powers et al., 2019a). According to the "person in environment" perspective, social workers need to have a holistic view in understanding and practicing social work, taking into account environmental (social, economic, political, ecological) considerations.

Ecosocial work has often been conceptualized as a holistic umbrella concept emphasizing the interrelated links between ecological issues and social work (Närhi & Matthies, 2018). In particular, ecosocial work requires a multidisciplinary and multidimensional approach to social work practice that incorporates concern for people and the biophysical environment at the micro, meso, and macro levels of social work practice (Boetto et al., 2022; Ramsay & Boddy, 2017). Ecosocial work has been discussed as a holistic perspective that sees human beings as part and parcel of a complex ecosystem. It promotes the equitable and sustainable use of ecosystem resources and services for the well-being of all. An ecosocial perspective allows social workers to consider the interdependent relationship between humans and the biophysical environment in a more coherent and holistic manner. For instance, in this book, Stamm and Dörfler discuss the intersectional link between human rights and ecosocial work approaches. Using a practical example, they demonstrate how two paradigms (human rights and ecosocial work) could be integrated in social work education. In a similar vein, Stamm et al. (2023) state, "ecosocial work, overlapping with concepts such as environmental, ecological, or green social work, is built on the premise that humans are part of, and dependent on, the web of life on Earth" (p. 179).

Ecosocial work can be considered an umbrella concept (see Fig. 1), closely linked to green social work, ecological social work, environmental social work, and eco-spiritual social work. Green social work is theorized as a concept relating environmental justice to social justice to enhance the well-being of people and the planet with a particular focus on transforming the sociopolitical and economic structures and forces that negatively impact the quality of life of the poor and marginalized population (Dominelli, 2012; Dominelli et al., 2018). Ecological social work and environmental social work have been conceptualized similarly. Ecological social work places both humanity and nature at the center of social work education and

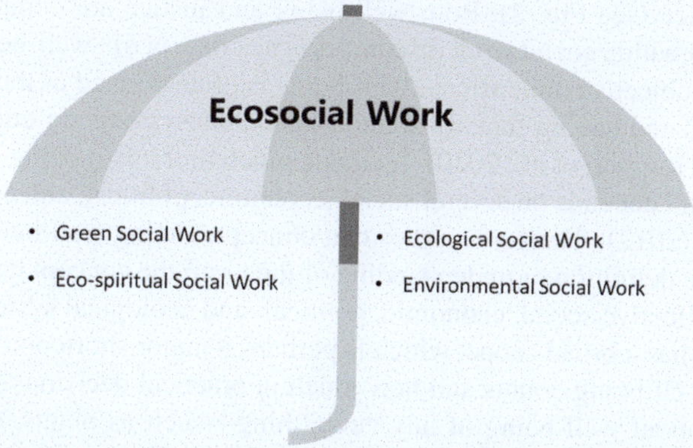

Fig. 1 Ecosocial work as an umbrella concept. (Source: author)

practice to ensure the sustainability of natural resources (Mckinnon & Alston, 2016). In a similar manner, environmental social work urges social workers to create sustainable conditions for the well-being of both human and the natural worlds (Gray et al., 2012). Finally, eco-spiritual social work proposes social work education and practice that is heavily influenced by the deep-ecological philosophy and indigenous values and knowledge in protecting and promoting human and ecological well-being (Besthorn, 2016; Ferreira, 2010). The common aspect of green social work, ecological social work, environmental social work, and eco-spiritual social work is integrating and promoting the biophysical environment within social work. The biophysical environment in promoting eco-spiritual social work includes forest, beaches, mountains, rivers, gorges, and so on.

Over the last few years, there have been several discussions on the gaps and concerns about ecosocial work in various discourses. For instance, Wang and Altanbulag (2022) state that it is crucial to clarify the concept of ecosocial work with its value concerns. In order to understand ecosocial work, one has to explore the concerns such as ecological injustices, climate crisis, and natural calamities that the concept is focused on. Moreover, Thysell and Cuadra (2022) mention that it is essential to clarify the conceptual understanding of the ecosocial on which social work bases its responses to the call for socio-ecological sustainability. Finally, Ranta-Tyrkkö and Närhi (2021) posit that there is still a need to continue the discussion on the basis and pathways through which an ecosocial work framework could be integrated into social work education and practice. In this connection, this chapter presents and discusses the values, vision, mission, goals, and practice strategies and orientations as a basis for integrating ecosocial work perspectives within the teaching and learning of social work.

Basic Values

Two central values put forward in various discourses on ecosocial work are well-being and justice (see Fig. 2). Both well-being and justice are complex and contested concepts within social work discourses. The concept of "well-being" has both subjective and objective dimensions. Well-being is often defined as the combination of feeling good and having functional and sustainable systems for enjoying a good quality of life (Ruggeri et al., 2020). It entails much more than being in a good and healthy state and depends on several social, economic, political, and ecological systems. Gamble (2012), therefore, argues that conceptualizing well-being should not require a single definition or understanding of the term; the concept largely depends on complex, effective social, economic, political, and ecological systems.

Well-being has a broad scope, which is perhaps a major shortcoming of the concept. Whose well-being counts and how much in practical decisions? For instance, one can talk about well-being of any being/thing—such as plants, organizations, nations, the economy, and the environment—without having a clear understanding of what are the precise indicators and what should be the final outcome/s of the

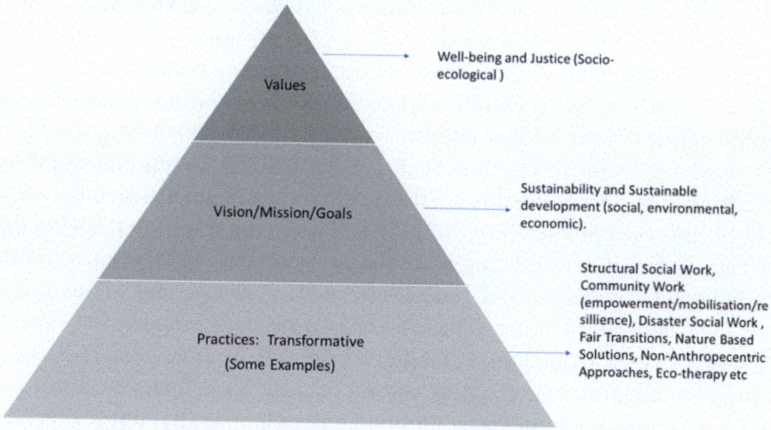

Fig. 2 Ecosocial work—values, missions, goals, and practice strategies. (Source: author)

concept (Campbell, 2016). However, there seems to be a consensus that well-being depends on a harmonious condition and relationship between people and the environment (including the biophysical environment and organizational environment). It is evident that people are happy when things around them function well and thrive.

Ecosocial work, therefore, puts greater emphasis on solving problems more holistically, looking at socioeconomic and, more importantly, ecological issues, concerns, and complexities through the lens of sustainability in achieving well-being for current and future generations of all beings (human and nonhumans) (Boetto, 2017; Rambaree et al., 2019; Salonen & Konkka, 2015). In this sense, ecosocial work focuses more on ecological well-being.

Ecological well-being is often defined as a well-balanced ecological system with a harmonious and thriving relationship between the planet and its habitants through sustainable social and economic conditions. The term recognizes the importance of the well-being of the Earth's natural systems alongside the well-being of humans and other species (Isham et al., 2022). An ecological system includes its inhabitants; therefore, ecological well-being encompasses the relationship between humans and the environment (Grouzet & Lee, 2014). Thus, ecosocial work calls for a paradigm shift, beginning with a revision of the traditional social work ethics by broadening its focus from human well-being only to other species and entire ecosystems, with the acknowledgment of the intrinsic value of the natural world through an ecocentric worldview (Rambaree et al., 2023; Stamm et al., 2023).

Another central value of ecosocial work is justice (see Fig. 2). Justice is one of the central virtues and principles of our society. It is commonly understood as fairness in dealing with humans and nonhuman beings. The conceptualization of justice originates in natural law and social contract theory. It has been conceptualized from different academic perspectives such as legal, political, social, economic, and environmental. Within social work, justice has been often discussed from a social perspective, as conditions for existence and solidarity and for promoting and maintaining stable and harmonious relationships through coexistence. In particular, social justice is commonly defined as an ideal condition for all human beings in a society having same fundamental rights, benefits, opportunities, obligations, and so on, with a focus on advocacy to eliminate power domination, discrimination, oppression, and inequities in the society (Barker as referred to in Besthorn, 2012, p.32).

Often, environmental justice has been conceptualized within the social justice frame (Beltrán et al., 2016; Besthorn, 2012). Similar to social justice, environmental justice has been conceptualized by Bell (2014) as having three overlapping dimensions: substantive or the overall quality of the environment; distributive or the extent of environmental equalities; and procedural or the fairness and inclusiveness of environmental decision-making. Thus, there is a need to broaden the concept of environmental justice to include ecological justice (Lin et al., 2021; Ross et al., 2020). Environmental justice remains anthropocentric as it focuses on the disparities among and impact of injustices on human populations. In contrast, ecological justice articulates justice for the ecosystem as a whole with the understanding that humans are only one part of this ecosystem (Rambaree et al., 2019). In this sense, it is important for ecosocial work to conceptualize justice within a holistic perspective and acknowledge justice as essential for all living organisms (Boetto, 2017). As Boetto (2017, p. 58) puts it, "Committing to ecological justice rather than environmental justice demarcates the difference between implementing transformative change within the profession or simply adding the natural environment to existing practice frameworks."

Within ecosocial work, there is, therefore, a need to go beyond the narrow focus on social justice and environmental justice to a broad focus on socio-ecological justice. Socio-ecological justice is based on a framework of repositioning the economy as a tool to assist human and ecological development rather than as a normative goal of societal development (Grossmann et al., 2021). Combining social justice with ecological justice provides better explanatory facts and a more coherent and holistic theorization of justice as a more profound and broader question within human and nonhuman relations. For instance, socio-ecological justice is argued to be a better concept/framework for sustainable development, especially in tackling the impediments of justice for all beings rooted in societal structures and functions (e.g., socioeconomic) (Grossmann et al., 2021; Kortetmäki, 2016). Justice for all (human and nonhuman) is however often not included in social work discourses. Rather, social work remains narrowly concerned about human justice.

According to Yaka (2020), socio-ecological justice "locates justice within a relational ontology that maintains an intrinsic relationship between social and ecological phenomena" (p. 176). Socio-ecological justice is therefore centrally founded on the right to cohabit with nonhuman beings and entities that are not merely resources but rather part of the social environment (ibid). The concept of socio-ecological justice, therefore, provides a broader (and more complex) perspective on rethinking justice through the relationality of human and nonhuman life (Davis et al., 2019; Pope et al., 2021; Yaka, 2019, 2020). For instance, socio-ecological justice provides a more suitable framework for capturing the growing concerns over climate injustices—in terms of the impact of the climate crisis on different population groups and places—requiring social work responses (Forbes & Smith, 2023).

Socio-ecological justice envisions true justice as a holistic, restorative, and inclusive concept that attends to the rights of all life and sees the good of humanity and the good of the planet as the same good (Lin et al., 2021). For example, socio-ecological justice is the value base in the preservation and conservational practical works, such as in protection of natural reserves and endangered species. In a similar manner to how social justice is used as a basis to defend the rights of disenfranchised and marginalized human populations, socio-ecological justice is used to defend interventions in favor of endangered species. It recognizes various forms of exploitation, oppression, domination, humiliation, and disrespect within the comprehensive framework of justice for all beings (Kortetmäki, 2016). Such comprehensiveness is vital in providing a more profound understanding of different complexities in creating necessary conditions for emancipatory or transformational change to deal effectively with current socio-ecological crises and urgencies. Pope et al. (2021, p. 10) posit that socio-ecological justice, first, recognizes the interconnections and differences between distinct demands and subjects of justice; second, is realized in a safe and just operating space for all forms of life, between two essential boundaries set by the promotion of ecological integrity (the ecological ceiling) and dignity of all life (the social foundation); and third, adopts ecocentric ethics that include human and nonhuman entities of present and future generations in a unified but not uniform community.

Vision, Mission, and Goals

The vision, mission, and goals of ecosocial work have been discussed in relation to many other contested concepts, of which two of the central ones are sustainability and sustainable development. Sustainability and sustainable development are broad and complex concepts that are interrelated with each other. The term sustainability raises concerns about the availability and adequacy of resources over time. The concept emerged into the mainstream discourses on development in the 1980s, although the term had existed in various discussions (such as in forestry management) long before. Before industrialization, the concept was mainly defined and discussed as a concern for responsible and long-term use of natural resources and therefore the preservation of natural capital due to its inherent worth (Purvis et al., 2019; Spindler, 2013). In the 1970s, debates about the *limits to growth* began, alongside discussions on "sustainable society" for the future with a main focus on equitable distribution of resources (Du Pisani, 2006; Purvis et al., 2019). Beyond ecological concerns, alongside scientific and technological advances, the need for social interventions requiring deep and enduring social and cultural changes became more apparent in various sustainability discourses (Boström et al., 2018).

Consequently, policymakers and advocates for international and national development started exploring the link between environmental sustainability and social factors such as inequalities, exclusion, and poverty. Sustainability was then gradually introduced into the discussion of "development." While development is a complex, contradictory, and multifaceted concept, its standard definition contains the central notion of transitioning from a less desirable to a more desirable kind of society in which the well-being of all is enhanced (Stiglitz, 1998). The term sustainable development was used in the year 1980, in a report on the World Conservation Strategy by the International Union for Conservation of Nature and Natural Resources (IUCN), where it was emphasized that for development to be sustainable, it must take account of social and ecological factors, as well as economic ones (Dylan, 2013; IUCN-UNEP-WWF, 1980). Thus, sustainable development emerged as a concept with three interconnected sustainability pillars—environment, economic, and social. Sustainable development became a popular and essential concept with the publication of the Brundtland Report in 1987 by the World Commission on Environment and Development, emphasizing the importance of the biophysical environment but with an equal concern for social and economic development. From a development perspective, sustainable development focuses on goals and processes for enhancing well-being and living standards without further damaging and jeopardizing the Earth's ecosystems (Mensah, 2017).

Sustainable development is a holistic concept that requires multidimensional and multidisciplinary knowledge and approaches. Social work, a discipline and profession with a long history and tradition of working on the mission of development in a holistic manner, has primarily been focusing on socioeconomic development. Despite the axiomatic interdependent link between humans and nonhumans, social work remained somewhat detached from consideration for the biophysical environment and nonhuman well-being during the 1970s, 1980s, and 1990s (Gray & Coates, 2012; Zapf, 2010).

During the beginning of the millennium, with growing concern for climate crisis, and the consequences of ecological destructions and calamities on human well-being, more and more attention began to be geared toward ecological sustainability within social work. A critical and holistic understanding of sustainable development in which the environmental, economic, cultural, and social are deeply interwoven became increasingly apparent in holistic approaches. Such holistic approaches offer potential pathways for social work to contribute to both human and nonhuman worlds (Buckingham & Kina, 2015). Sustainable development became a key concept in promoting holistic social work education and practice in several parts of the world. Peeters (2012), for instance, argued that the concept of development used within social work education and practice had to be reviewed considering the principle of respect for the Earth's ecological capacity to sustain, which is obviously a basic precondition for sustainable development.

The United Nations' resolution *Transforming Our World: The 2030 Agenda for Sustainable Development* adopted by the UN General Assembly of September 25, 2015, presented and prioritized 17 sustainable development goals and 169 targets to stimulate action over the next 15 years with the focus on 5Ps—people, planet, prosperity, peace, and partnership (United Nations, 2015). The United Nations' Sustainable Development Goals (UNSDGs) became an opportunity for social work to redefine its role and functions concerning socio-ecological issues and concerns (Jayasooria, 2016). For instance, Närhi and Matthies (2016) argue that an ecosocial framework provides structural social work (requiring interventions for structural changes in societies) with new holistic perspectives and opportunities to follow the principles of sustainable development. However, it is worth noting that even though the sustainable development goals are widely supported, many of them are far from being reached by the year 2030. National and international calls for social workers to consider complex social, economic, and environmental issues led several universities in different parts of the world to embed UNSDGs within social work curricula (Cordoba & Bando, 2022; Drolet et al., 2015; Matutini et al., 2023; Nhapi & Pinto, 2023; Powers et al., 2019b; Rambaree, 2020; Shatberashvili & Sadzaglishvili, 2022). In the USA, the American Academy of Social Work and Social Welfare presented *Create Social Responses to a Changing Environment* as 1 of the 12 grand challenges to be addressed by social workers (Forbes & Smith, 2023). Ecosocial work became a central concept in the 2020–2030 Global Agenda for Social Work and Social Development Framework (IASSW/ICSW/IFSW, 2020; Truell & Crompton, 2020). With the COVID-19 pandemic that began in the year 2020, the growing fear and concern for inequality, crisis, and a broken social contract toward meeting the UNSDGs amplified the importance of ecosocial work concept through the document entitled *A New Eco-Social Contract Vital to Deliver the 2030 Agenda for Sustainable Development* (UNRISD, 2021, 2022).

Ecosocial work is therefore becoming more and more evident and apparent in contemporary social work education, practice, and research. The vision and mission of ecosocial work therefore require the adoption and promotion of transformative practices and education.

Practices: Transformative

For attaining the UNSDGs, an in-depth and comprehensive understanding of how societal transformation in terms of strategies, interventions, and practices must be initiated, fostered, and governed toward sustainability is essential (Boström et al., 2018). Even if ecosocial work is much more than just working toward UNDSDGs, societal transformation remains a central focus in its strategies, interventions, and practices. Societal transformation needs a complete or even drastic positive change in society to achieve certain specific goals, such as the UNSDGs. Societal transition means incremental steps in strategies, interventions, and practices toward achieving transformational goals. Ecosocial transitions and transformations are focused on strategies, interventions, and practices for changing different areas of conditions and functioning in order to improve the planet for all (Matthies & Närhi, 2016; Stamm et al., 2023). In other words, ecosocial transitions and transformations require sustainability solutions in working with people (social work)—individually or collectively—to tackle the triple planetary crisis (climate change, nature and biodiversity loss and pollution) to ensure a viable future on Earth.

Ecosocial work practices are designed and implemented at micro, mezzo, and macro level. Micro level ecosocial work includes therapeutical interventions such as animal-assisted social work, horticulture therapy, eco-therapy, and nature-based clinical social work interventions with individuals, families, and groups (e.g., see Gomes, Borga, and Luís and Larocque, this book). Veterinary social work that incorporates animal-assisted social work and social work promoting animal welfare can also be grouped as micro level ecosocial work (Holcombe et al., 2016; Loue & Vincent, 2021). For instance, Boetto (2016) provides an ecosocial work framework for integrating theory with micro level practice, including comprehensive strategies for clinical casework. Mezzo level ecosocial work occurs in the community—geographical places and with specific groups such as marginalized, disenfranchised, and oppressed groups. It can include, for example, upcycling projects, bio-farming support, and community-commons (shared duties, responsibilities, and benefits in managing ecosystem resources) in income generation activities for enhancing livelihoods and tackling poverty. Mezzo level ecosocial work has been mainly focused on addressing socio-ecological injustices at community level with interventions and practices geared toward community mobilization, engagement, empowerment, resilience building, and meeting community ecological needs such as conservation and restoration. For examples of such initiatives, refer to chapters by Skinner and Lynch and Marques et al. in this book. A collection of such interventions and practices can be found in the book *Ecosocial Work in Community Practice: Embracing Ecosocial Worldviews and Promoting Sustainability* (Rambaree et al., 2023). The macro level of ecosocial work includes national and global policy level strategies, interventions, and practices. As the chapter by Anand et al. in this book attests, national and global mobilization and advocacy for socio-ecological justice, climate actions, biodiversity conservation, responding to energy crisis, and ending pollution are some of the areas in which ecosocial workers are engaged. *The People's Global Summit on Co-building a New Eco-Social World: Leaving No One Behind* (The Summit Partners, 2022) provides some examples of ecosocial work at macro level.

Ecosocial work practices require a transformational shift by distancing from social work's association with ecologically unsustainable paradigms—such as the growth ideology maintained and promoted within certain of welfare systems in social work (Hirvilammi & Helne, 2014; Powers et al., 2019a; Smith-Carrier & Macarthur, 2023). Indeed, growth-oriented economic systems have been identified as one of the main drivers of planetary crisis including global human miseries (Kallis et al., 2020). For decades, social workers from all around the world have been tackling the psychosocial problems emanating from development that promotes prosperity for a few at the expense of the other within the growth ideology (Powers et al., 2019a). Economic systems therefore have to be reorientated toward a post-growth model for promoting sustainability not only in the economic sphere but also other spheres such as social and ecological.

Peeters (2012) posits that ecosocial work promotes a radical dematerialization of the economy, the just distribution of wealth through a new perspective on well-being for every world citizen while preserving our biophysical environment. In this sense, Boetto (2016, p. 52) presents a transformative ecosocial model having an ontological base representing "being" as an identity of self (values, beliefs, and attitudes of social work practitioners) in social work that is interconnected with nature (the biophysical environment). She argues that "if we are consciously aware of the interconnected relationship between humanity and the natural world, then it is more likely the natural environment will be integrated into practice" (ibid). In a similar vein, Peter (2012) opines that to reorient social work toward addressing the ongoing ecological crises would require transformation of the very foundations of professional education and to rethink and revise the goals and purposes of social work education. In particular, ecosocial work requires ecological consciousness and application of transformative pedagogies through ecosocial work teaching and learning (Smith-Carrier & Macarthur, 2023).

There is a misconception that ecological well-being only refers to the well-being of the natural environment or the planet and that it is opposed to economic well-being through the "degrowth" ideology. What ecosocial work promotes and calls for is a sustainable economic system that works for the well-being for all (including nonhuman beings) and not a growth ideology that operates for the benefits and profit of only for small group of elites. As unsustainable growth ideology cannot be tackled by social workers alone; social work education requires knowledge and skills development in cross-disciplinary and multidisciplinary approaches and inter-professional collaborative works. Social work has a long history of interprofessional teamwork, and over the years, social workers have demonstrated successfully their role in bringing different stakeholders together (Ambrose-Miller & Ashcroft, 2016).

Ecosocial Work: Transformative Teaching and Learning

Social constructivists argue that knowledge is created through interactions involving diversity in how individuals understand and interpret a phenomenon/problem. When diversity, mutuality, and intersubjectivity among educators and learners are promoted in education, transformative teaching and learning take place. Through education, transformational teaching and learning imply a profound change in existing beliefs, assumptions, values, understanding, discourses, and actions. This kind of education requires in-depth critical reflection through which educators and learners consciously construct meanings of phenomena, creating conditions for transformations in perspectives (Simsek, 2012).

Transformative teaching and learning can be considered as a metatheory with a collection of several theories of learning that are classified together based on their commonalities regarding human nature (Hoggan & Higgins, 2023). It is based on the basic nature of the human being as an active agent with rationality, critical reflectivity, and emancipatory power. It recognizes the need and necessity for human beings to be critical, reflective, and innovative in thinking and doing. It is a form of learning that avoids sets of fixed assumptions and expectations by being more inclusive, discriminating, open, reflective, and ready to enable change and create new meanings and understandings through creativity, imagination, problematizing, and questioning (Erisen et al., 2016; Jones, 2011; Witkin, 2014).

Transformative teaching and learning are heavily influenced by critical theory and pedagogies. Critical theory draws on a wide range of disciplines to critique society with the purpose of not only to understand it but also to advocate for social, economic, ecological, and political changes aimed at achieving justice, fairness, and freedom in society (Fleming, 2016; Fuchs, 2017). Morley et al. (2020) posit that education informed by critical theory is transformative by valorizing critique, fostering democratic citizenship, and developing critical consciousness capable of identifying power relations between dominant groups and the oppressed, linking the personal with the political and highlighting the need for change. Knowledge about the condition, approaches, and actions for change is cocreated through democratic dialog and interactions with different stakeholders and members of society. Therefore, a theory about transformation in society is abstracted from the everyday lived experiences of various community members. Where it does not, there is a need to change approaches in how knowledge construction takes place within the education system (Seal & Smith, 2021). Social work education systems are being criticized for being subordinate to the neoliberal market demand, whereby universities are aggressively turning more attention toward pulling in external funding rather than focusing on creating sustainable societal impacts through transformative teaching and learning (Morley et al., 2020). In this sense, it is important to emphasize that the central purpose of education is to create and foster a transformative learning environment for emancipation of both educators and learners.

Walker (2018) has validated four essential environments that foster transformative learning. The environments are categorized as different phases that do not necessarily occur in chronological order. The first environment is where a "disorienting dilemma" takes place. This is the environment where problems are identified and existing beliefs, values, assumptions, and knowledge are questioned and challenged. The dilemma arises during the process with a disconnect or discomfort between old inadequate frames of reference and the possibility offered by new ones (Fleming, 2016; Mezirow, 2009). The second environment is where "self-reflection" occurs (Walker, 2018). Self-reflection is the process through which own thoughts are examined by observing, reviewing, and analyzing cognitive and emotional contents that influence own knowledge construction (Savicki & Price, 2021). It is an environment that promotes better self-awareness on own assumptions and limitations within the process of knowledge construction. The third environment is referred to as "meaning perspective and critical discourse" where perspective and discourses are critically examined (Walker, 2018). Meaning perspective is an essential aspect in transformative learning. In this sense, educators have the responsibility to help learners to become aware and critical of their own and others' meaning perspectives by recognizing the assumptions and frames of reference through critical reflections on discourses. Discourses convey meanings and interpretations of problems and concepts from different perspectives based on up-to-date and relevant information and context (Mezirow, 2003). The final environment is called "acting," where actions are formulated based on new knowledge and behavior (Walker, 2018). Such an environment facilitates critiquing and appreciating others' viewpoints and discourses and creates conditions for a shift in perspectives. A shift in perspectives within transformative teaching and learning requires action to enact the altered perspective in the social world (Jones, 2016; Mezirow, 1997). The newly formed knowledge and behavior are tested and validated through actions accompanied by continuous critical reflections.

Particularly, social interactions in educational settings provide opportunities to explore, understand, and reflect autonomously as an individual and collectively with others. Such options allow everyone to articulate and argue about phenomena and life experiences using different perspectives. In this connection, educators need to promote positive attitude, equity, diversity, and justice in bringing learners' existing frames of reference into awareness and critically reflecting on how different frames restrict or facilitate thoughts, concept formations, and actions in transforming the world for the betterment of all (Mezirow, 2009; Witkin, 2014). In a sense, transformational learning requires challenging meaning structures in concept formation. Meaning structures are perspectives and schemes formed on our prior knowledge which is based on existing beliefs and values of individuals, depending on what they experience (Şahin & Doğantay, 2018).

A common understanding in ecosocial work is that ecological sustainability is an essential aspect in social work education and practice needs to be environmentally proactive. Given that ecosocial work is principally founded on the premise that humans are an integral part of a complex interlinked ecosystem, social work education therefore needs to be primarily grounded on concerns for ecological sustainability. However, the Introduction to this book emphasizes that the strong modernist and anthropocentric base of Western countries poses a strong challenge for social work to adopt an ecological perspective. In several social work discourses, ecosocial work has remained anthropocentric and anchored on a social sustainability domain, making little contribution to the area of ecological sustainability.

Anthropocentrism is based on the belief that value (based on human judgements) is human-centered and that all other beings are means to human ends (Kopnina et al., 2018). Anthropocentrism is often justified on the basis that humans are superior as they are the beings who can think, reflect, and take rational decisions not only for humans but also for the welfare of nonhumans. The argument for anthropocentric social work is influenced by a modernist, humanist, and human rights framework which defines "social" within social work as inclusively comprising of human beings (Bozalek & Pease, 2021; Thysell & Cuadra, 2022). Bell (2020) states that non-anthropocentric social work is seen as being too radical by some thinkers.

Given the urgency related to the triple planetary crisis (climate change, loss of biodiversity, and pollution), ecosocial work teaching and learning (including practices) *must* be radical and transformative (Powers et al., 2019a). In this book, Lievens presents an interesting chapter on the need for an ecocentric worldview in social work education in tackling global challenges and related psychosocial problems such as eco-anxiety. The context of growing crisis and urgency requires a social work transformational shift away from meeting only human needs and pursuit of only human rights as its raison d'être (Woods & Hölscher, 2021). In this sense, the International Federation of Social Workers (IFSW) urges for the inclusion of ecosocial work perspectives in social work by promoting a holistic rights framework and recognition of the interconnectedness of all life in our ecosystem as integral in the guiding ethics of social workers (IFSW, 2022).

In particular, ecosocial work is much more than adding ecological perspectives to prevailing social work approaches, and it requires a shift from and a critical challenge to the anthropocentric and modernist assumptions through transformative learning and actions (Närhi & Matthies, 2016; Smith-Carrier & Macarthur, 2023). In the same sense, in this book, Mattocks et al. argue for reenvisioning the social work curriculum from a transformative philosophical perspective. Within transformative philosophy, the nature of ecosocial work knowledge is not only created by and for humans but with significant contributions from integrity and creativity through holistic learning from both humans and nonhumans (Bell, 2020). The teaching and learning of ecosocial work in higher education institutions requires transformative pedagogies targeted at graduate attributes such as active global citizenship, multidisciplinary knowledge, broader system thinking, and sustainable solution-focused problem-solving.

Graduate attributes are "the skills, knowledge and abilities of university graduates, beyond disciplinary content knowledge, which are applicable to a range of contexts" (Barrie, as quoted in Wald & Harland, 2019, p.361). Within the discipline of social work, graduates are expected to have a range of capabilities to integrate and adapt to different contexts and practice situations, as part of work-readiness (Howard et al., 2015). Given that sustainability and sustainable development are a global concern and require international and global commitments, social work graduates need to promote active global citizenship as one of the key graduate attributes. Active global citizenship refers to the application of one's "moral responsibility" based on a cosmopolitan mindset or a feeling of being at home in the world, with a willingness to take on obligations and responsibilities to meet the needs and welfare of others in one's global community (Lough & Mcbride, 2014). Within the context of globalization and global sustainability challenges, social work is uniquely positioned to promote global (and national) sustainability curricula that promote active engagement of the educators and learners in the global arena.

Another key graduate attribute of ecosocial work needs to be based on multidisciplinary (or rather pluri-disciplinary) knowledge. As an applied social science, social work requires knowledge from other subject areas or disciplines such as sociology, psychology, economics, politics, and geography. Given that sustainability and sustainable development cut across different disciplines, social work learners need to have opportunities for multidisciplinary education (see Whelan and Hurley, Forde, and Ó Suilleabháin in this book). Multidisciplinary education will allow them to develop broader perspectives that can promote broader system thinking. Sustainability and sustainable development challenges are immensely complex, complicated, and intricate problems, interlinked in processes of rapid, dynamic change that are fundamentally systemic (Voulvoulis et al., 2022). Broader system thinking is therefore essential in sustainable solution-focused problem-solving. Social work has a long history of using a problem-solving perspective. In particular, solution-focused approaches to problem-solving have commonly been used in clinical social work. Within the teaching and learning of ecosocial work, a sustainable solution focus is, therefore, a must, given the complexities and complications surrounding sustainability and sustainable development challenges.

Conclusion

All in all, ecosocial work requires teachers, students, and others to navigate several complex and contested concepts through critical reflections for meaning-making and enabling positive changes in society in tackling one of the most significant challenges for the planet Earth and the survival of all beings. In this sense, ecosocial work embraces and associates with several other complex and contested concepts such as development, justice, and sustainability.

In particular, ecosocial work teaching and learning in higher education institutions requires transformative pedagogy in promoting graduate attributes such as active global citizenship, multidisciplinary knowledge, broader system thinking, and sustainable solution-focused problem-solving. From a transformational learning perspective, conceptualization of terms such as ecosocial work needs to be critically examined by looking at different individuals' and groups' values, beliefs, assumptions, and discourses. In addition, the conceptualization of ecosocial work requires the promotion of the self-reflections of everyone—teachers, students, practitioners, policymakers, service users, etc. Moreover, a "good" and "broad" conceptualization of ecosocial work is based on considering and critically analyzing the different meanings and discourses gathered from various individuals and groups. Finally, ecosocial work must focus on bringing a shift in perspective, behavior, action, and practices.

References

Ambrose-Miller, W., & Ashcroft, R. (2016). Challenges faced by social workers as members of Interprofessional Collaborative Health Care Teams. *Health and Social Work, 41*(2), 101–109. Available at: https://doi.org/10.1093/hsw/hlw006

Bell, K. (2014). *Achieving environmental justice a cross-national analysis*. Polity Press.

Bell, K. (2020). A philosophy of social work beyond the anthropocene. In V. Bozalek & B. Pease (Eds.), *Post-anthropocentric social work* (pp. 58–67). Routledge. https://doi.org/10.4324/9780429329982-6

Beltrán, R., Hacker, A., & Begun, S. (2016). Environmental justice is a social justice issue: Incorporating environmental justice into social work practice curricula. *Journal of Social Work Education, 52*(4), 493–502. https://doi.org/10.1080/10437797.2016.1215277

Besthorn, F. H. (2012). Radical equalitarian ecological justice a social work call to action. In *Environmental social work* (pp. 31–45). Wichita State University. https://doi.org/10.4324/9780203095300

Besthorn, F. H. (2016). Deep ecology ' s contributions to social work: A ten - year retrospective. *International Journal of Social Welfare, 21*(3), 248–259. https://doi.org/10.1111/j.1468-2397.2011.00850.x

Boetto, H. (2016). Developing ecological social work for micro level practice. In J. McKinnon, & M. Alston (Eds.), *Ecological social work: Towards sustainability (1st ed., pp. 59–77). Palgrave Macmillan.*

Boetto, H. (2017). A transformative eco-social model: Challenging modernist assumptions in social work. *British Journal of Social Work, 47(1), 48–67.* https://doi.org/10.1093/bjsw/bcw149

Boetto, H. et al. (2020). Raising awareness of transformative ecosocial work: Participatory action research with Australian practitioners. *International Journal of Social Welfare, 29(4), 300–309.* https://doi.org/10.1111/ijsw.12445

Boetto, H., Närhi, K., & Bowles, W. (2022). Creating 'Communities of Practice' to enhance ecosocial work: A comparison between Finland and Australia. *British Journal of Social Work, 52*(8), 4815–4835. https://doi.org/10.1093/bjsw/bcac092

Boström, M., Andersson, E., Berg, M., Gustafsson, K., Gustavsson, E., Hysing, E., et al. (2018). Conditions for transformative learning for sustainable development: A theoretical review and approach. *Sustainability (Switzerland), 10*(12). https://doi.org/10.3390/su10124479

Bozalek, V., & Pease, B. (2021). Towards post-anthropocentric social work. In V. Bozaleck & B. Pease (Eds.), *Post-anthropocentric social work: Critical posthuman and new materialist perspectives* (pp. 1–17). Routledge.

Buckingham, S., & Kina, V. J. (2015). Sustainable Development and Social Work. In *International Encyclopedia of the Social & Behavioral Sciences: Second Edition (pp. 817–822). Brunel University, Uxbridge, United Kingdom.* https://doi.org/10.1016/B978-0-08-097086-8.28095-1

Campbell, S. M. (2016). The concept of well-being. In G. Fletcher (Ed.), *The Routledge handbook of the philosophy of well-being* (pp. 402–413). Routledge.

Cordoba, P. S., & Bando, L. (2022). Social work and the sustainable development goals: A student placement model. *Australian Social Work, 75*(4), 519–526. https://doi.org/10.1080/0312407X.2021.1947335

Davis, J. et al. (2019). Anthropocene, Capitalocene, … Plantationocene?: A manifesto for ecological justice in an age of global crises. *Geography Compass, 13(5), 1–15.* https://doi.org/10.1111/gec3.12438

Dominelli, L. (2012). Green social work. In *International Encyclopedia of the social & behavioral sciences* (2nd ed.). Polity. https://doi.org/10.1016/B978-0-08-097086-8.28037-9

Dominelli, L., Nikku, B. R., & Ku, H. B. (2018). The routledge handbook of green social work. In L. Dominelli, B. R. Nikku, & H. B. Ku (Eds.), *The Routledge handbook of green social work.* Routledge Taylor & Francis Group. https://doi.org/10.4324/9781315183213

Drolet, J., Wu, H., Taylor, M., & Dennehy, A. (2015). Social work and sustainable social development: Teaching and learning strategies for 'Green Social Work' curriculum. *Social Work Education, 34*(5), 528–543. https://doi.org/10.1080/02615479.2015.1065808

Du Pisani, J. A. (2006). Sustainable development – Historical roots of the concept. *Environmental Sciences, 3*(2), 83–96. https://doi.org/10.1080/15693430600688831

Dylan, A. (2013). Environmental sustainability, sustainable development, and social work. In J. Mel Gray, Coates, & T. Hetherington (Eds.), *Environmental social work* (pp. 62–87). Routledge. https://doi.org/10.4324/9780203095300-3

Ebbesen, S., & Gregoric, P. (2022). Forms of representation in the Aristotelian tradition. Volume three: Concept formation. In C. T. Thörnqvist & J. Toivanen (Eds.), *Forms of representation in the Aristotelian tradition. Volume three: Concept formation* (pp. 1–33). Brill. https://doi.org/10.1163/9789004506114

Engeström, Y. (2020). Concept formation in the wild: Towards a research agen-da. *Education et Didactique, 14*(2), 99–113. https://doi.org/10.4000/educationdidactique.6816

Erisen, Y., Çeliköz, N., & Şahin, M. (2016). The transformational learning of three adult academicians. Mediterranean. *Journal of Humanities, 6*(1), 299–299. https://doi.org/10.13114/mjh.2016119306

Ferreira, S. B. (2010). Eco-spiritual social work as a precondition for social development. *Ethics and Social Welfare, 4*(1), 3–23. https://doi.org/10.1080/17496531003607891

Fleming, T. (2016). Rethinking the critical theory influences on transformative learning: In dialogue with Honneth. In D. Andritsakou & E. Kostara (Eds.), *The role, nature and difficulties of dialogue in transformative learning* (pp. 88–96). ESREA & Hellenic Adult Education Association.

Forbes, R., & Smith, K. (2023). Introduction: Foundations of ecosocial work practice. In *Ecosocial work: Environmental practice and advocacy* (pp. ix–xvi).

Fuchs, C. (2017). Critical social theory and sustainable development: The role of class, capitalism and domination in a dialectical analysis of Un/Sustainability. *Sustainable Development, 25(5), 443–458.* https://doi.org/10.1002/sd.1673

Gamble, D. N. (2012). Well-being in a globalized world: Does social work know how to make it happen? *Journal of Social Work Education, 48*(4 SPL. ISS), 669–689. https://doi.org/10.5175/JSWE.2012.201100125

Gerring, J. (1999). What makes a concept good? A criterial framework for un-derstanding concept formation in the social sciences. *Polity, 31*(3), 357–393.

Gray, M., & Coates, J. (2012). Environmental ethics for social work: Social work's responsibility to the non-human world. *International Journal of Social Welfare, 21*(3), 239–247. https://doi.org/10.1111/j.1468-2397.2011.00852.x

Gray, M., Coates, J., & Hetherington, T. (2012). *Environmental social work. Environmental social work* (Vol. 9780203095). School of Humanities and Social Science, University of NewcastleNSW, Australia. https://doi.org/10.4324/9780203095300

Grossmann, K. et al. (2021). *From sustainable development to social-ecological justice: Addressing taboos and naturalizations in order to shift perspective, 0(0), 1–23.* https://doi.org/10.1177/25148486211029427

Grouzet, F., & Lee, E. (2014). Ecological well-being. In A. C. Michalos (Ed.), *Encyclopedia of quality of life and well-being research* (pp. 1784–1787). Springer.

Hirvilammi, T., & Helne, T. (2014). Changing paradigms: A sketch for sustainable wellbeing and ecosocial policy. *Sustainability (Switzerland), 6*(4), 2160–2175. https://doi.org/10.3390/su6042160

Hoggan, C., & Higgins, K. (2023). Understanding transformative learning through metatheory, new directions for adult and continuing. *Education, 2023*(177). https://doi.org/10.1002/ace.20475

Holcombe, T. M., Strand, E. B., Nugent, W. R., & Ng, Z. Y. (2016). Veterinary social work: Practice within veterinary settings. *Journal of Human Behavior in the Social Environment, 26*(1), 69–80. https://doi.org/10.1080/10911359.2015.1059170

Howard, A., Johnston, L., & Agllias, K. (2015). Ready or not: Workplace perspectives on work-readiness indicators in social work graduates. *Advances in Social Work, 17*(2), 7. https://search.informit.com.au/documentSummary%0Adn=609917178610392%0Ares=IELHSS%0Ahttps://lens.org/188-874-878-080-762

IASSW/ICSW/IFSW. (2020). *Global agenda for social work and social development October 2020.* https://www.ifsw.org/wp-content/uploads/2020/11/GlobalAgenda-Press-Release.pdf

IFSW. (2022). *The people's charter for an eco-social world.* https://newecosocialworld.com/the-peoples-charter-for-an-eco-social-world/. Accessed on 17 Mar 2023.

Isham, A., Elf, P., & Jackson, T. (2022). Self-transcendent experiences as promoters of ecological wellbeing? Exploration of the evidence and hypotheses to be tested. *Frontiers in Psychology, 13*(November), 1–20. https://doi.org/10.3389/fpsyg.2022.1051478

IUCN-UNEP-WWF. (1980). *World conservation strategy: Living resource conservation for sustainable development*. Geneva.

Jayasooria, D. (2016). Sustainable development goals and social work: Opportunities and challenges for social work practice in Malaysia. *Journal of Human Rights and Social Work, 1*(1), 19–29. https://doi.org/10.1007/s41134-016-0007-y

Jones, P. (2011). Transformative learning, ecopedagogy and learning ecology: Pathways for social work education. In *Eco-social justice: Issues, challenges and ways.*

Jones, P. (2016). *Shift happens: Transformative learning in social work education*, 355.
Kallis, G. et al. (2020). *The case for degrowth. Polity Press.*
Kopnina, H., Washington, H., Taylor, B., & Piccolo, J. (2018). Anthropocentrism: More than just a misunderstood problem. *Journal of Agricultural and Environmental Ethics, 31*(1), 109–127. https://doi.org/10.1007/s10806-018-9711-1
Kortetmäki, T. (2016). Is broad the new deep in environmental ethics? A comparison of broad ecological justice and deep ecology. *Ethics and the Environment, 21(1), 89–108.* https://doi.org/10.2979/ethicsenviro.21.1.04
Lalumera, E. (2014). On the explanatory value of the concept-conception distinc-tion. *Rivista Italiana di Filosofia del Linguaggio, 8*(2), 73–81. https://doi.org/10.4396/sfl1316
Lin, J., Hiltebrand, G., Stoltz, A., & Rappeport, A. (2021). Environmental justice must include the rights of all species to life and respect: Integrating indigenous knowledge into education. *International Studies in Sociology of Education, 30*(1–2), 93–112. https://doi.org/10.1080/09620214.2020.1854827
Loue, S., & Vincent, A. L. (2021). Directions in veterinary social work. *Revista Romaneasca pentru Educatie Multidimensionala, 13*(4), 633–651. https://doi.org/10.18662/rrem/13.4/502
Lough, B. J., & Mcbride, A. M. (2014). Navigating the boundaries of active global citizenship. *Transactions of the Institute of British Geographers, 39*(3), 457–469. https://doi.org/10.1111/tran.12035
Matthies, A. L., & Närhi, K. (2016). The ecosocial transition of societies: The contribution of social work and social policy. In *The Ecosocial transition of societies: The contribution of social work and social policy*. https://doi.org/10.4324/9781315615912
Matutini, E., Busacca, M., & Da Roit, B. (2023). Ready, steady, go? Obstacles to the spread of eco-social work approaches: An Italian case study. *Sustainability (Switzerland), 15*(4). https://doi.org/10.3390/su15043050
Mckinnon, J., & Alston, M. (2016). In J & A. M. McKinnon (Eds.), *Ecological social work towards sustainability*. Palgrave.
Meadowcroft, J., & Fiorino, J. D. (2017). Conceptual innovation in environmental policy. In J. Meadowcroft & J. D. Fiorino (Eds.), *Conceptual innovation in environmental policy*. Massachusetts Institute of Technology. https://doi.org/10.1080/23251042.2016.1259865
Mensah, J. (2017). Sustainable development: Meaning, history, principles, pillars, and implications for human action: Literature review Justice Mensah | Sustainable development: Meaning, history, principles, pillars, and implications for human action: Literature review PUBLIC INTEREST STATEMENT. https://doi.org/10.1080/23311886.2019.1653531
Mezirow, J. (1997). Transformative learning: Theory to practice. *New Directions for Adult and Continuing education, 74*, 5–12.
Mezirow, J. (2003). Transformative learning as discourse. *Journal of Transformative Education, 1*(1), 58–63. https://doi.org/10.1177/1541344603252172
Mezirow, J. (2009). An overview of transformative leaning. In *Contemporary theories of learning: Learning theorists in their own words* (pp. 90–105).
Morley, C., Ablett, P., & Noble, C. (2020). Introduction: The imperative of critical pedagogies for social work. In C. Morley, P. Ablett, C. Noble, & S. Cowden (Eds.), *The Routledge handbook for critical social work* (pp. 1–16). Routledge. https://doi.org/10.4324/9781351264402
Närhi, K., & Matthies, A.-L. (2016). The ecosocial approach in social work as a framework for structural social work. *International Social Work, 47*(September 2016), 1–13. https://doi.org/10.1177/0020872816644663
Närhi, K. and Matthies, A.-L.A.-L.L. (2018) . The ecosocial approach in social work as a framework for structural social work. *International Social Work, 61(4), 490–502.* https://doi.org/10.1177/0020872816644663
Nhapi, T., & Pinto, C. (2023). Embedding SDGs in social work education: Insights from Zimbabwe and Portugal BSW and MSW programs. *Social Work Education, 00*(00), 1–18. https://doi.org/10.1080/02615479.2023.2221284
Nylén, E.-J. (2023). *Theorizing ideological steering*. Tampere University.

Peter, J. (2012). Transforming the curriculum social work education and ecological consciousness. In J. Mel Gray, Coates, & T. Hetherington (Eds.), *Environmental social work* (pp. 213–230). Routledge. https://doi.org/10.4324/9780203095300

Pope, K., Bonatti, M., & Sieber, S. (2021). The what, who and how of socio-ecological justice: Tailoring a new justice model for earth system law. *Earth System Governance, 10, 100124.* https://doi.org/10.1016/j.esg.2021.100124

Powers, M. C. F. M. C. F., Rambaree, K., & Peeters, J. (2019a). Degrowth for transformational alternatives as radical social work practice. *Critical and Radical Social Work, 7*(3), 417–433. https://doi.org/10.1332/204986019X15688881497178

Powers, M., Schmitz, C., & Beckwith Moritz, M. (2019b). Preparing social workers for ecosocial work practice and community building. *Journal of Community Practice, 27*(3–4), 446–459. https://doi.org/10.1080/10705422.2019.1657217

Purvis, B., Mao, Y., & Robinson, D. (2019). Three pillars of sustainability: In search of conceptual origins. *Sustainability Science, 14*(3), 681–695. https://doi.org/10.1007/s11625-018-0627-5

Rambaree, K. (2020). Environmental justice in the case of the chagos marine protected area: Implications for international social work. *Sustainability (Switzerland), 12*(20). https://doi.org/10.3390/su12208349

Rambaree, K., Powers, M. C. F. M. C. F., & Smith, R. J. R. J. (2019). Ecosocial work and social change in community practice. *Journal of Community Practice, 27*(3–4), 205–212. https://doi.org/10.1080/10705422.2019.1660516

Rambaree, K., Powers, M. C. F., & Smith, R. J. (2023). *Ecosocial work and social change in community practice.* Routledge Taylor & Francis Group.

Ramsay, S., & Boddy, J. (2017). Environmental social work: A concept analysis. *British Journal of Social Work, 47*(1), 68–86. https://doi.org/10.1093/bjsw/bcw078

Ranta-Tyrkkö, S., & Närhi, K. (2021). Striving to strengthen the ecosocial framework in social work in Finland. *Community Development Journal, 56*(4), 608–625. https://doi.org/10.1093/cdj/bsab030

Ross, D., Bennett, B., & Menyweather, N. (2020). Towards a critical posthumanist social work. In V. Bozalek & B. Pease (Eds.), *Post-anthropocentric social work* (pp. 175–186). Routledge. https://doi.org/10.4324/9780429329982-17

Ruggeri, K., Garcia-Garzon, E., Maguire, Á., Matz, S., & Huppert, F. (2020). *Well-being is more than happiness and life satisfaction: A multidimensional analysis of 21 countries.* Health and Quality of Life OutcomesHealth and Quality of Life Outcomes [revista en Internet] 2020 [acceso 4 de julio de 2021]; 18(1):1–16. Health and Quality of Life Outcomes, 1–16. https://hqlo.biomedcentral.com/track/pdf/10.1186/s12955-020-01423-y.pdf

Şahin, M., & Doğantay, H. (2018). Critical thinking and transformative learning. *Journal of Innovation in Psychology, Education and Didactics, 22*(1), 103–114.

Salonen, A. O., & Konkka, J. (2015). An Ecosocial approach to well-being: A solution to the wicked problems in the Era of Anthropocene. *Foro de Educación, 13(19), 19–34.* https://doi.org/10.14516/fde.2015.013.019.002

Savicki, V., & Price, M. V. (2021). Re fl ection in transformative learning: The challenge of measurement. *Journal of Transformative Education, 19*(4), 366–382. https://doi.org/10.1177/15413446211045161

Seal, M., & Smith, A. (2021). An introduction to critical pedagogy. In J. Jarvis & K. Smith (Eds.), *Critical pedagogy in higher education* (pp. 1–20). Criitical Publishing. https://www.iso.org/news/ref2271.html

Shatberashvili, N., & Sadzaglishvili, S. (2022). SDG-focused social work education in Georgia. *Social Work Education, 41*(7), 1507–1524. https://doi.org/10.1080/02615479.2022.2103111

Simsek, A. (2012). Transformational learning. In N. M. Seel (Ed.), *Encyclopedia of the sciences of learning* (pp. 3341–3344). Springer. https://doi.org/10.1007/978-1-4419-1428-6_373

Smith-Carrier, T., & Macarthur, J. (2023). The state of eco-social work training in Canada : Transformative praxis for climate constrained futures. *International Social Work, 00*(0), 1–17. https://doi.org/10.1177/00208728231196366

Solonchak, T., & Pesina, S. (2015). Concept and its structure. *Procedia - Social and Behavioral Sciences, 192*(3519), 352–358. https://doi.org/10.1016/j.sbspro.2015.06.050

Spindler, E. (2013). The history of sustainability the origins and effects of a popular concept. In I. Jenkins & R. Schröder (Eds.), *Sustainability in tourism: A multi-disciplinary approach* (pp. 9–31). Springer Gabler. https://doi.org/10.1007/978-3-8349-7043-5

Stamm, I., Ranta-Tyrkkö, S., Matthies, A. L., & Närhi, K. (2023). THE ECOSOCIAL PARADIGM IN SOCIAL WORK: Striving for planetary well-being. In M. Elo, J. Hytönen, S. Karkulehto, T. Kortetmäki, J. S. Kotiaho, M. Puurtinen, & M. Salo (Eds.), *Interdisciplinary perspectives on planetary well-being* (pp. 177–188). https://doi.org/10.4324/9781003334002-18

Stiglitz, J. (1998). *Towards a new paradigm for development*. United Nations Conference on trade and development. Geneva.

Taylor, S. D., & Vosgerau, G. (2021). The explanatory role of concepts. *Erkenntnis, 86*(5), 1045–1070. https://doi.org/10.1007/s10670-019-00143-0

The Summit Partners. (2022). *Co-building a new eco-social world: Leaving no one behind. Contribution book*. IFSW.

Thysell, M., & Cuadra, C. B. (2022). Imagining the ecosocial within social work. *International Journal of Social Welfare*, 1–18. https://doi.org/10.1111/ijsw.12571

Truell, B. R., & Crompton, S. (2020). *To the top of the cliff how social work changed with COVID-19*. IFSW.

United Nations (2015). *Transforming Our World The 2030 Agenda for Sustainable Development. Resolution Adopted by the General Assembly on 25 September 2015, 42809, 1–13*

UNRISD. (2021). *A new eco-social contract: Vital to deliver the 2030 agenda for sustainable development*.

UNRISD. (2022). *Crises of inequality: Shifting power for a new eco-social contract. Nature Structural & Molecular Biology* (Vol. 14). UNRISD.

Voulvoulis, N., Giakoumis, T., Hunt, C., Kioupi, V., Petrou, N., Souliotis, I., et al. (2022). Systems thinking as a paradigm shift for sustainability transformation. *Global Environmental Change, 75*(May). https://doi.org/10.1016/j.gloenvcha.2022.102544

Yaka, Ö. (2019). Rethinking justice: Struggles for environmental commons and the notion of socio-ecological Justice. *Antipode, 51(1), 353–372*. https://doi.org/10.1111/anti.12422

Yaka, Ö. (2020). Justice as relationality: Socio-ecological justice in the context of anti-hydropower movements in Turkey. *DIE ERDE, 151(2–3), 167–180. Available at:* https://doi.org/10.12854/erde-2020-481

Wald, N., & Harland, T. (2019). Graduate attributes frameworks or powerful knowledge? *Journal of Higher Education Policy and Management, 41*(4), 361–374. https://doi.org/10.1080/1360080X.2019.1613310

Walker, S. L. (2018). Development and validation of an instrument for assessing transformative learning: The Transformative Learning Environments Survey (TLES). *Journal of Transformative Learning, 5*(1), 23–46. https://jotl.uco.edu/index.php/jotl/issue/view/23/showToc

Wang, P., & Altanbulag, A. (2022). A concern for eco-social sustainability: Background, concept, values, and perspectives of eco-social work. *Cogent Social Sciences, 8*(1). https://doi.org/10.1080/23311886.2022.2093035

Witkin, S. L. (2014). Change and deeper change: Transforming social work education. *Journal of Social Work Education, 50*(4), 587–598. https://doi.org/10.1080/10437797.2014.947897

Woods, G., & Hölscher, D. (2021). Return of the posthuman: Developing indigenist perspectives for social work at a time of environmental crisis. In: Bozalek V., & Pease, B. (Eds.), *Post-anthropocentric social work: critical posthuman and new materialist perspectives (pp 121–133). Routledge.*

Zapf, M. (2010). Social work and the environment: Understanding people and place. *Critical Social Work, 11*(3), 366–367. https://doi.org/10.1080/0312407x.2010.500654

Chapter 2: How to Approach the Materiality of the Ecosocial Transition in Social Work Education

Satu Ranta-Tyrkkö

Introduction

When social work is commonly perceived to consist of activities that take place in the social realm between people, these activities are in multiple ways material. The materiality of social work is tangible, for example, in needs-based forms of material support for people whose other societal safety nets have failed. When and if not a starting point, social work usually aspires to achieve a holistic and balanced understanding of the issue or phenomena behind the need for social work intervention, including its causalities and complexity. Moreover, in addition to the core social work principle to secure the dignity and worth of every person (e.g., NASW, 2021), ensuring service users' basic material needs is usually a precondition for more in-depth forms of work, which have greater chances of success when the participants' time and energy are not consumed to mere survival. But if historically social work has been commonly associated with help in situations of material and other forms of deprivation, nowadays the apparent cause of intervention can sometimes be excess, as in cases of compulsive hoarding or inability to get rid of anything, even trash. Amidst current material excess typical especially to consumer societies, it is possible to end up with amounts of stuff the managing of which is beyond one's capacity, stuff here meaning things that we have and are unwilling or unbale to dispose of even when they complicate and compromise our lives (Boscagli, 2014). At any rate, the material and the social influence constitute each other: materiality, including nonhuman organisms, is part of society, and human existence is tied to material practices and arrangements (Schatzki, 2010).

While dire material deprivation continues to be a serious problem falling on the most disadvantaged people, in this chapter, the focus is on the problem of overt and

S. Ranta-Tyrkkö (✉)
Department of Social Sciences and Philosophy, University of Jyväskylä, Jyväskylä, Finland
e-mail: satu.ranta-tyrkko@jyu.fi

C. Forde et al. (eds.), *Teaching and Learning in Ecosocial Work*,
https://doi.org/10.1007/978-3-031-58708-5_3

taken for granted material excess that features contemporary consumer societies and consumerist lifestyles. From this starting point, this chapter contemplates issues of matter and materiality in and for social work. In so doing, its point of departure is that while matter is self-evidently "what things are made of" (Ingold, 2011, p. 16), and life, including human life, "itself undergoes continual generation in currents of materials" (ibid., p. 31.), in closer examination, materiality as a ubiquitous aspect of human and nonhuman existence and experience is, despite its obviousness, somewhat hard to grasp. Nevertheless, understanding and addressing the current environmental challenges requires engaging with their material underpinnings, as well as with the very materiality of our own practices and being.

This chapter is built on the ecosocial (ecological/environmental/green social work) premise that humans are part of nature and dependent on the delicate balance of the ecosystems and the climatic system of the Earth (Dominelli, 2012; McKinnon & Alston, 2016; Boetto, 2017; Närhi & Matthies, 2018;) and that societies overall need to go through a profound and exhaustive sustainability transition. This includes the reality that there is no way out of the fact that societies and individuals must construct new material systems to produce the essentials—food, energy, housing, and transportation (e.g., Schlosberg & Craven, 2019, 148; BIOS, 2019). At the minimum, these systems must be far less harmful than present ones and ideally be climate cooling and life (in its multitudes) preserving by their nature (Ibid.)

Renewing understandings of materiality, sufficiency, and affluence brings along epistemological and ontological processing. While this kind of thinking and doing is already taking place in various forms and layers of the society, its spread and establishment is complicated by the fact that it challenges deep etched common "truths" and taken for granted ways to perceive, think of, and speak about the world. These include, for example, at the course of history hierarchy-laden binaries of nature/culture, animal/human, body/soul, material/spiritual, woman/man, or inanimate/animate, which potentially shakes up not only how we see the world but ourselves and our professional practice (Wilson, 2021, p. 41).

This chapter is particularly interested in the material underpinnings of consumerist relationships to "stuff" as one of the key fronts of the unresolved climate and environmental emergency, providing thus only a limited view on the multifaceted conceptual discussion on materiality. This framing notwithstanding, the chapter argues that social work at large, and ecosocial work by default, must reexamine and reformulate how it relates with and is embedded in the material world. This, in my view, is not only professional but a deeply personal and political challenge especially in affluent countries and population groups. Living in the world, we are part of its materiality, among others, by being entangled in exploitative and destructive relationships with the nature (Blühdorn & Deflorian, 2021). To learn from the manifold meanings of stuff, the easy and concrete way ahead is to start from ourselves, for example, along the lines of the exercises proposed later in this chapter.

As writing and thinking always takes place from some place and material foundation, here are a few words about my own background. My interest in the topic has grown sporadically both in- and outside the academy. Although from an economically modest background, as a citizen of an affluent country, I have been able to lead my life relatively securely, bolstered with many publicly funded services. Against this background, my opportunities to spend time in the Gandhian (after Mohandas Gandhi) rural development organization Gravis in northwest India in 1990s, and with the cultural action organization Natya Chetana in eastern India during 2000s, have been formative experiences. The Gandhians taught me that the essence of social work is "simple living and high thinking," which could be a motto for ecosocial work, as is also the other Gandhian dictum, "the Earth can sustain everyone's need, but not anyone's greed." Moreover, I had to travel to India to realize that I'm a habituated consumer, the ability to be a customer making consumption choices providing me a sense of being a doer, brushing up my mood and sense of autonomy amidst otherwise a highly collective mode of living (Ranta-Tyrkkö, 2010, p. 53–57). Working as a social work educator and researcher since early 2000s, I have gradually made my way to ecological and ecosocial issues in social work, realizing with hindsight that materiality, in different guises, has been one of the topics that has long intrigued me, although the scope to go into the matter has been scarce in the context of the tightly packed social work curriculum.

This chapter draws specifically from the learning gained as part of my postdoctoral research project on the consequences of the mining industry for disadvantaged groups and their resultant social work needs in northern Finland and eastern India and as teacher at the University of Jyväskylä social work curricula course Social Work in Ecosocial Transition, as well as in numerous sustainability trainings (Polku 2.0 project, 2021–2023). Moreover, several predecessors and contemporaries have developed useful exercises, which I unhesitatingly borrow and re-introduce. This chapter pays homage especially to the US-based professor emeritus of social work Fred Besthorn, utilizing exercises that he has developed, tested, and disseminated for others to try two decades prior to the writing of this chapter (Besthorn, 2003), as well as to the work of Finnish theologian Panu Pihkala on climate and environmental emotions (e.g., Pihkala, 2020a, b, 2022).

In what follows, I first provide a brief overview on diverse materialisms, turning thereafter to discuss our everyday excess and consumerist relationship to the world as one of the key underlying causes of the global environmental crises at hand. As consumerism is largely about stuff, I start from our personal relationship to stuff, thereafter proceeding to the need to work also the emotional and political side of things.

Multiple Materialisms and the Material Turn

From earlier to later times, humans have aspired to make sense of the world and their own place in it, resulting in diverse explanations about the material, biophysical world, and its beings, humans included. There are and have been, for example, relational indigenous cosmologies which do not differentiate entities into animate or non-animate (Ingold, 2011, p. 67), or traditions, including Chinese medicine and

Ayurveda, which perceive the world, including humans, as a dynamic energetic whole (e.g., Svoboda & Lade, 1995).

Theories of matter are commonly divided into old/ancient/premodern and new/modern. One old form of materialism is the version of atomism (with various contemporaries; see Berryman (2022)) developed by ancient Greek philosophers Democritus and Epicurus, which posits that everything in the world consists of atoms “careening and colliding in the void” (Gamble et al., 2019, p. 113). Later, the modern scientific and rationalist worldview has been associated with a mechanistic vision of the world that functions according to “the laws of nature” (ibid., p. 115), as well as tight dualism between matter and spirit, and thus body and mind, a view explicated in the modernizing seventeenth-century Europe by the philosopher and mathematician René Descartes. This “cartesian” separation of mind and matter has been influential in science ever since. Furthermore, it has often been interpreted in a hierarchical manner, giving higher value to qualities assumed more rational and civilized and lesser value to qualities claimed more corporeal, mundane, and animallike, the interests behind these judgements being often politically, patriarchally, anthropocentrically, and so on, vested.

From the late nineteenth century onward, historical materialist theories, first and famously outlined by Karl Marx and Friedrich Engels, have elaborated on how human history and social relations reflect the material foundations and productive forces of societies. Wolff and Cullenberg (1986, p. 128) argue that unlike in earlier class theories, Marx’s originality lays in his definition of class as the process of producing and distributing surplus labor as separate from the related matters of property and distribution of power. Since then, a diverse cadre of post- and neo-Marxist scholars have developed a web of theories that reflects the original formulations of Marx and Engels to varying degrees (ibid., p. 130), paying attention how production and reproduction of life has been shaped by the struggle of successive social classes to mold society in their own interests. Of late, for example, environmental historian Jason W. Moore (2016, p. 11) has pointed that capitalism has throughout its existence relied on its capacity to extract cheap nature, especially cheap energy, raw materials, food, and human life. Moreover, it has been a way to organize nature, and humans as part of it, into “a world-ecology that joins the accumulation of capital, the pursuit of power, and the co-production of nature in successive historical configurations” (ibid., p. 7).

Although matter and materiality have thus been long thought of, a more recent realization is that in a sense, both premodern and modern materialisms have neglected matter by conceiving it “as essentially passive” (Gamble et al., 2019, p. 116) while presuming humans as uniquely external to nature among beings (Lettow, 2017, p. 113). Social sciences and humanities have taken this differentiation between society and nature largely for granted, treating materiality and biophysicality mainly as a backdrop of human activities and focusing extensively on relations and interactions between people while ignoring the role materiality plays in these relations (e.g., Schatzki, 2010, 126–127). Inherently a product of modernity, also academic professional social work has been until very recently axiomatically human-centric and generally a prime example of modern faith in progress and development (McDonald, 2006). An emblematic and often mentioned example of social work's anthropocentrism is that even the person-in-environment (PIE) framework that strives toward holistic understanding of individual client situations has been commonly confined to social, cultural, and economic environments but excluded the natural environment (Besthorn, 2003).

After the turn of the millennium, articulations of another kind of "new" materialism—discussed under umbrella terms such new materialism, postmaterialism, and posthumanism (see Lievens in this volume, as well as Bozalek & Pease, 2021)—have started to emerge. These overlapping discussions challenge the heretofore human centeredness of social sciences and humanities and argue for alternative theoretical understandings of materiality. Besides just physicality, materiality can be understood, for example, to play a role in social life and thought or "compose social phenomena" (Schatzki, 2010, p. 134), "changing configurations of materiality," characterizing social life (Ibid., p. 125–141).

Distinguished most of all with a turn to the agency of matter, this emerging material and thereby often also (neo)metaphysical turn in social sciences and humanities attempt to "overcome anthropocentrism and the neglect of the nonhuman world" (Lettow, 2017, p. 107), with specific interest in human-nature relations, materiality, and agency as a feature that not only humans but also nonhuman entities have (ibid; Gamble et al., 2019, p. 111). Rather than a consistent doctrine, for the time being, new materialism consists of a bundle of evolving theorizations that approach matter most of all as *doing*, having active and agential roles in the constantly evolving world (e.g., Barad, 2003). New materialism has been further distinguished, for example, into vital (Bennett, 2010/2020) and performative (Barad, 2003) strands, which critique "the presumption of matter as inherently passive" (Gamble et al., 2019, p.12) and point that the performances of humans are not "external to those of the rest of the material world" (ibid.).

While the new materialist arguments have been treated with interest and a positive challenge to rethink human-nature relationships in social sciences and humanities, they have been criticized for unnecessary and at times overt obscurity (Monforte, 2018, p. 381–382) and for several theoretical-political problems. A key criticism is that the new materialist discussion largely ignores and is reluctant to acknowledge the situatedness of knowledge production and thus "relations of dominance, power and difference within and among human societies" (Lettow, 2017, p. 107; Schatzki, 2010). Especially in equating all forms of material agency—be it agency of persons, worms, or hurricanes (as in Bennett, 2010, p. 107, quoted in Lettow, 2017, p. 111)—new materialism is accused of submerging relations of difference in favor of a false universalism that employs "a 'view from nowhere'" (Lettow, 2017, p. 110). Overall, so far new materialism provides very few tools to analyze historically constituted social structures or to conceptualize the differences between humans and nonhumans, of just human forms of agency (ibid., p. 111). Further, new materialist conceptualization, such as the commonplace differentiation of actors into human and non- or other-than-human, is too general for nuanced discussion on the complex and diverse realities that these concepts strive to delineate (Ibid.).

The criticisms notwithstanding, new materialist arguments have, in a short time span, significantly influenced the discussion of human-nature relations in social sciences and humanities. Meanwhile, also natural scientific research has challenged heretofore commonplace assumptions of the separateness or difference of humans and matter or humans and animals. On one hand, the human organism is found to be literally "a hive of activity" (Ingold, 2011, p. 17) of the microbiota that live in and on humans, while on the other, many animals have been found to be far more sentient and intelligent than previously thought (Rowan et al., 2021) and plants capable of intra- and interspecies communication and (in human terms) strategic action (Moreira & Abdala-Roberts, 2019).

In sum, it seems that understandings of human-nature relations are in a process of paradigm change and that also social theory must learn to take materiality systematically into account (Schatzki, 2010). Even so, the newest and older forms of materialism are not on every aspect different from each other. As Lettow (2017, p. 108) points out, even when epistemologically dissimilar, especially from the perspective of critical social theory (foundational, e.g., for critical social work), new materialism and historical materialism share similar interests and questions. Moreover, open to a broad understanding of matter and materiality, historical materialism could provide new materialism with tools for social critique and thus perhaps pave the way for a critical materialism that combines critical social theory with critical epistemology (Lettow, 2017, p. 118; Boscagli, 2014). For the purposes of this chapter, combining insights from both historical and new materialism appears a good match. Leaning primarily on perspectives founded on historical materialist premises, but open also to new materialist insights, this chapter is built on a somewhat eclectic body of literature while focusing on stuff, consumerism, and their multifaceted significance in and beyond social work.

Our Everyday Excess

As Hansen and Nielsen (2023, p. 3) posit, "The way people lead their lives and the goods and services that are incorporated in their mundane practices - in particular related to housing, mobility and food (Tukker et al., 2010) can have dramatic environmental consequences." Most of us (majority of this book's readers supposedly belonging to relatively affluent classes and population groups globally) consume way too much of the Earth's material resources. When a decent living standard is achievable with around 6 tons of raw material consumption per person per year (Vélez-Henao & Pauliuk, 2023), in 2022 an average EU citizen consumed 14.5 tons. In the author's home country Finland, the same year, the domestic consumption of raw materials—metal ores, fossil energy materials, biomass, and nonmetallic minerals—was 43.7 tons per capita, the highest among 35 European countries (Eurostat, 2023, Domestic material consumption by main material category, 2022), a situation caused by the resource and energy intensive industrial sector and high household consumption levels in Finland.

Whether rich or poor, the way we lead our lives is entwined with the economic system(s) in place. Globalization together with capitalism's evolution into its current neoliberal, globally ubiquitous stage has been greatly fortified by "the Fossil Machine" (Salminen & Vadén, 2015, p. 19) and the superpowers provided by oil and other fossil fuels. As condensed, highly efficient, and transportable form of energy, particularly oil has fed immense amounts of energy work in the systems of production globally. Fossil fuels have rendered possible ever accelerating volumes of production, transportation, and consumption of goods, fossil-time economic and political interests thus shaping our societies, culture, and behavior (Salminen & Vadén, 2015; Rosa, 2019). At the same time, the growth based on large scale extraction of natural resources has brought along productivism and a tendency to see everything, even fellow humans, as resources to utilize (Salminen & Vadén, 2015, p. 43).

Capitalism has been highly successful in enabling and promoting material prosperity on a scale unparalleled in history. Within a short period of time, living with excess, or rather "imperial mode of living" (Brand & Wissen, 2021), has become a normal, taken for granted situation in affluent societies and social classes and a desired condition for many who are not yet quite there. Nonmaterial wealth, on the other hand, is more difficult to comprehend. While there have been societies at the course of history where the ideal amount of material property has been not more than what one can comfortably carry along (e.g., Sveiby & Skuthorpe, 2006), in the present moment, such lifestyles may appear as disturbing anomalies that societies should get rid of. When and if well-being, progress, and equality are envisioned merely as access to modern infrastructures, commodities, and monetary economy, for example, indigenous people who are displaced to make way for infrastructure or resource extraction projects may realize that their nuanced understanding of and belonging to a particular land is considered insignificant (e.g., Padel & Das, 2010). Overall, lifestyles that are anti- or external to consumerism are, from a consumerist perspective, easily perceived as other-worldly, freakish, and even pitiful. When many low-income social work service users lead their lives with relatively limited material assets out of necessity, their ability to do so is seldom turned into a virtue that also others should aspire to, although from an ecological sustainability perspective that should be the case.

Consuming on the scale that is mainstream in consumer societies sucks dry the planet, including many humans at various points of the chains of production, and generates enormous amounts of waste. Illustrative of the scale of human influence on Earth, Elhacham and associates (2020) argue that the amount of material produced by humans (themselves 0.01% of biomass) exceeded the biomass of the Earth at some point in 2020. This was just one step in the longer-term development since the agricultural revolution. Since then, the plant mass has reduced by half (from 2 to 1 Tera, i.e., 1,000,000,000,000 tons), while the amount of human made material has increased in accelerated speed (to circa 1.1 Tera tons in 2020).

The estimate is that in 2020, the cattle of humans (cows, pigs, etc.) made nearly 60% of mammals, and poultry 70% of all birds; nowadays, only 4% of all mammals are wild animals, and the world's most common bird is broiler chicken. While the mass of plastic alone is already twice the mass of animals, human made material output (buildings, roads, smaller artifacts, etc.) keeps doubling up every two decades (Elhacham et al., 2020; Bar-On et al., 2018). In short, the space and scope humans (and some humans more than others) with their varied systems of production take from the rest of life on Earth is simply outrageous, a situation unlikely to change very soon. Globally, the human population is estimated to grow from 8 billion in 2023 to nearly 10 billion in 2060s, which implies increased demand for resources. Much of the harm that humans cause to the planetary system of the Earth thus has to do with the massive scale of human influence.

Shady Sides of Stuff

When the previous subchapter was a grand narrative about the massive scale of human influence on Earth, a more tangible way to explore consumerism and the importance of material possessions is to turn the gaze to our own material practices. One way to start can be the following exercise, originally formulated by Fred Besthorn, known as the developer of deep ecological social work:

Exercise 1 Exploring and challenging consumerism (adopted from Besthorn, 2003)

1. Make an exhaustive list of everything you own.
2. How are your individual, emotional, cultural identities and activities reflected in your possessions? Write a short interpretive essay and attach to your property list.
3. Collective discussion: how does property provide us with/keep us from developing spiritual meaning and collective accountability?

As none of us is an outsider to the material world and the current environmental crises, the aim of this exercise is not to moralize to anyone on the amount of their property but basically to stop to grasp what one's material possessions are and explore the varied meanings and significance they have for us. As listing one's possessions may take some time, depending on the role of this assignment in each teaching situation/course, it can be given as homework, or if a course listing is fit for the purpose, with a time limit for completion by students.

People today have more stuff than during any preceding civilization (Valkonen et al., 2019). To provide some scale and background info for the previous exercise, in Finland, for example, the ordinary amount of stuff in a household is between 10,000 and 50,000 items, whereas in the USA, the amount goes often up to 300,000 (Kinnunen, 2017). Yet, just a few hundred items—even when one counts each one—can be enough to lead a decent life. Those who have lived for a while out of suitcase may have experienced this; similar number of belongings are common also to many of my Indian social worker friends working in the Indian NGO/voluntary sector. A century or two back the property of ordinary lower-class people in the global north was usually no more, a fact that one can marvel at in historic museums displaying labor class lives or basically any other than affluent lifestyles before the early twentieth century. But times change. Besides the "traditional" basic needs of food, water, clothing, and shelter, nowadays also access to Internet, communication means, transportation, electricity, healthcare, and modern waste management are necessary for participation and safety (Watkins, 2022).

In the acquiring of new things, their materiality is rarely the uppermost criteria. In view of Janet Bennett (2020, 31), excessive consumption is, in a way, *anti*materiality indifferent to the vitality of matter. While our homes get filled with stuff almost without our noticing, our purchasing decisions stem largely from the visions of a good and pleasant life that we associate with the things in question (Salonen et al., 2021, p. 77–82). Moreover, many of our purchases are driven by the desire to look acceptable in the eyes of others, our social aspirations thereby determining our consumption. In proportion, poor people sometimes avoid social gatherings because they feel they have nothing decent to wear, but do not want to stand out as shabby either.

In a similar vein, Hartmut Rosa (2019) theorizes that people in late modern societies keep acquiring goods and streaming services to have them available for their potential good life moments. Such moments are however scarce, for irrespective of social class, many suffer from a constant lack of time and energy to lead the kind of life they desire (ibid.). In an increasingly uncertain world (Kantola et al. 2022), much time and energy go into gaining social, economic, cultural, and symbolic capital to secure desired positions, relationships, experiences, and even peace of mind, as if the name of the game is either incessant status optimization or reversal. For Rosa (2019), this kind of racing is symptomatic of the acceleration and the need to have the world *available*, *accessible*, and *attainable* that features contemporary societies and neoliberal capitalism. The flip sides include that constant pursuing something worth having easily turns the focus onto what one might be lacking (Salonen et al., 2021). Moreover, availability and accessibility may equal disposability. Multitude diminishes the value of an individual item or person, as one can be substituted with another. What's more, when a thing or device breaks down, buying new is often easier and cheaper than repairing the old, and many items are planned to be disposable in the first place (Valkonen et al., 2019; Salminen & Vadén, 2015, 77). However, having everything constantly available is not necessarily amusing for long; the tighter we grasp everything, the more distanced we risk becoming from ourselves, others, and nature (Rosa, 2019).

Each good we purchase either for their symbolic value or to keep established everyday practices going (which alone encompass significant share of resource use globally) contains multiple streams of material flows (Valkonen et al., 2019). In the end, as Valkonen et al. (2019) explicate, the consumerist mode of living is one kind of way to live from the land. Intensely coupled with the nature and the nonhuman, consumer societies rely heavily on science and technology and production of waste. Indeed, waste is integral to societies in which newness is highly valued and the economy is founded on constant acquisition of new goods, requiring getting rid of old ones, respectively (there are limits to storing and living with stuff, even with the increasingly popular options to rent storages, employ personal organizers, etc.).

(Ibid.) Or else the more we have, the more we must maintain, look after, insure, and guard.

Carbon emissions can be considered a form of waste, to the extent that the carbon in the atmosphere and oceans is a central aspect of the Wasteocene, the rest of the related concepts for the current era, such as Anthropocene and Capitalocene notwithstanding (Moore, 2022). However, even when the consequences of consumption scale up, for example, as the Great Pacific Garbage Patch of marine debris in the North Pacific Ocean (BBC, 2023) or as the numerous former mining sites and regions become contaminated by toxic waste, modern waste disposal infrastructures are highly efficient in dispelling waste from vicinity (Valkonen et al., 2019, p. 135). There is a "disassembly of material feedback loops, experienced as a lack of a general view and as alienation" (…) "when carbon dioxide emissions, waste, and the production of raw materials are removed far enough from consumption, they vanish into blind spots" (Salminen & Vadén, 2015, p. 37).

Globally and locally, power works through materials and labor (Latimer, 2018; Moore, 2022). Market prices do not fully convey the ecological and social costs involved in production (Padel & Das, 2010). Many of the products we consume are produced in globally fractured production chains, which makes identification and monitoring of each work and material contribution difficult. While companies increasingly invest in improving the social and ecological sustainability of their production chains, the situation is prone to unequal power relations, affluent centers gaining advantage over politically and economically marginalized locations and people while materially dependent on them. Meanwhile, consumption increasingly relies on diverse background infrastructures from Internet to logistic, communications, and educational systems that back up the use of single products. Even seemingly immaterial products are often based on or hidden streams of raw materials, as in the case of search engines or various cloud services (Valkonen et al., 2019, p. 45).

All this boils in that although we do not pursue forced labor, species extinction, sea level rise, or other social and ecological disasters, we likely further them. While ecological sustainability would require absolute reduction in material consumption, reaching a political agreement and effective and compelling regulation is highly challenging, our political and economic systems being founded on the necessity of economic growth (Valkonen et al., 2019, p. 43; Blühdorn, 2014). Even in circular economy, which in principle is about more sparing and reduced use of raw materials, much effort has been put on creating "zero waste" products and closed production circles, as if elimination of waste and solving the problems at the heart of overconsumption would be a matter of better planning only (Valkonen et al., 2019, 49). While that is absolutely needed, tackling the problems inherent to consumerism also requires thoroughgoing ecological reconstruction (BIOS, 2019), aka ecosocial or green sustainability transition of societies. In so doing, it is however important to take care of the function—and response—ability (Haraway, 2016) of ourselves and others.

Coming to Terms with Emotions

The multiple environmental, social, and political challenges at stake are apt to increase feelings of hopelessness, powerlessness, alienation, and anxiety. Environmental and social justice issues evoke political contestations, which alone can cause confusion and uncertainty. Moreover, living the end times of fossil fuel-based civilization, whose infrastructures, technologies, and worldviews are still largely in place, is in many ways ambivalent—urgent and profound change already looming but not yet here in its entirety. Many old practices and habits are known to be problematic, and developing new or alternative ones is in progress (Pelttari, 2021).

According to the agential realist (new materialist) view of Karen Barad (2003, p. 828), the "knower" is not external to the natural world, but part of nature s/he seeks to understand or feels for. Living in a porous relationship with our environments, the misery of the planet influences us in diverse forms, such as in psychological or somatic symptoms, or both, as the two are connected. Our emotional reactions to the gloomy environmental and climate situation and related societal and economic scenarios matter, for they influence our reasoning and action. Emotions come and go and are not as such "right" or "wrong." While the eco- or climate anxiety of some stems from the realization that things are far off the mark, others may experience sustainability transition demands as criticism for their lifestyle and identity (Virtanen, 2022). Further, emotions being often connected to our values and identity, emotions perceived as negative, such as anger or fear, can sometimes convey important messages, while positive emotions can be problematic, for example, if originating from environmentally harmful activities (Pihkala, 2017, 2022).

Common emotions related to environmental issues include astonishment, confusion, disappointment, shock, sorrow, melancholy, fear, guilt, anger, anxiety, and sense of inadequacy. For example, in the Youth Barometer conducted in Finland 2021, 76% of young people aged 15–29 years responded feeling sad because of species extinction and loss of biodiversity (Kiilakoski, 2021). On the other hand, environmental issues can also evoke feelings of pleasure, satisfaction, pride, capability, and togetherness, especially when one acts to safeguard nature. Also feeling good, joyful, relieved, and hopeful (also overoptimistic) are common among environment related feelings (Pihkala, 2017, 2020a, b; Toivoa ja toimintaa, n.d.).

The following exercise is adopted, with slight modifications, from the educational resource website *Toivoa ja toimintaa* (Hope and Activities, in Finnish only), developed by similarly named project of the Finnish Association of Biology and Geography Teachers, BMOL, since 2019, focusing on global education. The website's climate emotions exercises, including this one, have been developed in collaboration with Panu Pihkala, who has worked extensively on climate and environmental emotions (e.g., Pihkala, 2020a, b, 2022). The exercise aims to acknowledge the varied emotions we have regarding the climate crisis and value the knowledge they contain, as well as rehearse talking about emotions. I have tried out varieties of this exercise with diverse participants, and while requiring some warmup and attentiveness to whatever emerges in the group, it has always been helpful in grasping and discussing the role of emotions and the importance of emotional work in climate change mitigation and adaptation and the sustainability transition at large. Moreover, as the exercise is about emotions and not the climate change as such, it has usually engaged also those participants, who for a reason or another trivialize climate change, enabling a non-polarizing change of tone in the discussion.

Exercise 2 Climate emotions (Toivoa ja toimintaa, n.d.)

1. Identify/invent words that depict your emotions related to the climate crisis (can be supported with a readymade list of common emotions).
2. Discuss your emotions with a pair or in small groups. While discussing, keep it safe for everyone: discuss personal issues only in ways which you and others can feel safe and comfortable with, don't be judgemental and avoid feelings of inferiority or superiority in relation to others.
3. Make a visual illustration (depending on the available/attainable materials, a drawing on a paper, making a picture collage, mind map, etc.) that somehow illustrates one or several of the emotions discussed (alternatively utilize some word cloud app to enlist all the emotions discussed). Display the illustrations.
4. Collective discussion: what kind of emotions can you identify at the visual illustrations (—what stands out)? How do you interpret these emotions—what kind of situations and why do they stem from?
5. Further discussion: what kind of ramifications climate emotions have in and for social and community work, and how could they be addressed? How can social and community work facilitate and support emotional work as part of the sustainability transition?

In my heretofore experience, the emotions that most of all come up include confusion, sorrow, inconsistency (such as between available livelihood and transportation options and what is known to be good for the climate and the environment), anxiety, fear, and shock, occasionally added with nature loving of various sorts, or utility, if the family livelihood is tied to natural resources, such as forest. (For those wanting to go deeper with emotions, Fred Besthorn has further developed an exercise called "Eco-Confessional Aiming for Ritualised Cleansing and Renewal"; see Besthorn, 2003).

The way we channel our emotions is of social and political as well as material significance. The transition toward profoundly sustainable worldview and actions is a comprehensive process for individuals and societies alike, requiring emotional reflection and processing. In other words, it is important to recognize and approve the diverse emotions that the situation evokes and deal with them (Pihkala, 2017). While neither overt optimism nor pessimism helps us forward (Haraway, 2016), the hopelessness caused by repeated and severing ecological crises can, at the worst, debilitate one's functional capacity and lead to apathy and inability to imagine a better world (Amsler, 2010). With difficult feelings, it is important to strengthen one's emotional skills and general mental health and well-being, maintain daily routines, have sufficient sleep and nutrition, and seek help from mental health professionals if need be. These are things that also social and community workers and educators have a mandate to promote.

To be able to act and work for better future, we need to be able to believe that positive change is possible (Virtanen, 2022)—an approach that in general features social work. It is thus vitally important to nurture and look after our own hopefulness and ability to act in our multiple roles as social workers, educators, and citizens and cherish and nurture the same qualities in the people and communities we work with. More than just a state of mind, hope is action and springs up from action, which regarding planetary scale environmental problems will be needed day after day for decades to come. Change that permeates our everyday values and worldviews is truly radical and can make miracles (Ibid.).

Here, an important and free flowing resource (to use oil-time utilitarian language) is nature itself, as well as spirituality inherent in each of us. Research shows that spending time outdoors, in the natural environment, is in many ways beneficial to human health and overall well-being (see Larocque in this volume). Time spent in nature relieves stress, calms down, and furthers self-reflection amidst the constant surge of various stimuli of the otherwise fast-moving world. Natural environments also provide opportunities for physical exercise and doing things together. With social work clients, going out changes the nature of the worker-client relationship, enabling to see each other beyond the customary roles, often to the benefit of the relationship and actual goals of the work (Malinen et al., 2020). Access to nature is an important part of participation, and depending on our background and personality, different natural environments, with connections to cultural identities, speak differently to each of us, although personal relation to nature can change over time (THL, 2023). Among others, these benefits can be promoted with various nature assisted services, which are increasingly available.

Related with the above, it is anyhow worth emphasizing that nature and thus various natural environments, ecosystems, and lineages of species have an intrinsic value beyond and above human needs and liking and thus beyond economic and other forms of utilization. Humans being a species totally dependent on the web of life on Earth, safeguarding the continuity of that life in its diverse forms is of utmost importance to the future of humanity. That requires, to start with, nothing less than getting rid of lifestyles corrupted by oil, including the current circles of resource extraction and habitat destruction turned to waste. As Salminen and Vadén point, a way ahead is to revive human abilities to experience the wild as sacred "separate from the technological totalisation," for "The sacred is unproductive and it prevails" (Salminen & Vadén, 2015, 108, 125).

Enjoying the gifts of nature does not require heading to consume the farthest national parks; stopping by and cherishing the local nature can make miracles as well. In view of Besthorn:

> As one develops an ever-widening identification with the whole, there is no need for self-sacrificing, moral altruism, (...) since the interests of those with whom we identify, nature included, may be seen as one's own interests as well. Self becomes most fully realised not solely when self-interest and wants are met, but rather when one begins to identify with more-than-one's self. (Besthorn, 2003)

At the end of the day, for most of us, the important denominators of good life include relatedness to our close ones and nature and the experience that one can be of service to greater goals than just personal gain (Virtanen, 2022, 147). Other important qualifiers for good life and emotional well-being are sense of sufficiency, contentment, trust, love, and peace with oneself and the surrounding reality (Salonen et al., 2021, p. 67–68), as well as personal growth and ability to make choices that strengthen one's virtues (Heinimäki, 2018, p. 49). Emphasizing these kinds of qualities of experiential well-being, that anyhow are at the heart of social work, can be a guiding light for ecosocial work seeking the way toward a more enchanted and sacralized humanity and understanding of the world.

Engaging Politically Toward More Sustainable Materialisms

As noted, this chapter is written in the understanding that the pending ecosocial or green sustainability transition process must involve massive social and cultural change, to which facilitating and carrying out social work has the potential and responsibility to contribute with its specific know-how on the social. The change implies political change, and social work is always political work, among others, in the sense that it cannot avoid having a political location and agenda in a society. Ecosocial work as a system-aware, community-based, and structural mode of work is a particularly political mode of work, conducting which eye and readiness for constructive political engagement are an advantage. This is pertinent also due to fact that one already existing and supposedly deepening dividing line within, and to an extent between societies, goes between those committed to proceed with the transition and those principledly antienvironmental or unconcerned of the environmental present and futures—both with material implications. As politics is about both conflicts of interests and values and finding compromises over disagreements, it is worthwhile to explore what one's own starting points are to partake such processes.

Exercise 3 Political genogram (adapted from Besthorn, 2003)
Aim: to articulate and understand the formation of one's own political identity (for issues of social and economic justice are inseparable from ecological justice and the distribution and use of power).

1. Construct a political genogram (a diagrammatical portrayal) of your family's political history, such as belief systems and ideologies that your family holds.
2. In so doing, pay attention to the kind of values, behaviors, and temperament it provides you in confrontational situations: how has your upbringing shaped your political identity and comfort/discomfort zones of political action?

The point is to understand the power and dynamics of one's political history to build a more constructive base for the ecological and social justice-oriented activities, *not* to dwell on pains, contradictions, anxieties, and so forth, of the past (Besthorn, 2003).

In my own teaching, I have modified this exercise into an assignment to construct a sociocultural autobiography. Depending on the group and the study unit in question, the autobiography has focused either on one's upbringing and inheritance what it comes to different dimensions of sustainability or the core legacies and value base from which one enters social work. As this exercise contains personal information, I have instructed the students to report on the level and style they themselves are comfortable with, with the possibility to use a primarily visual illustration added with some explanation. In my view, stopping to reflect on one's past and path to the present moment without being judgmental is useful for self-knowledge as a (future) social worker to clarify one's sociopolitical location and reflect its meaning for one's practice, goals, and what one takes as given in life.

Regarding the transition imperative, liberal democracies have been slow, if not incapable, to make radical enough political changes required for a more sustainable course; for example, carbon emissions have kept growing. As truly effective sustainability reforms would hit most political interest groups in a way or another, inefficient politics stem, among others, from fear of political backlash that effective reforms would likely cause and the continued political support for prioritizing growth, competitiveness, and (although not formulated as such) overconsuming mode of living (Blühdorn, 2014; Brand & Wissen, 2021). Moreover, off late liberal democracies have been marked by so called post-political condition, i.e., reduction of politics into consensual governance driven by market liberalism, continuation of the prevailing order, and management of the ever-rising complexity of societies. While interest in politics among citizens has been on the decline, significant repoliticization has taken place especially on two fronts. These include, on one hand, new social movements and forms of activism that push forward social and environmental reorganization with participatory and egalitarian agendas. On the other hand, right-wing anti-globalization, anti-liberal, and antienvironmental mobilization has in a short time altered the political space and discourse in many countries, as, for example, Donald Trump has done in the USA. While on many aspects oppositional, these two fronts share widespread lack of confidence on established political institutions and inclination for lifestyle activism (Blühdorn & Deflorian, 2021).

Though there are still today many for whom pursuing ecological sustainability is not a priority (Blühdorn, 2014), there have long been individuals and movements embodying dissent to mainstream consumerism and experimenting and practicing socially, politically, and ecologically alternative lifeforms. These include, among others, the few still active Gandhian rural development organizations in South Asia, eco-villages globally (see GEN, n.d.), and the political activation of relational and communal ways of life and cultural and political autonomy under diverse pluriversal (as opposed to universal) discourses of transition (Escobar, 2018), including the paradigm for radical ecological democracy (Radical Ecological Democracy, n.d.). From the margins, such approaches are capable to question current transition politics, when seeing it reduced into greenwashing or replacing one form of energy with another without deeper de- and reconstruction of the modes of thinking and living behind the current socioecological crises.

Ecosocial work shares many interests and values with the new social movements. Driven often with lack of satisfaction with mainstream politics and disillusionment with contemporary material life, these movements are increasingly "working on environmental *practices* and *systems*" to produce, for example, food, energy, and sustainable fashion themselves, in a community-based manner. In other words, beyond or instead of just voting in elections and/or practicing political consumerism, such as boycotting of "buycotting" of certain products, the political choice is embodied action to develop "new collectively organised *systems* of material flows" (Each emphasis original, Schlosberg & Craven, 2019, p. 23–34, 47). According to Forno and Graziano (2014, p. 143, quoted in Schlosberg & Craven, 2019, p. 44), common characteristics of sustainable consumption movement organizing include linking consumption to environmental damage and social injustice, seeing artisanal production more sustainable than mass production, being concerned of transnational justice and the sustainability of global supply chains, and encouraging direct relationships between producers and consumers. While thus perceiving (sustainable) consumption as the way ahead, these movements also take steps beyond the dominance of capitalism with active detachment from the systems of economic and material power they disagree with (Schlosberg & Craven, 2019, p. 46, 82).

However, when the previously described material and circulatory systems can increase local and community-based ownership over processes of production and provide political meaning and livelihood opportunities, they are also highly vulnerable to co-optation (Schlosberg & Craven, 2019). As Blühdorn and Deflorian (2021) argue, neoliberal capitalism is highly capable to mobilize alternative visions, creative energies, and organizational capacities for its own purposes. However, political support for alternatives is often selective and partial, and activism that challenges the logics of privatization, growth, and social inequality is largely ignored or repressed. Collaborating social movements may find themselves in a situation where they are expected to provide solutions for the pending ecological and social problems with inadequate resources and support to do so. (Ibid.) Yet, even if temporary, movements nurture alternative imaginaries and critique the neoliberal status quo and have increased public awareness of the unsustainability and fragility of capitalist consumer societies.

Concluding Thoughts

Reflecting to some extent also my own frustrations and perplexity but also hopefulness amidst the current Earthly challenges, this chapter turned out to be a compilation of perspectives and exercises broadly associated with materiality, especially stuff, and generally addressing environmental issues in social work education. As has become clear, I wrote the chapter with the kind of societies and social classes in mind, in which consumerism is presently the norm and getting one's stomach full is mostly not an issue. Further, the exercises that I included, starting from one's material possessions, proceeding to emotions, and finally to political or sociocultural milieus and heritage, explore ownership, emotions, and political views and their socio-material and ecological impacts primarily from an individualist perspective, which, as Pichle et al. (2020, p. 200) point out, is typical to liberal democracies "structurally entangled with the capitalist mode of production and living." Notwithstanding the global spread of late capitalism, this is a clear limitation, which I apologize, for as a gesture it repeats northern, affluent self-centeredness and perpetual and continuous tendency to omit or fade out unequal relations and disparities among humans globally, not to mention the unequal relations with other-than-humans.

Either way, I do think that especially affluent population groups and countries must urgently find ways to come down to Earth from their bubbles of imperial mode of living (Brand & Wissen, 2021), i.e., lifestyle structurally premised on global access to resources and externalization of negative economic, social, and environmental consequences. While this certainly requires structural and political change, and thus widespread and active support for governments and political interest groups enforcing such changes, it also comes down to individuals, including individual (eco)social workers, and whether and how they are, do, and promote the change in their everyday. If the upper-level politics fails or does not proceed fast enough, the alteration of the course must be supported and can emerge also from below, in many voices and with diverse (pluriversal) practices. Either or, it is important to pay attention to the power relations and economic structures involved. As Rosi Braidotti (2019, p. 153) reminds, while we are all together in this, "we are not One and the same." For the sustainability transition to proceed, especially the over-consuming classes globally must learn fairer and ecologically less destructive meaning and content into their lives.

Stemming from the notion that personal is political, the exercises I delivered are somewhat elementary but have, in my experience, potential to shed light on how each of us, although embodied and positioned in diverse material locations, is implicated in materiality (Braidotti, 2019, p. 153) and thus ultimately part of the world one may want to change. Each of us thus not only undergoes the continual generation in currents of materials (Ingold, 2011, p. 31) during (and after) one's life course but also influences the flow of materials in the world. It is especially at this latter level that we can personally and collectively work on politics and cultures of materiality in ecosocial work and beyond.

References

Amsler, S. (2010). Bringing Hope to 'Crisis'. Crisis thinking, ethical action and social change. In S. Skirmshire (Ed.), *Future ethics. Climate change and apocalyptic imagination* (pp. 129–152). Continuum.

Barad, K. (2003). Posthumanist performativity: Towards an understanding of how matter comes to matter. *Signs, 28*(3), 801–831.

Bar-On, Y. M., Phillips, R., & Milo, R. (2018). The biomass distribution on earth. *Proceedings of the National Academy of Sciences -PNAS, 115*(25), 6506–6651.

BBC. (2023). Victoria Road beach: More than 70 volunteers help clean up coastline (*News* England/ Regions/Liverpool 18th September, 2023. https://www.bbc.com/news/uk-england-merseyside-66842296

Bennett, J. (2010/2020). *Vibrant matter. A political ecology of things*. Duke University Press. Translated into Finnish in 2020 by Tapani Kilpeläinen as *Materian väre: Olioiden poliittinen ekologia*. Tampere, Niin & näin.

Berryman, S. (2022). Ancient Atomism. In E. N. Zalta & U. Nodelman (Eds.), *The Stanford Encyclopedia of philosophy* (Winter 2022 ed.). Metaphysics Research Lab. Stanford University. https://plato.stanford.edu/archives/win2022/entries/atomism-ancient/, retrieved 28.11.2023.

Besthorn, F. (2003). Radical ecologisms: Insights for educating social workers in ecological activism and social justice. *Critical Social Work, 3*(1), 66–106.

BIOS. (2019). *Ecological Reconstruction*. Ecological reconstruction » BIOS: Ecological reconstruction.

Blühdorn, I. (2014). Post-ecologist governmentality: Post-democracy, post-politics and the politics of unsustainability. In E. Swyngedouw & J. Wilson (Eds.), *The post-political and its discontents: Spaces of depoliticisation, spectres of radical politics* (pp. 146–166). Edinburgh University Press.

Blühdorn, I., & Deflorian, M. (2021). Politicisation beyond post-politics: New social activism and the reconfiguration of political discourse. *Social Movement Studies, 20*(3), 259–275. https://doi.org/10.1080/14742837.2021.1872375

Boetto, H. (2017). A transformative eco-social model: Challenging modernist assumptions in social work. *British Journal of Social Work, 47*, 48–67.

Boscagli, M. (2014). Introduction. Of Jena Glassware and potatoes: Matter in the moment. In M. Boscagli (Ed.), *Stuff theory: Everyday objects, radical materialism* (pp. 1–36). Bloomsbury.

Bozalek, V., & Pease, B. (2021). *Post-anthropocentric social work. Critical Posthuman and new materialist perspectives*. Routledge.

Braidotti, R. (2019). *Posthuman knowledge*. Polity.

Brand, U., & Wissen, M. (2021). *The Imperial mode of living. Everyday life and the ecological crisis of capitalism*. Verso.

https://ec.europa.eu/eurostat/web/products-eurostat-news/w/ddn-20230731-1 through which access to excel file Eurostat 2023, Domestic material consumption by main material category, 2022.

Dominelli, L. (2012). *Green social work. From environmental crises to environmental justice*. Polity.

Elhacham, E., Ben-Uri, L., Grozovski, J., Bar-On, Y. M., & Milo, R. (2020). Global human-made mass exceeds all living biomass. *Nature, 588*, 442–444.

Escobar, A. (2018). *Designs for the Pluriverse: Radical interdependence, autonomy, and the making of worlds*. Duke University Press.

Eurostat. (2023). EU's domestic material consumption remained stable in 2022. News articles, 31 July 2023.

Forno, F., & Graziano, P. (2014). Sustainable community movement organizations. *Journal of Consumer Culture, 14*(2). https://doi.org/10.1177/1469540514526

Gamble, C. N., Hanan, J. S., & Nail, T. (2019). What is new materialism? *Angelaki, 24*(6), 111–134.

GEN. (n.d.). About GEN - The Global Ecovillage Network (GEN) bridges countries, cultures, and communities!. Retrieved 14th Dec 2023.

Hansen, A., & Nielsen, K. B. (2023). Introduction. Consumption, sustainability, and everyday life. In A. Hansen & K. B. Nielsen (Eds.), *Consumption, sustainability, and everyday life*. Springer.

Haraway, D. (2016). *Staying with the trouble. Making Kin in the Chthulucene*. Duke University Press.

Heinimäki, J. (2018). *Arvon Porukka*. Kauppakamari.

Ingold, T. (2011). *Being alive. Essays on movement, knowledge and description*. Routledge.

Kantola, A., Aaltonen, S., Haikkola, L., Junnilainen, L., Luhtakallio, E., Patana, P., Timonen, J., & Tuominen, P. (Eds.). (2022). *Kahdeksan kuplan Suomi. Yhteiskunnan muutosten syvät tarinat*. Gaudeamus.

Kiilakoski, T. (Ed.). (2021). *Kestävää tekoa. Nuorisbarometri 2021*. Valtion nuorisoneuvosto, Nuorisotutkimusseura/Nuorisotutkimusverkosto, & Opetus- ja kulttuuriministeriö.
Kinnunen. V. (2017). Tavarat tiellä. Sosiologinen tutkimus esinesuhteista muutos. Acta Universitatis Lappoensis 362. Rovaniemi, University of Lapland. https://lauda.ulapland.fi/bitstream/handle/10024/63026/Kinnunen_Veera_ActaE_229_pdfA.pdf?sequence=1
Latimer, J. (2018). Afterword: materialities, care, 'ordinary affects', power and politics. *Sociol Health Illn., 40(2), 379–391.*
Lettow, S. (2017). Turning the turn: New materialism, historical materialism and critical theory. *Thesis Eleven, 140*(1), 106–121. https://doi-org.ezproxy.jyu.fi/10.1177/0725513616683853
Malinen, H., Finel, N., Tiitu, M., Vierikko, K., Tuhkanen, E., Sinkkonen, A., Matila, A., Lahti, E., Hannonen, P., & Hapuoja, A. (2020). *Elämänmittainen lähivihreäpolku Tietopaketti lähiluonnon hyvinvointivaikutuksista*. Ympäristöministeriö ja Suomen ympäristökeskus. https://julkaisut.valtioneuvosto.fi/handle/10024/162679
McDonald, C. (2006). *Challenging social work: The institutional context of practice*. Palgrave.
McKinnon, J., & Alston, M. (Eds.). (2016). *Ecological social work. Towards sustainability*. Palgrave.
Monforte, J. (2018). What is new in new materialism for a newcomer? *Qualitative Research in Sport, Exercise, and Health, 10*(3), 378–390.
Moore, J. W. (2016). Introduction. Anthropocene or Capitalocene? In J. W. Moore (Ed.), *Anthropocene or Capitalocene? Nature, history and the crisis of capitalism* (pp. 1–11). PM-Press.
Moore, J. W. (2022). Waste in the limits of capital: How capitalism lays waste to the web of life, and why it can't stop. *Emancipations, 2*(1), article 4.
Moreira, X., & Abdala-Roberts, L. (2019). Specificity and context-dependency of plant-plant communication in response to insect herbivory. *Current Opinion in Insect Science, 32*(April), 15–21. https://doi.org/10.1016/j.cois.2018.09.003
Närhi, K., & Matthies, A. (2018). The Ecosocial approach in social work as a framework for structural social work. *International Social Work, 61*(4), 490–502.
NASW. (2021). *Code of Ethics of the National Association of Social Workers*. https://www.socialworkers.org/About/Ethics/Code-of-Ethics/Code-of-Ethics-English
Padel, F., & Das, S. (2010). *Out of this earth. East India Adivasis and the Aluminium Cartel*. Orient BlackSwan.
Pelttari, M. (2021). *Lämpenevä Maa – ilmastolukutaidon käsikirja*. Vastapaino.
Pichler, M., Brand, U., & Görg, C. (2020). The double materiality of democracy in capitalist societies: Challenges for socio-ecological transformations. *Environmental Politics, 29*(2), 193–213. https://doi.org/10.1080/09644016.2018.1547260
Pihkala, P. (2017). *Päin Helvettiä? Ympäristöahdistus ja toivo*. Kirjapaja.
Pihkala, P. (2020a). Anxiety and the ecological crisis: An analysis of eco-anxiety and climate anxiety. *Sustainability, 12*(19), 7836. https://doi.org/10.3390/su12197836
Pihkala, P. (2020b). Eco-anxiety and environmental education. *Sustainability, 12*(23), 10149. https://doi.org/10.3390/su122310149
Pihkala, P. (2022). Towards a taxonomy of climate emotions. *Frontiers in Climate, 3*(January). https://doi.org/10.3389/fclim.2021.738154
Polku 2.0 Project. (2021–2023). Kestävän tulevaisuuden Polkukartta — Polkukartta (jyu.fi).
Radical Ecological Democracy. (n.d.). About – Radical Ecological Democracy. Retrieved 14th December 2023.
Ranta-Tyrkkö, S. (2010). *At the intersection of theatre and social work in Orissa, India. Natya Chetana and its theatre. Acta Unviersitatis Tamperensis 1503*. Tampere University Press. http://urn.fi/urn:isbn:978-951-44-8003-4
Rosa, H. (2019). Available. Accessible, attainable: The mindset of growth and the resonance conception of the good life. In H. Rosa & C. Henning (Eds.), *The good life beyond growth: New perspectives* (pp. 39–53). Routledge.

Rowan, A., D'Silva, J., Duncan, I., & Palmer, N. (2021). Animal sentience: History, science, and politics. *Animal Sentience, 31*(1). https://doi.org/10.51291/2377-7478.1697

Salminen, A., & Vadén, T. (2015). *Energy and experience: An essay in Nafthology*. MCM. Energy and Experience.indb (mcmprime.com).

Salonen, A., Kurenlahti, M., & Jaaksi, A. (2021). *Rakkautta ja valoa: Tie hyvään tulevaisuuteen*. Docendo.

Schatzki, T. (2010). Materiality and social life. *Nature and Culture, 5*(2), 123–149.

Schlosberg, D., & Craven, L. (2019). *Sustainable materialism: Environmental movements and the politics of everyday life*. Oxford University Press.

Sveiby, K.-E., & Skuthorpe, T. (2006). *Treading lightly: The hidden wisdom of the world's oldest people*. Crow's Nest, Allen & Unwin.

Svoboda, R., & Lade, A. (1995). *Chinese medicine and Ayrveda*. Lotus Press.

THL. (2023). Luonto ja osallisuus. Luonto ja osallisuus - THL, retrieved 14th December 2023.

Toivoa ja toimintaa. (n.d.). https://toivoajatoimintaa.fi/

Tukker, A., Cohen, M. J., Hubacek, K., & Mont, O. (2010). The impacts of household consumption and options for change. *Journal of Industrial Ecology, 14*(1), 13–30.

Valkonen, J., Pyyhtinen, O., Lehtonen, T., Kinnunen, V., & Huilaja, H. (2019). *Tervetuloa jäteyhteiskuntaan! Aineellisen ylijäämän kanssa eläminen*. Vastapaino.

Vélez-Henao, J., & Pauliuk, S. (2023). Material requirements of decent living standards. *Environmental Science & Technology, 57*, 14206–14217.

Virtanen, H. (2022). *Trauma ja luonto eli kuinka selviytyä ekoahdistuksesta*. SKS.

Watkins, A. (2022). How Have People's Basic Needs Changed Over Time. *Popped Blog* January 25, 2022, https://populationeducation.org/how-have-peoples-basic-needs-changed-over-time/ (1.12.2023).

Wilson, T. (2021). An invitation to the trouble with humanism for social work. In V. Bozalek & B. Pease (Eds.), *Post-anthropocentric social work. Critical Posthuman and new materialist perspectives*. Routledge.

Wolff, R. D., & Cullenberg, S. (1986). Marxism and post-Marxism. *Social Text, 15*(Autumn), 126–135.

Case Study 1: Ecosocial Experiments in Finnish Adult Social Work

Elisa Rainerma

Finding Ecosocial Social Work?

Having already completed degrees in environmental engineering and theology and work experience with refugees and immigrants, I turned to social work studies in search of means to help people more comprehensively. My move was motivated by my experience of the current (Finnish) welfare system as too bureaucratic and public services not meeting the needs of clients. I wanted to be able to bring about change in this.

My interest in ecosocial work was sparked even before my social work studies began. Due to my previous studies and work experience, I had a good understanding about the global environmental crises and their impacts on disadvantaged people globally. Moreover, I strongly felt that social work, being committed to tackle inequality and increase people's wellbeing, should and can work to ensure that the negative effects of environmental crises are more equitably distributed and that efforts are made to prevent the expansion of the crises. While for me ecosocial work was a welcome and desired subject area to start with, in what follows, I recall my own experiences of social work education in Finland and resultant views on teaching and learning ecosocial work. My master's degree contained three ecosocial elements: a 5 ECTS (credits as per the European Credit Transfer and Accumulation System) introductory course Social Work in Ecosocial Transition, a 10 ECTS internship in project, and eventually my master's thesis, which explored the foundations of ecosocial work and how ecosocial work could be introduced into Finnish adult social work.

E. Rainerma (✉)
University of Jyväskylä, Jyväskylä, Finland
e-mail: elisa.h.rainerma@jyu.fi

C. Forde et al. (eds.), *Teaching and Learning in Ecosocial Work*,
https://doi.org/10.1007/978-3-031-58708-5_4

The Ecosocial Framework: A Huge Transformation for Social Work with Adults

In 2020–2021, I did my internship in the ESF (European Social Fund) funded project called *TASOS (Taloudellinen toimintakyky ja sosiaalinen osallisuus) (Financial capability and social participation)*, carried out in collaboration with the University of Jyväskylä and five medium-sized municipalities in Central Finland. In TASOS we tried out nature-based and community-based methods in social work with adults. In the project team, which consisted of both social work researchers and practitioners, my task was to assist in compiling and sharing research-based knowledge on ecosocial social work, brainstorming with the team on what kind of nature-based and community-based methods could be implemented, and participating in these experiments. I also collected the empirical data for my master's thesis in this project, interviewing social workers about ecosocial social work.

Even if the ecosocial framework was natural for me, I soon realised that for my fellow students or the project participants it was anything but a given. First, ecosocial work is not yet very well known. Most social work practitioners are still quite unfamiliar with the ecosocial framework. Even if they know the concept, its actual content and practice applications remain hazy, and carrying out ecosocial work in practice is rare. Also in social work education, ecosocial work is still a marginal, though emerging theme. Secondly, ecosocial work is often difficult to grasp. I realised that both social work students and practitioners had difficulty incorporating the ecosocial framework into the existing service system and ways of working and envisioning its actual content and realisation. The ecosocial approach changes accustomed understandings of the role and goals of social work, including typical social worker and service user roles. In other words, rather than just a method, in the Finnish context the ecosocial framework presents and provides a completely new perspective on social work with adults, thus implying a huge transformation.

In the project I found that currently, social work with adults is highly system-oriented, which the social workers perceived as problematic. Due to obligations set by the Finnish labour administration, the work focuses largely on client employment issues and measures, aiming to spur the client's employability and labour market status, which besides necessary financial assistance leaves little room for anything else. Even when the social workers themselves view that the key problems of their clients do not enable a return to unemployment, employment measures define the work. The work takes usually place in the social welfare office and the time per client is short. Social workers do not have enough time to get to know their clients and their situations holistically, or to tackle the root causes behind the problems. The relationships between social workers and clients remains thin, which diminishes the opportunities and likelihood of promoting sustained changes in the lives of the clients. However, not all social workers even see promotion of change as the primary objective of social work with adults but appear content with provision of financial assistance and employment as the core of the work.

The ecosocial approach shakes up the previously described setting and can thus be mind-blowing for service users and social workers alike. In the TASOS project, social workers found internalising and adopting ecosocial social work as part of the practice difficult. In their experience the system-oriented approach was so pervasive that even the occasional nature-based or community-based events were difficult to carry out. The priority was to solve acute problems of clients as quickly as possible, usually by financial assistance. What is pursued through nature-based or community-based activities remained obscure, as these activities did not seem to relate directly to clients' problems, or at least didn't appear to solve them. It was also difficult to convince the superiors of the importance of ecosocial work. Social workers themselves perceived that in ecosocial work the client is at the centre of the work and the goals of the work are determined from the client's point of view, and this was new in social work with adults. Instead of mere financial support, ecosocial work aims to improve the overall wellbeing of the clients and thus extend the work beyond financial assistance.

During the project only a few social workers tried out some nature- or community-based activities. They realised that meeting clients outside the office, for example, taking a walk in the forest together, getting to know the clients in their own living environments instead of clients visiting the social welfare office, and generally working in a community-based manner had several benefits. First, it improved the quality of life for clients, many of whom made new friends and found meaningful content for days. Secondly, the relationship between the social worker and the client became more equal. Strengths and weaknesses of both were present in the relationship and this built trust in a new way. Clients also talked more openly about their lives and problems, which enabled the social workers to help them better. These benefits notwithstanding, however, social workers were also concerned that the time spent in this type of work would be somehow detrimental to their office-based duties and efficiency. In my view, in part at least, such inner conflicts highlight that the leap from mainstream bureaucratic social work practice to nature- and community-based ecosocial practice is huge and requires not only habituation but also profound rethinking and reorganisation of the work.

How Can the Ecosocial Change Happen?

The project TASOS was essentially about collaborative learning and trying out new ways of working and promoting change in the lives of the service users. The social workers involved emphasised that for them the key realisation in the project was the importance of their own transformation process as social workers, enabling them to work in new ways. In their case, transformation meant reflecting on their own way of working, as well as re-examining their personal and work community values and attitudes considering the ecosocial framework. To start off, this process required both knowledge of ecosocial work and the opportunity to experiment with it in practice, which the project facilitated.

Because the ecosocial approach is quite new in the Finnish context (e.g. Ranta-Tyrkkö & Närhi, 2021; Rainerma, 2021; Nöjd et al., 2023), and it significantly challenges taken for granted ways of doing adult social work, both social workers and social work students need strong research-based knowledge of ecosocial social work. They need to know what ecosocial social work as a concept means, why and in what context it arose, and what it aims to achieve. In general, there is still need for explication of why social work should address environmental issues and crises as part of its practice. In social work education, most examples of ecosocial work, for the time being, are about work done elsewhere, not in Finland. Often these descriptions are from communities and societies where concrete environmental disasters already directly affect people's lives. The problem of these kinds of examples is that they are easy to ignore in societies like Finland where environmental disasters are not yet commonplace.

In discussions with my fellow social workers and social work students, we have reflected on the fact that ecosocial work is specific to society and place and thus also the problems and content of the work varies. In Finland, for example, people do not live amid severe environmental damage, nor do social professionals have to tackle the concrete effects of environmental damage on clients' living conditions or lives. Yet, this does not mean that ecosocial work is not needed in Finland. In the TASOS, we gathered to learn about ecosocial work, reflecting on the acquired knowledge and brainstorming on what ecosocial work could be like in practice. Joint workshops and joint reflection paved the way for everyone's learning process and personal change. Amongst others, this process helped the social workers to realise that in social work with adults, the consequences of the ecological crisis are reflected in the lives of clients as experience of deprivation, meaninglessness, and exclusion, as well as poverty and unmet basic needs.

In the project, experimenting with nature-based and community-based methods was important for the service users as a collaborative learning process and for the social workers' personal professional development. In conclusion, rather than mere information delivery, the ecosocial framework is easier to grasp experientially, learning by doing with others – in this case as a team of social workers, service users, and researchers together. Based on the project experience, collaborative learning modes can enable the creation of new ideas and develop both ecosocial practice and theory further. Moreover, such processes, when successful, spread and facilitate the adoption of the ecosocial framework, supporting social workers' professional and personal learning processes and broadening of perspective. While reflecting on these experiences together creates opportunities for change in practical work, it is also important that the connection between theory and practice remains strong. Otherwise, ecosocial social work risks being seen as pointless tinkering and may fail to bring about any lasting change.

Bibliography

Nöjd, T., Kannasoja, S., Niemi, P., Ranta-Tyrkkö, S., & Närhi, K. (2023). Social welfare professionals' views on addressing environmental issues in social work in Finland. *Nordic Social Work Research.* https://doi.org/10.1080/2156857X.2023.2253456

Rainerma, E. (2021). *Ekososiaalisen työorientaation mahdollisuudet suomalaisessa aikuissosiaalityössä* [Possibilities of ecosocial work orientation in Finnish adult social work]. Master's thesis, University of Jyväskylä. http://urn.fi/URN:NBN:fi:jyu-202201311325

Ranta-Tyrkkö, S., & Närhi, K. (2021). Striving to strengthen the ecosocial framework in social work in Finland. *Community Development Journal, 56*(4), 608–625. https://doi.org/10.1093/cdj/bsab030

Chapter 3: Teaching Crises: The Uncomfortable Place of Ecosocial Work Within Social Work Education

Pieter Lievens

Introduction

This chapter is based on my experience in teaching "Dealing with Global Challenges" at Thomas More University of Applied Sciences (from 2008 till 2019) in Flanders and "Sustainability" (from 2016 till 2018) and "Global Challenges" (from 2014 till 2022) at KdG University of Applied Sciences and Arts, Belgium. The course "Global Challenges" was a three ECTS course (workload of 75 hours) of one semester (12 weeks, 2 hours per week teaching), compulsory for every Social Work student in their third year bachelor's degree, which include in Flanders General Social Work, Legal Counseling, Sociocultural Work, Arts and Cultural Mediation, and Human Resource Management. "Global Challenges" took place in an auditorium setting with 200 students. The aims were mainly awareness of the macro-level, deeper understanding of social situations, looking at relations and connections with other areas and sectors of society, holistic thinking and systems thinking, critical reflection, and radical understanding of developments in society including awareness of the future impact of these "Global Challenges" on the welfare of people. Basically, it was an introduction to global perspectives in Social Work.

Since 2005 I have taught an elective minor "International Perspectives on Social Work" which included an extensive chapter on global challenges. In the beginning, the wording *climate change* was kind of self-evident. Around 2016 I realized that this framing understated the ongoing events, so I changed the title of the chapter to *climate crisis*. Then, I realized that with a diplomatic crisis or a government in crisis, the "crisis" fades away over time. So since 2018 I speak about *climate*

P. Lievens (✉)
KdG University of Applied Sciences and Arts, Antwerp, Belgium

Thomas More University of Applied Sciences, Geel, Belgium
e-mail: pieter@ecosocialwork.eu

C. Forde et al. (eds.), *Teaching and Learning in Ecosocial Work*,
https://doi.org/10.1007/978-3-031-58708-5_5

catastrophe and "global heating" rather than "global warming," because this articulates more accurately what it is.

Using the word "catastrophe" feels a bit unfair because it refers to something that already happened and the framing somehow deprives young people of a hopeful future. Over the years, more and more, I have come to relate to what dystopian information does to engaged young people. In an auditorium setting (and certainly with online teaching during the COVID period), there is no room for emotional aftercare when facts and figures on global challenges lead to fear, anger, sadness, or feeling betrayed by older generations (Hickman et al., 2021). This is especially true for social workers as on the one hand they long to feel that efforts for a "more beautiful world" are not futile and that there is a contribution to be made to more positive progress. On the other hand, there is the negativism of "destruction of the planet" which creates a context of doubt about the long-term goal and the meaningfulness of their efforts. In this context, social workers can cling to the belief that there are always multiple "futures" possible, no matter how the climate catastrophe and ecological degradation will unfold. Like their profession always did, social workers can continue to work and collaborate on the most desirable scenario by promoting solidarity, humanity, and resilience, come what may.

Global challenges have always made a reference to the future of global society and the planet Earth. Next, let us take a look at the main elements of global challenges before we go into detail about teaching ecosocial work and the existential risk of climate catastrophe.

The Content of Global Challenges

In 2008, in the fall semester with the banking crisis, I decided somewhat arbitrarily to teach six main global challenges.

The first challenge is the knowledge crisis, containing limits to knowledge, limits of technology, and limits of science. For me, this is the most important and the most influential crisis regarding the other five global challenges. As people are often told that they are incapable of understanding the complex challenges, this often leads them to feel inferior to scientists, politicians, or opinion makers, and consider these experts as the ones who know "what is best." In this chapter, we question such common sense, look critically at crucial concepts, and challenge the framing of ideas. The challenge is getting out of information bubbles and empowering people to promote human rights, free speech, and democratic values.

The second challenge is the economic crisis, or rather *the economy as a system in crisis*. From facts and figures on growth, employment, private debt, public debt, and pollution, it is concluded that growth equals (and continues to equal!) degradation of ecosystems and ecosystem services on a global level. This degradation is twofold: on the one hand, *the sources*, being the input for the economic system, get emptied (see challenge 3 below: resource depletion), while on the other hand, *the sinks*, the output of the economic system, fill up (see challenge 4 below: pollution, greenhouse gases, and ocean acidification). So, there is a "lock-in": growth is needed for prosperity, for paying back private and public debt, and to give people work and an identity. Growth cannot be sustained into the future. It undermines welfare already today, and further growth undermines well-being in a fundamental way in the long run and will lead to an ecological disaster. Economics is a system in crisis and the crisis won't go away. The course elaborates on the alter-globalization movement, covers interesting experiments with change and other economic models, and touches upon neoliberalism, solidarity, basic income, *buen vivir*, gift economy, and LETS (or time banking). Each concept is linked to Social Work practices.

The third challenge is the commodity crisis. Most future social workers in Flanders (Western Europe) don't think in material terms. However, many human rights, like housing or food, have a material basis (see Ranta-Tyrkkö in this volume). This basis is fairly predictable from which the evolution of society and future well-being can be derived, especially in regard to access to basic rights, welfare and health. Fossil fuels like coal, oil, and natural gas are used for energy and petrochemical products like plastics, synthetic fibers (clothing and textiles), cosmetics, paint, electronics, and (automotive) components. They are subjected to depletion. Minerals and metals can be reused but cause a human rights problem during extraction and cause damage to local environments and communities. They are equally subjected to depletion. Biotic commodities are regrowable and renewable. All food crops like wheat, corn, coffee, rice, soya, fruit, vegetables, and meat and other agricultural crops like palm oil, wood, rubber, and cotton are not subjected to depletion. These crops come under pressure when phasing out fossil fuels, especially because chemical fertilizers and crop treatment are made of fossil fuels. Most studies project net adverse impacts on crop yields due to climate change. In oceans, a peak in fisheries has already been observed, and the world capture of fish has been in decline since 2000. Oceans are ecosystems that give free services. Fish is one of these. The ecosystem service "fish from oceans" has been replaced by "men made" aquaculture. In this part of the course, we touch upon the agroecological paradigm and compare it to the economical paradigm in agriculture, with some reference to the role of an ecosocial worker in green spaces, biophilic spaces, or community gardens. The course elaborates on the demand side by discussing nutritional patterns (fat and sugar), overconsumption of food (overweight and obesity), the evolution toward animal proteins, and the impact of meat consumption. Hunger and agrofuels, land grabbing, and deforestation are part of this learning unit. The commodity crisis makes the need for an ecosocial transition very clear. Social Work can initiate strategic experiments, build resilience by solidarity, and envision societal innovation in a circular economy or closed loop production. In this chapter, the basics of nuclear and renewable energy, rebound effect, greenwashing, cradle to cradle, and services instead of products are touched upon.

The fourth challenge is formed by the ecological crisis. Strictly considered, the commodity crisis ("the sources") is also an ecological crisis but currently less problematic than "the sinks." The atmosphere is a sink that is filling up with greenhouse gases causing catastrophic, irreversible climate disruption which leads to floods, heat (waves), drought, storms and cyclones, and deaths as a direct result of weather phenomena. This leads to indirect climate-related consequences like failed harvests, infectious diseases, food and water scarcity, sea-level rise, erosion, biodiversity loss, air quality, and forest fires. People suffer or die from "indirect effects of weather phenomena" (IPCC, 2019, p. 564). Consequently, humans react to direct or indirect climate impacts with conflict, looting, revolution, poverty, vulnerability, migration, suicide, price increases, unemployment, and economic crisis. Most of these processes are a consequence of linear global heating and risks resulting from dynamic interactions (IPCC, 2023, p. 128). Next to linear or step by step change, a risk of abrupt, nonlinear, large-scale climate disturbance exists due to looming global tipping points such as sudden sea-level rise due to ice-sheet instability, massive methane emission from permafrost melting or from deposits stored deep in the seafloor, leading to a runaway heating resulting in a 'hothouse' planet (Lenton et al., 2023, p.63, 64, 72, 73 and 153).

Next to the climate catastrophe, there is a historically unseen loss of biodiversity due to pollution, material excess leading to overexploitation (see Ranta-Tyrkkö in this volume), industrial agriculture, deforestation, urbanization, and habitat loss. This crisis has been occurring independently of the climate catastrophe, and unlike climate, it cannot be fixed, managed, or manipulated by technical means. Loss of biodiversity implies smaller ecosystem services (crop yield, fisheries, and forestry), less water and air quality, loss of viability, and resilience of ecosystems. Although human-driven extinction can never be overstated (being not only the death of species but also the end of their evolutionary destinies), this ongoing depletion remains a non-topic in society.

The course devotes attention to oceans, a global common that suffers from acidification, heavily disrupted marine life, dead zones, and deep-sea mining. Oceans used to absorb around 35% of the anthropogenic CO2 emitted in the atmosphere. The free service of capturing and storing CO2 by oceans is in decline. As nitrogen pollution has been very much in the media in Belgium since 2019, this issue is explained in detail.

The facts and figures on the ecological crisis are existential and call for need satisfaction and well-being while preserving natural capital, lower ecological footprints for companies and organizations, and a good life within planetary boundaries for families. This requires significant gains in energy efficiency but primarily in energy sufficiency (collective abstinence), a dematerialization of the economy leading to equitable distribution of material wealth, and applying a vision of well-being which is less connected to economic growth (see Rambaree in this volume). Local production and self-sufficiency are part of the ecosocial transformation. Other chapters of this book discuss the impact of ecological crises on vulnerability and justice and the role of social workers in applying normative principles, approaches, and actions for sustainable development. These are therefore not elaborated on here but are also part of the course.

The fifth challenge is the "inequality crisis." We look to inequality within countries by means of Pen's parade (referring to the Dutch economist Jan Pen), a visual representation of income distribution and inequality by means of a symbolic procession of 1 hour in which the whole population comes along, and the length of a person represents his income, the Gini coefficient compared over time and place and the evolution of the highest incomes. We give facts and figures on inequality between countries and inequality in the world. For example, we explore how much every family would earn if the global GDP would be equally distributed after a general taxation of 30% for public goods and services[1] like global ocean management, ecosystem restoration, global basic healthcare, and education. Based on 2017 figures, net available monthly income per family (on average two adults between 23 and 63 years old, 1.42 children below 23, and 0.42 elderly above 63) would be €3.185 in purchasing power parity. The course mentions the growth in global billionaires (e.g., from 15 in 1990 to 177 in 2023 in the UK). There is attention to *The Spirit Level* showing that health and social problems in a society are not related to income but to income inequality. Equal societies do better in terms of education, life expectancy, trust, violence, mental illness, drug abuse, obesity, and so on. Students learn about human development indicators and pathways to development. Inequality is linked with the ecological crisis by the observation that the super richest 1% (just 77 million people) emit as much carbon as the 5 billion poorest people on Earth (Khalfan et al., 2023). The course discusses how social workers can dismantle structural inequalities and touches upon advocacy and working at the structural and policy level.

The sixth and final global challenge is the so-called demographic crisis with three elements: population growth, aging, and urbanization. From past figures on world population, the current growth is explained and projections toward the future are made based on the evolution of the birth rate (UN DESA, 2022). We problematize this by referring to Thomas Malthus (1766–1834) and "deproblematize" this by explaining the concept of population momentum, the connection with life expectancy, and the level of schooling. Although in the course of the lifetime of my father, born in 1945 and died in 2009, the world population tripled from roughly 2.5 billion to 7.5 billion people, the concept of "overpopulation" is questioned by the observation that—roughly spoken—the planet is able to sustain 1.5 billion people living an "American lifestyle" and 13 billion people having an "Ethiopian lifestyle" (Earth Overshoot, Global Footprint Network). The concept of "population explosion" is contested by the observation that the means of production (GDP) grew 12-fold between 1945 and 2009 (population only 3-fold) and global food production each year grew faster than the population, so the food per capita increased by 50% since World War 2. In the context of population growth in the so-called developing countries, one should be careful with "birth reductions" being a form of "blaming the victim" since poverty and hunger have structural causes (systemic drivers) like discrimination, economic policies, corruption, political instability, displacement of people, violence, conflict, and war (HLPE, 2023, p. 57). Thinking starting with "There are too many people" leads to "Who or whom is too much?" People exist and hence have the right to a decent life. Empowerment and secondary education, especially for girls, are more effective than birth control and population or family

[1] Belgium spends around 24% of its GDP on public goods and services.

planning. "Each single person of the whole current world population can have a house and a garden of 100m^2 in the surface of Namibia" is a quote from the course. So basically in regard to resource scarcity and ecological considerations, consumption patterns and the economy can be more easily and faster adapted than the course of the population that is roughly fixed for next three decades.

Regarding aging, global life expectancy has almost doubled since 1945. Getting older is not a crisis, because it means that people remain healthy longer. Elderly people are a growing and super diverse target group for social workers with huge potential for (social) action, arts, and cultural inspiration.

Regarding urbanization, the course speaks about the growth of megacities, land usage, and population density. From a Social Work perspective (in Western Europe), attention should be given to urban ecology, livability of cities, air quality, heat-island effect, and urban planning like adding green and blue (public waters) to living conditions, especially in neighborhoods that need that the most. The relationship with the *hinterland*, referring to the place where water, energy, food, and other materials come from, is worth looking at from a Social Work perspective.

Basically, the sixth crisis is "deproblematized." There is no demographic "crisis." The other five global crises are directly linked to well-being and are likely to cause expanded human suffering in the future. That is why there is a "duty to act" for social workers, in the same way as they would act in light of witnessing child abuse. Social workers cannot pretend not to see or not to know about global challenges. Speaking up (*Parrèsia*) and questioning are the first steps for breaking away from the dominant culture that largely defaults on the obligation to act.

As previously described, the challenges may seem separated from each other. During the lectures, however, links are made between challenges. In-depth exploration of underlying factors, complexities, and nuances makes connections and relations clear that allow students to come to systems thinking and (multiple) futures thinking.

The didactical approach is the 5E model for teaching global challenges[2] (Lievens, 2018) is as follows:

1. **E**xperiencing the challenge and **E**xploring the solution: each lecture starts with experiencing the global challenge in everyday life with questions, pictures, or teasers that spark an interest and explore solutions or options to think differently (see Rambaree "disorienting dilemma" in this volume).
2. **E**xplaining and **E**laborating the challenge: analyzing profoundly the figures and trends for understanding of the status of the world in regard to that challenge. In this step, we elaborate on statistical data, theories, causes, consequences, and underlying assumptions to understand each challenge (see Rambaree "meaning perspective and critical discourse" in this volume).

---

[2] Due to the reduction in ECTS credits and changes in workload, this model couldn't be fully applied at KdG. Get in touch with the author for study guides, assignments, concrete exams, and more detailed descriptions.

3. Enlightening the challenge and Envisioning alternative views or solutions: we give examples and initiatives that answer the challenge. We discuss conditions and resources for action. We make reference to "transformative organizing" and not just "issue-based organizing" in Social Work. This step elaborates on visions, passion, and creativity to show what is possible when we use our full human potential.
4. Engaging in global change and Exemplifying local practices: the last step shows actions in practice or in the student's placement. Students are encouraged to go from understanding to professional roles and concrete behavior in practice (See Rambaree 'action to enact the altered perspective' in this volume) by means of an individual blog assignment and a collaboration in a collective wiki. In the written exam, there is an assimilation question given in advance. *What inspires action? What resources are needed for these achievable steps? How can you collaborate with existing initiatives from your position as a starting professional? How will you involve target groups and colleagues? How are these steps addressing structural injustices or ecological degradation?*
5. Evaluating the effect of actions and Evolving in thinking and understanding: to what extent has the idea, suggestion, action, or practice of the student contributed to approaching a more social just and ecologically sustainable world? Ecosocial work practices are seldom evaluated in terms of impact on global challenges as most of the goals and actions don't go beyond the local and regional society. Impact on future generations and communities elsewhere are however a key element in this context. A few weeks after the examination, students are offered the possibility to discuss their answers with the lecturer. Evolving refers to adapting thinking, strategies, or interventions based on the feedback, outcomes, or changing circumstances identified during the evaluation process. Ecosocial workers need flexibility and continuous improvement in response to new experiences, better understanding, and deeper reflection on the complex nature of ecosocial issues and global challenges. E-mails and informal discussions constitute a crucial aspect of this stage in the 5E model, particularly in the teacher's understanding and learning process. Students' questions, reflections, and challenges especially during informal interactions sparked by serendipity and events on campus or in society are essential elements in the didactic approach and should be encouraged throughout the entire curriculum.

Discussion of Teaching Ecosocial Work

A course on global issues is demanding for most students. As their main motivation is mostly direct help of specific target groups, dealing with global challenges is outside their comfort zone. The evaluation of this course is based on the taxonomy of Bloom. That causes great differences in marks: varying from very low to very high. Especially weak students tend to respond to assignments in a vague and normative way. But let us first discuss the content.

Merits of "Global Challenges" as a form of teaching ecosocial work

Firstly, the main added value of the course is the inclusion of the international and global dimension in the curriculum. Social workers have an international definition and a global agenda. These need to be an integrated part of their future job as it changes the identity of the Social Work profession from a local issue-based struggle to a global fight for justice and sustainability.

Secondly, the course gives a language for clearly stating what is going on and for formulating alternatives. Putting things in a larger perspective helps and motivates. The IFSW calls for *urgent global action* in a policy paper stating that "social workers have an important role in working with these complex yet interconnected issues on local, national, and global levels" in the context of "the world facing unparalleled levels of social and environmental crises" (IFSW, 2022).

Thirdly, insight into global challenges is liberating for students, as it helps them to understand the future. A student formulated it like this: "the course was a mental rollercoaster. It was as if I needed it for really becoming adult." Particularly for students who are hungry for understanding the world they live in, global challenges and the 5E model are a very powerful learning environment. Fundamental understanding of world-dynamics is often a surprise and challenges their existing worldview.

Fourthly, the course has the potential to organize incoherent information abundantly present in the media, enabling students to deepen their understanding and think more independently from mainstream media and social media algorithms that filter information and create "knowledge-bubbles."

Shortcomings of the Global Challenges course as a form of teaching ecosocial work

Firstly, ambiguity about whether this content is part of the "Body of Knowledge" in Social Work makes the link to practice difficult. Many practitioners believe that global heating or biodiversity loss is not an element of the profession and hence not part of the role of Social Work to act upon, especially in the Flemish context where Legal Counseling and Human Resource Management are part of Social Work. That is why the course is a bit beyond the everyday understanding of Social Work. Fact-based teaching, with numbers and tables, is not so common in social science. That is why "Experiencing and Exploring" and "Engaging and Exemplifying" in the 5E model are important. It is crucial not to leave it with local environmental or social issues, but really "upscaling" these experiences toward a global agenda and making links with global injustice and ecological sustainability.

Secondly, on a more theoretical level, many students fail to see in a holistic way how the social, economic, ecological challenges influence each other. This is explained by a lack of teaching interconnectedness and futures thinking in Flemish education. That makes it difficult to address the complexity of the macro-level and draws attention to actions at the microlevel, where some issues are tackled without understanding the systems they belong to and depend on.

Thirdly, the main shortcoming of the course is the lack of a fundamental choice to integrate the ecosocial perspective in the curriculum from the first day to the last day of education (see Section "Social Accountability of Schools of Social Work").

Fourthly, the most provoking in teaching global challenges is the fact that it brings the polarization of society into the auditorium. Students often fail to see the difference between facts or figures and opinions or beliefs they hear daily, especially when these are not grounded in observations. For example, when teaching the course "General Welfare" and making a comment to the WHO definition on health, no single student questions my expertise. A comment to the IPCC assumption on the linearity of climate change or the feasibility of carbon removal was questioned by some students who labelled it as "radicalistic" or "activistic" or made comments like "Oh no not again."

Last and not least, Global Challenges create discomfort for some students and faculty members because the root causes of social and environmental problems question the status quo and the underlying premises of society in a very fundamental way. As such, this is not new to Social Work. However, the preference of a group of students to a more conservative or traditionalist understanding and approach of Social Work creates polarization. An ecosocial perspective moves Social Work away from therapeutic and controlling practices and creates a strong ethical orientation and appeal. A lot of actors in society aim for sustainability but stay away from root causes and ethical orientations. When ecosocial work sharply questions certainties, like, for example, care for people and environment above income or "return on investment," it gives an uneasy feeling. A future with degrowth, more redistribution, and care for the Global South and future generations is not supported by everyone.

Experiences among students

In April 2015, we conducted a small piece of research on how the course was experienced by students. Some students found the topics very interesting and inviting to think critically about what is going on in our world e.g., "It makes us reflect on events, which actually affect us as well"; "It broadens your view on all kinds of aspects"; "It makes you dwell on important things that are happening." Others expressed that the content is very current and has a strong social relevance. Some said that there were many slides that were not significant until they got into the class and heard the explanation. Pictures and charts made everything very visual.

Some students stated that there were many numbers and data with too much information. The topics and challenges made the content too broad. Some students advised bringing down the level and using less graphs, charts and tables. Several students stated that they were forced to follow the teacher's opinion.

In essence, the course was controversial with both strong negative and strong positive feedback. Maybe this is in line with the marks, as students already knew their results when filling in the questionnaire.

Experiences as a teacher, the case of global heating

When I started teaching Global Challenges in 2007, based on the available literature with the existing scenarios, I concluded an estimate of between +3.5 °C and +4.5 °C global average temperature increases by 2100 (Fisher et al., 2007, p. 228). That estimate, around +4 °C by 2100, still stands. The continuation of policies implemented, leading to +3.2 °C, are currently not granted, and climate sensitivity or carbon cycle feedbacks could be higher than the best estimate (IPCC, 2023, p. 68). The small difference between +1.5° and +2° makes a world of difference (Carbon Brief, 2018; CASW, 2020, p. 4&5). Studies on the consequences of a world around 2.5 °C warmer in 2100 than today, already +1.4 °C warmer than preindustrial levels (WMO, 2023), are incredibly inhuman and truly pessimistic (Steffen et al., 2018; Lynas, 2021). Despite all actions of environmental movements, policy documents, and corporate promises, this trajectory toward "hothouse earth," posing an existential risk, still is the most likely scenario for me. This likelihood is based on genuine assessment of the underlying assumptions of the climate models like taking carbon out of the atmosphere after 2050 (87% of the IPCC scenarios assume large-scale technology that removes CO2 from the air although currently unproven, expensive, technically challenging and unrealistic to be scaled up), the current impact of global dimming due to human-made aerosols (Hansen et al., 2023), and the inability of models to take tipping points into account, like permafrost thawing not accounted for in climate projections (Glikson, 2018). There is a significant statistical likelihood that global heating will exceed 5 °C by 2100, especially in scenarios with strong economic growth. Some publications outline frightening scenarios and sobering ecological, social, and economic consequences of a significant warmer planet, like "Reasons for Concern" and the "Burning Embers Diagram" filtered out of the IPCC report in April 2007 (Smith et al., 2009), and "risks of large-scale discontinuities." Students feel pessimism when I teach. That is why I spent only one single hour on the climate catastrophe.

It is highly likely that my daughter will be alive in 2100. She will be living in ecological conditions and profound functional shifts in "System Earth" that haven't been experienced by any Hominin species during the last 6 million years and beyond.[3] Two elements of explanation: firstly I fail to see this conclusion as an opinion on climate. For me this is pure physical science like an ice cube that will melt in a room of 21 °C. Even if that process has never been observed before, it can be deduced that further melting will take place. Having said that, secondly, this does not mean that one should passively watch this happening. Given the injustices global heating creates, one should try to contribute as much as possible to a desired scenario for the future.

How to deal with this pessimism hanging over the future and gain power over the system? Why would we call behavior of clients or anybody as "acting well" or "acting badly" when the 77 million super rich emit as much as carbon as the 5 billion poorest people on Earth (Khalfan et al., 2023)? How do we deal as a professional in "rich countries" with jeopardizing well-being in the very long run, by trying to achieve well-being in the short run? Should we choose to contribute to strategies for mitigation or adaptation? Mitigation means avoiding further heating by reduction of

[3] Sapiens evolved from around 200,000 years before the present.

greenhouse gases and degrowth in certain economic sectors, like petrochemicals, cement, and military, and in regions with a high ecological footprint. Adaptation means accepting change and taking protection measures by building higher dykes, larger food stocks, more air conditioners, or emergency capacity for hospitals.

Perspectives on Social Workers Dealing with Global Heating

Why would Social Work combat further global heating (= mitigation)? First, runaway global warming threatens the well-being of billions of people. Advocating against further heating safeguards the current ecological conditions that provide quality of life. Securing quality of life, advocacy for future generations, and people now already suffering from a warmer planet can be a role of Social Work.

Second, certain impacts such as reduced agricultural yields, biodiversity loss, and loss of ecosystems and their ecosystem services cannot be addressed curatively. The damage is irreparable and affects first and mostly people in the global south, whose life depends on these ecosystems. When Social Work advocates for prevention of further heating (*every 0.1 °C less counts*), it might lead to protection and continuity of daily livelihood.

Third, only choosing for adaptation would leave the injustice and root causes of suffering in place. People who have no responsibility for global heating will continue to be victims. It is the job of Social Work to address injustice. Countering overconsumption and opposing neocolonial practices, for example, in resource extraction by large *fossil* corporations[4] can help to create a more just society. The international definition of Social Work and the global agenda invite social workers to engage in transformative practices toward a new, sustainable, and equitable society.

Fourth, the climate catastrophe being just one single element of the ecological crises, focusing on climate adaptation alone, risks keeping the other ecological crisis out of vision. Advocating against further global heating has the potential to address other ecological risks such as resource scarcity, pollution, endocrine disruption, and particulate matter. Addressing these simultaneously improves people's health, quality of life, and well-being already today. Improving well-being is at the core of Social Work.

Why would Social Work focus on the effects of global warming (=adaptation)? First, people are affected and displaced by climate. They suffer or leave their country due to increased risk of flooding, persistent drought, or heat waves. It is the task of Social Work to support and take initiatives to help affected communities and

[4] What make governments choose to steer the behavior of billions families, businesses, and organizations with a jumble of measures to promote "sustainability," instead of imposing some simple quotas and rules around pricing on—probably less than 200 worldwide—coal, oil, and gas companies to keep global heating limited?

integrate newcomers in the short run by offering assistance and in the long run by advocating for a legal status for climate refugees.

Second, in Flanders, like in many other regions, there are already three times more heat waves and more days of intense rain than in the 1970s, affecting particularly people who live in urban centers (heat islands), in poorly insulated or ventilated houses, with poor health (asthma) and elderly people less able to protect themselves. Social Work can set up a service for the elderly during heat waves or advocate places where people of disadvantaged neighborhoods can cool off.

Third, mitigating the climate catastrophe is the responsibility of the entire society and requires involvement of numerous actors. If all sectors are involved, social work can focus on its core mission: caring for the most vulnerable and recognizing that socioeconomic status, age, gender, and race intersect with climate impacts. As social work is already short of resources due to austerity, it should not "lose time" with mitigation and focus on target groups to solve their immediate problems and communities adapting to the consequences of global heating.

Fourth, due to the Paris Agreement, money might go to the wealthy, for example, as subsidies for electric cars. Yet there are already climate victims, such as people at risk due to poor health or living in poverty, who cannot adequately resource themselves due to lack of resources. Social work needs to give the already suffering a voice and demand resources to meet their needs and ensure that the transition overall is just, reducing inequalities and securing social and basic rights for all within planetary boundaries. By highlighting the inequality of the climate risks and—when pertinent—of the mitigation and adaptation measures, social work puts social sustainability on the political agenda, facilitates participation, and promotes solidarity within current generations.

Social Work in times of crisis

It is important to see vulnerability to global heating as an outcome of historical, *purposely executed* socioeconomic processes[5] that reflect power inequalities. Looking to "vulnerable communities" from an apolitical lens can lead to blaming those who cannot "adapt." Adaptation to changing climate is in this sense a very neoliberal idea: it expects individuals or communities to adapt. Those who can't lack adaptive capacity (due to their own shortcomings) and need to be taught resilience (as those in power sometimes claim having provided opportunities and knowhow). Ecosocial workers should therefore carefully disentangle concepts like "vulnerability" (as it is not just "unluck" to be at the wrong place) or "resilience" (as it is not just a capacity to be learned) as these concepts are heavily affected by power relations and the lack of access to basic resources like land, water, and food and basic rights like participation in decision-making.

Moreover, ecosocial work should not overly focus on global mean temperature increase. Heat is only one of the many risks and exposures of weaker groups in society, precipitation, flooding, wind, storms, drought, and fire each causing serious loss of well-being. In line with Xu and Ramanathan (2017), mitigation goals should

[5] See Climate Summit of Noordwijk, Netherlands, November 1989.

be rather structured in terms of risk categories instead of mere temperature thresholds. In this way, the distribution of risks from dangerous over catastrophic to unknown, meaning existential, over a certain period of time becomes visible. Global heating is not only about the world getting few degrees warmer, but a chain of social, economic, and environmental phenomena that interact and form an existential threat.

Ecological degradations, both on the side of the sources and the sinks, aggravate existing injustices at present and toward people born after 2000. These are likely to worsen in the coming decades. Social work therefore must combat further ecological and social degradation and deploy resources and methodologies to deal with earthly challenges. Global heating is one of them and should not be left solely in the hands of technicians and engineers.

Why are social workers more needed than engineers or technicians?
First, while producing smaller, more energy-efficient products reduces their environmental impact, simultaneous economic growth increases consumption: more sales, more usage, more miles, and continued overall global ecological degradation. This is the "rebound effect" or "Jevons paradox" and observed many times. That is why education, prevention, and regulation are needed. Ecosocial workers can tutor, promote collective *refuse* or less use, and play a role in the formulation of legislation that takes vulnerable people and rebound into account.

Second, fundamental change is about social and political struggle over resources. "Leaving the oil in the soil, the coal in the hole, the tar sands in the land" confronts the massive economic value at stake. The existing stocks, assets in balance sheets, and means of production will not be easily abandoned. Engineers and technicians mostly work for the owners of resources and for goals defined by them. In such a conflict of interests, social work needs to build power and to mobilize for change. Often technologists claim that something is "inevitable." That is why social workers also need understanding of the system and technical knowledge.

Third, oceans, climate, biodiversity, and Earth's resources are in essence commons. Therefore, responding to ecological crises can be done within the framework and existing infrastructure of social work dealing with commons: from a neighborhood center where people can borrow drill or lawnmower, over community owned energy production, to countrywide citizens science mapping air pollution. Ecosocial work can help to establish an inclusive governance structure, equitable sharing of benefits and responsibilities or duties, legal protection, and conflict resolution around commons from the community level to the global level.

Four, resource scarcity increases inequality. The gap between people having access to raw materials (such as neodymium for electric cars) and the people not having access will increase. At the same time, some *technical* "solutions" like the allegedly sustainable aviation fuels have a negative impact on food security and land use. Many engineered solutions can thus create new injustices and are often not equally accessible to everyone.

Five, social workers need to bring in perspectives that technicians and engineers do not think about. The social and environmental impact of an economy powered by solar panels, wind turbines, batteries, and fuel cells may turn out to be as negative as an economy with oil from tar sands and gas from shale rock. Moreover, it may be a conscious choice to shift pollution in Canadian tar sands and the risks of fracking in the USA to devastating mining in Mongolia or detested metallurgy in Malawi under the guise of "switching to sustainable energy," resulting in a "green" economy having negative effects on biodiversity, arable land, water availability and quality, human rights, child labor, and heavy, dangerous, and unhealthy working conditions. Developed nations should not further externalize their environmental impacts to less developed countries (Jorgenson, 2016).

All this boils down to is that social workers need a holistic view. More than technical or economic feasibility, combating global heating is about social feasibility. If around 2030 it appears that the +2 °C target is no longer achievable and some tipping points are already observed, technical solutions like reflecting sunlight away from Earth's surface by deploying reflective aerosols in the stratosphere, using space-based mirrors to cool the planet, or building the Northern European Enclosure Dam to protect against the inevitable sea-level rise should be democratically discussed and decided on a global level as there is no way back afterward. As engineering solutions require further ecological degradation, the crucial role of ecosocial work working is crystal clear.

After 2030, trade-offs between mitigation and adaptation will become sharper. With its global agenda, ecosocial work must promote mitigation that avoids further injustice, whereas adaptation often covers smaller sections of people, being limited to national or regional contexts. Affluent countries should see mitigation as a fundamental obligation based on global solidarity and understanding of the colonial injustices that have been created by systems of economic growth. This brings this chapter to the core: how to look at global challenges and examine the underlying assumptions of our worldview and our position in global change.

Perspectives on Ecosocial Transformation

Naming perspectives, each reflecting different attitudes and approaches, can help to understand positions toward underlying assumptions—with each position being to some extent true or visible when teaching ecosocial transformation. Elaborating and adapting on four narratives of the Anthropocene from Bonneuil (2015), we suggest the following perspectives:

First, there are forms of neglect or disregard for environmental issues. Global challenges are underestimated, possibly due to lack of awareness or a belief that the Earth is resilient enough to withstand human activities. The nonhuman world is largely ignored. Even in a context of allegedly "growing awareness," this perspective minimizes the ecological crises. Individuals and societies downplay the significance of environmental problems and focus more on the short term without considering longer term or the global scale which leads to unsustainable practices and further ecological degradation. In community work, this position is visible in cultural and art-related work with the goal to "build community," divert, or amuse. In deep ecosocial work, this neglect is observed in a belief in an inherent balance of nature.

Second, there is fatalism or powerlessness regarding ecological degradation. Climate disruption and biodiversity loss are seen as inevitable, resulting in disbelief that the course of the planet can be altered. Individuals and societies fail to take proactive measures. Apathy contributes to a self-fulfilling prophecy. This perspective hinders efforts to implement positive change, and inaction exacerbates ecological degradation. In social work, this position is visible in individual practices that relieve eco-anxiety (see Ahlin, this volume), ecological grief, and solastalgia.

Third, there is the ecomodernist position or belief in techno-fix. Further economic growth, scientific advancements, and human innovation will solve global challenges by decoupling economic growth from environmental impact through progress. This "green growth" position is observed in some manifestations and narratives of Green Social Work. However, when individuals and societies do what they have always done, they will receive what they always got: ecological degradation, inequality, and unsustainability. Ecomodernism may overlook potential risks, unintended consequences, further destruction and social objection, undesired power relationships, growing injustices, and existing inequalities.

Fourth, there are posthumanist and post-sustainability positions that challenge the anthropocentric view that places (a small subset of) sapiens at the center of the environmental discourse. Based on the interconnectedness of all living beings and ecosystems and the deep history of the Living Planet, a more inclusive perspective is taken: "We are nature." This position is an invitation to reconsider what and who we are as human beings on this planet. Environmental crises are in the deepest sense a crisis of the mind and the soul. What do we need to become for a liveable future? This is difficult to translate in practical terms and faces resistance from those who are deeply entrenched in anthropocentric perspectives. That is why the course leads to critical learning regarding the prevailing mental infrastructures and dominant narratives.

Taking Different Perspectives in Ecosocial Teaching and Learning

Along with the previously described positions, values, attitudes, missions, and goals also shape the discourse and decision-making and hence the teaching and learning of ecosocial work (see Rambaree in this volume), which makes teaching global challenges politically sensitive. Social workers with a neoliberal view start from the premise "We can do this!". By working together, humans can find solutions for both ecological and social problems. Social workers with a liberal view start from mutual understanding, finding consensus, and looking for steps "in the right direction." Radical social workers with a critical view start from structural causes, take postcolonial positions, and try to act upon the root cause of both ecological collapse and social injustices. Ecosocial workers have a post-critical view and aspire to go beyond human-centric thinking. Ecosocial workers see symbiosis with the Living Planet as a legitimate goal for intervention and they recognise the political existence of "the nonhumans" to whom *we* are bound and who deserve *our* care by demanding more humility and caution.

It is important here to recognize that there is no single position nor one "right" position. People have pluralistic positions, and depending on the context, the one or the other perspective, or rather narrative, is more or less suited. Environmental and sustainability education in essence should be pluralistic. From any of these perspectives, Social Work can propose similar alternatives such as a 4-day working week, a wealth tax (on luxury consumption, energy or resource use), a maximum income, banning investing in or subsidizing of weapons technology and arms, fossil sectors and mining, limiting marketing in public places, seabed mining moratorium, and so on.

Over the years of teaching global challenges, I have become increasingly critical toward solutions embedded in modernity and Western-centric perspectives. Many techno-fix solutions introduced over the years proved nonfunctionable or non-scalable. In line with Eversberg et al. (2023), I have realized that often technocratic promises and ecomodernist solutions function themselves as barriers to the needed ecosocial transformations. As economic growth has always had a very strong correlation with ecological damage, why would we believe that we now, unlike in the past, can decouple growth from resource depletion, climate disruption, biodiversity loss, or other challenges? Notwithstanding that, overall, things are getting better for humanity and there are more opportunities for people available in general, further growth of advanced economies and rich regions undermines the welfare of individuals and communities already today.

While the academic freedom and open mindedness toward different views in higher education are in principle cherished, profound degrowth or broader de-civilization thinking is sometimes marginalized, and people with this position risk being stigmatized. Thinking freely within a certain paradigm is welcomed, but thinking from a posthumanism position is often socially not accepted (Kopnina & Cherniak, 2016). Part of teaching and learning ecosocial work is about exploring and changing connotations of words, narratives, and mental infrastructures, as modernity is not a neutral starting point. Ecosocial transformation requires employing new wordings and frames in proportion, including inviting students to de-learn and to think about "ex-novation" (instead of in-novation). Transformation is thus not only about solidarity with the Global South or future generations. It is also about thinking outside our current culture and progress-oriented Western worldview (see Anand in this volume). Posthumanism is often obscured by eco-fascism that puts our "nature" above "human animals." Yet, saving "nature" or taking care of the planet is no more than securing the essence of our humanity.

While for some degrowth might evoke the connotation of replacing paid social work with voluntary or informal practices, it does not mean giving less value to social integration through paid labor. For the billion poorest people on the planet becoming richer and hence striving for economic growth is a legitimate way to improve quality of life or protect against climate disasters. Working for the betterment of the lives of the vulnerable and excluded remains, even in a posthumanist position, which is just one of the many pluralistic positions that education should incorporate. In an overall context of social work education emphasizing that well-being, empowerment, and social protection are dependent on economic growth, the course on global challenges leads to cognitive dissonance among students.

Dealing with Cognitive Dissonance When Teaching Ecosocial Work

Students feel uneasiness and discomfort when they must deal with information that is incompatible with their current worldview and often deploy mechanisms to reduce cognitive dissonance both during and after class.

When new knowledge is incompatible with existing knowledge, students' reactions vary from questioning, neglect or adapting the new knowledge to their existing belief system, rationalizing various ethical positions, or questioning the expertise or legitimacy of the person who brings this information. The same applies to teachers and researchers. Let me give a personal example of information that could cause cognitive dissonance.

My family and I use our car only occasionally. When we drive, I limit speed to reduce the environmental impact. During the fall of 2018, we went on holiday to Texel in the Netherlands by car. By driving only 90 km/h, our travel takes 1.5 hours longer, but this saves 6 liters of fuel compared to driving at maximum speed. Two days later, I am sitting on an A350-900 to Hanoi on behalf of an institute of higher education. One seat in this "state-of-the-art aircraft" for that distance consumes 400 liters of kerosine (120,000 liters in total), or 800 liters of CO2 emission for the entire trip, not even accounting the net warming effect of contrail cirrus and NOx, being twice as high as CO2 emission. What do the climate efforts of my family mean in relation to these emissions? A one-way plane trip consumes as much energy just by one seat as our family of two adults and two teenagers in 4 years and 2 months. With a lower seating capacity or a stopover, this would be even more. That academic year, I made two more intercontinental and one European trips by plane for an organization that has sustainability in its mission statement almost fully subsidized by a government that aims to keep global warming well below +2 °C.

Education on climate brings about existential concerns, which students feel often inhibited about talking openly about in class and which manifests and unfolds itself in a diversity of expressions (Vandenplas et al., 2023). As one teacher preparing his teaching expressed, "Climate Change is emotionally, it is existentially so overwhelming that it is almost impossible to live with" (Vandenplas et al., 2023). Teaching and learning ecosocial work can therefore be confronting and confusing. A genuine learning experience can be both triggered and hindered by the existential dimension and emotions related to it. These dynamics can affect educational relationships and have psychological effects for many stakeholders. Institutions of higher education should provide support both on the level of the content and on the level of the didactical setting to help teachers to provide the right education in the right circumstances.

Global challenges, such as the climate catastrophe, are framed in the media as "now or never" or last chance strategic decisions. "Act Now!" is a clip released in 2008, in which a child explains tipping points and says that "we have between 4 and 10 years to make sure we don't reach this fatal point," based on science of the day (Friends of the Earth International, 2008). Writing this chapter 15 years later, the growth of greenhouse gases in the atmosphere continues. The conclusion available to students is that disaster is inevitable, although the prediction of temperatures by 2100 in the "with current policies" scenario has gone down by 1 °C, but is still +3.2 °C compared to well below +2 °C in the Paris Agreement. How to deal with both a sense of global failure and a sense of urgency in educational contexts?

Ecosocial workers or educators confronted with cognitive dissonance, climate distress, or eco-anxiety should see these emotions *not* simply as caused by awareness as such. Feelings of injury, grief, rage, and so on are caused by more powerful others: governments, companies, or scientists failing to act (Hickman et al., 2021). For people born after 2000, the destruction of the planet is personal. Current narratives largely risk individualizing cognitive dissonance and the absence of hope suggesting a personal response to work on it. Yet, action needs to be taken by those in power by placing the emotions of young people at the center of decision-making (Hickman et al., 2021).

Solutions for "saving the climate" are problematic in relation to other ecological, social, and political boundaries (Smil, 2016). That awareness takes hope away. While dystopia is to be avoided in social work, facts and figures on global challenges bring along anger, sadness, hopelessness, and frustration. A way of alleviating cognitive dissonance and navigating political intricacies is teaching less pessimistic content and orienting to near future (2030–2040) rather than far future and to nearby rather than the global level. Is this adhering to a perspective that downplays the significance of global challenges, hence minimizing the need for radical change and contributing to further ecological degradation? What about students keen on understanding the world they live in? As some posthumanist, de- and postcolonial perspectives have a huge potential for a leap in competencies like systems thinking for students who are open to question mental and cognitive positions that are taken for granted.

Discussing hope itself during teaching could be instructive: "Naming what we hope for" rather than hoping that things will work out (Macy & Johnstone, 2012). Ecosocial workers not only do crisis management or therapeutic work but also use visions, passion, and creativity to collaborate for the most desirable future and contribute to global change. This requires safety (smaller groups), time (profound process is possible within other educational duties), and space (teacher is available and can give what is needed) (see Section "Social Accountability of Schools of Social Work").

Regarding cognitive dissonance, a sensitive content warning in the study guide or in the course description stating that *facts and figures and statements of the lecturer may be upsetting* may be useful. *As this content is directly related to the learning outcomes, it cannot be omitted. Contact study counseling or the lecturer if you have concerns about this.* In contexts like Flanders, where the ecomodernist perspective is dominant and disregard for environmental issues common, an informed consent on positions of the lecturer or the institute might be suited as well as an explicitly stated learning outcome that students will be capable of discussing ecosocial practices or expressing opinions from a posthumanist, decolonial, and postcolonial perspective among other positions. This requires agreement that these positions are part of teaching and learning ecosocial work and the curriculum enables us to build toward this goal.

Social Accountability of Schools of Social Work

Teaching and learning ecosocial work enables future social workers to deal with the complexities of global challenges without despair, to develop understanding that connects their local practice and context with the global ecological and social systems, to experience a holistic sense of belonging and interconnectedness with all living beings and all ecosystems, to advocate an ethical and sustainable relationship in harmony with nature, to rebuild symbiosis, and to work together in ways that open up sustainable possibilities for ecological regeneration and just coexistence with all living beings.

Schools of Social Work that choose to offer this kind of content to students are open to fundamental alternatives to development, welcome different views, and are open to a collective process of learning and unlearning that may lead to better and viable futures currently difficult to imagine. They educate social workers who are capable of transformative practices toward a new, sustainable, and equitable model of society even when they are dismissed as radical or extreme. By doing this, schools of Social Work take responsibility for the desired impact: *How do we contribute to a just and sustainable society? What do we want our students to be (as human beings)?* They establish a teaching and learning culture where values, facts, and political positions of ecosocial workers are jointly explored.

Faculties which do this should provide training and support for scholars to enhance their understanding and professional development. Enough time should be dedicated to exchange between faculty members about evolving content and pedagogy. Adequate funding, in terms of allocated work time, is needed to follow research on global challenges, which is a fast moving and broad field and for developing the pedagogy in ecosocial work or sustainability education in general.

Schools could formally pronounce their position, scope, and standards for teaching, research, and service delivery on the ecosocial transformation (including on economic growth). Such a position paper could be an integral part of the managerial loop justifying curriculum, action and evaluation programs, and incremental progress in excellence toward the anticipated impacts for the community and global society.

Heads of departments, program directors, or any staff member should explicitly provide support when posthumanist or postcolonial perspectives are marginalized or attacked due to polarization in society or post-truth skepticism. This culture of support and regeneration is needed to guarantee educational practices of inclusive pluralism. Students and lecturers should be encouraged to participate in outreach activities that share and nourish these with the broader community.

Conclusion on *Teaching Crises*

One can disagree on normative questions but not on the facts and figures that describe the state of the world as it is. In a context of decades of ongoing transformative rhetoric of policy bodies like the UN or the European Commission (Eversberg et al. 2023), teaching global challenges has been an endeavor toward clarity and truth. It is a privilege to engage with young people around solid science and understanding of systems dynamics. Over the years, this area of expertise has become a sobering and confronting story about what will likely happen during the students' lifetime. Notwithstanding the human resilience and the restorative power of ecosystems, I have come to understand that global society is entering a post-sustainability area in which life-supporting services, resources, food, and habitable space will be under pressure and the number of displaced people, feelings of existential risk, and the need for global redistribution and solidarity are greater. This has direct consequences for social work. Teaching ecosocial work therefore should be not mere "as preparation" for certain job competencies and responsibilities, but "as separation," questioning our dominant culture. Educating can be understood in the meaning of the ancient Greek word *scholè*: a free place where students can be liberated from what determines them. Today schools can help learners to become free from unsustainable culture and habits by fundamentally rethinking teaching starting from renewing content and finding partners and frameworks toward global justice. Taking responsibility for Earthly challenges means questioning the status quo and linking social work education again to values, ethics, social justice, and ecological sustainability. That is why unmasking claims that a certain response to a global challenge is "inevitable" and formulating political alternative standpoints and activism for justice are crucial elements of teaching and learning ecosocial work. In a Western European context, a lot of barriers need to be removed for this. A few of these on the institutional level are lack of expertise and experience in the field of activism and advocacy, curriculum constraints, market-oriented recruitment, historical patterns of institutional logics like short-term focus, and dealing with the institution's own cognitive dissonance when bringing up root causes. Teaching and learning ecosocial work imply removal of mental barriers, such as being able to critically assess social work's own premises and assumptions in the context of a growth-oriented consumerist and extractivist society in which global challenges, for example, are reduced to climate "change" and climate "change" is reduced to "temperature." By endorsing such reductions, social workers themselves become agents that confirm the dominant relationship with nature and reconfirm all inequalities that came and still come along. Education is not neutral but deeply entrenched with a worldview. Posthumanist teaching of ecosocial work inspires and encourages practitioners, teachers, and students alike all over the globe to swim against the current, especially in "advanced economies."

References

Bonneuil, C. (2015). The geological turn, narratives of the Anthropocene. In C. Hamilton, F. Gemenne, & C. Bonneuil (Eds.), *The Anthropocene and the global environmental crisis: Rethinking modernity in a new Epoch* (Vol. 2015, pp. 15–31). Routledge.

Carbon Brief. (2018). *The impacts of climate change at 1.5C, 2C and beyond,* retrieved 17 December 2023 from https://interactive.carbonbrief.org/impacts-climate-change-one-point-five-degrees-two-degrees/

CASW. (2020). Canadian Association of Social Workers. *Position Statement "Climate Change and Social Work"* by Schibli K.

Eversberg, D., Koch P., Lehmann R., Saltelli A., Ramcilovic-Suominen S., & Kovacic, Z. (2023). Promises of bioeconomic change as a strategy for avoiding socio-ecological transformation. *Sustainability Science, 18*, 557–568. https://doi.org/10.1007/s11625-023-01321-4

Fisher, et al. (2007). Issues related to mitigation in the long term context. In *Climate change 2007: Mitigation. Contribution of Working Group III to the fourth assessment report of the intergovernmental panel on climate change*. Cambridge University Press.

Friends of the Earth International, Act Now!, 3 dec 2008, Director: Nic Balthazar. Music: Hooverphonic. Retrieved on 17 December 2023 from https://www.youtube.com/watch?v=T7mETr1SqXk

Glikson, A. (2018). The methane time bomb. *Energy Procedia, 146, 23–29. ISSN 1876-6102.* https://doi.org/10.1016/j.egypro.2018.07.004

Hansen, et al. (2023). Global warming in the pipeline. *Oxford Open Climate Change, 3*(1). https://doi.org/10.1093/oxfclm/kgad008

Hickman, et al. (2021). Young people's voices on climate anxiety, government betrayal and moral injury: A global phenomenon. *The Lancet*. https://doi.org/10.2139/ssrn.3918955

HLPE. (2023). *Reducing inequalities for food security and nutrition*. CFS, HLPE-FSN.

IFSW. (2022).*The role of social workers in advancing a new eco-social world,* Policy UN May 24, 2022 retrieved 17 December 2023 from https://www.ifsw.org/the-role-of-social-workers-in-advancing-a-new-eco-social-world/

IPCC. (2019). *Climate Change and Land*, Cambridge, UK and NY. https://doi.org/10.1017/9781009157988

IPCC. (2023). *Climate change 2023: Synthesis report*. Contribution of Working Groups I, II and III to the Sixth Assessment Report of the Intergovernmental Panel on Climate Change, Geneva, Switzerland, p. 184, https://doi.org/10.59327/IPCC/AR6-9789291691647

Jorgenson, A. (2016). The sociology of ecologically unequal exchange, foreign investment dependence and environmental load displacement: Summary of the literature and implications for sustainability. *Journal of Political Ecology, 32*, 335–349.

Khalfan, et al. (2023). *Climate equality: A planet for the 99%*. Oxfam International. https://doi.org/10.21201/2023.000001

Kopnina, H., & Cherniak, B. (2016). Neoliberalism and justice in education for sustainable development: A call for inclusive pluralism. *Environmental Education Research, 22*(6), 827–841. https://doi.org/10.1080/13504622.2016.1149550
Lenton, et al. (2023). *The global tipping points report 2023*. University of Exeter.
Lievens, P. (2018). *Study guide dealing with global challenges*. Thomas More.
Lynas, M. (2021). *Our final warning, six degrees of climate emergency*. Harper Collins UK.
Macy, J., & Johnstone, C. (2012). *Active Hope: How to face the mess we're in without going crazy*. New World Library.
Smil, V. (2016). *Energy transitions: Global and national perspectives*. Bloomsbury Publishing.
Smith, et al. (2009). Assessing dangerous climate change through an update of the Intergovernmental Panel on Climate Change (IPCC) "reasons for concern". *PNAS, 106*(11), 4133–4137. https://doi.org/10.1073/pnas.0812355106
Steffen, et al. (2018). Trajectories of the earth system in the Anthropocene. *PNAS, 115*(33). https://doi.org/10.1073/pnas.1810141115
United Nations Department of Economic and Social Affairs, Population Division. (2022). *World population prospects 2022: Summary of results*. UN DESA/POP/2022/TR/NO. 3.
Vandenplas, E., Van Poeck, K., & Block, T. (2023). 'The existential tendency' in climate change education: An empirically informed typology. *Environmental Education Research*. https://doi.org/10.1080/13504622.2023.2246694
WMO. (2023). *World Meteorological Organisation, Provisional State of the Global Climate.*
Xu, Y., & Ramanathan, V. (2017). Well below 2 °C: Mitigation strategies for avoiding dangerous to catastrophic climate changes. *PNAS, 114*(39), 10315–10323.

Part II
Practices in Ecosocial Work Teaching and Learning

Chapter 4: Reenvisioning Social Work Curricula: Ecosocial Work Is *Everybody's Thing*

Nicole Mattocks, Melissa Indera Singh, Lauryn Smith, Georgianna L. Dolan-Reilly, Meredith C. F. Powers, and Kelly Smith

Introduction

The impacts of climate change and environmental degradation are experienced and recognized globally. Vulnerable populations (e.g., racial and ethnic minoritized populations, women, the global poor, and individuals with disabilities) are disproportionately impacted, creating environmental injustices (IPCC, 2021; Bullard et al., 2008) by exacerbating existing inequalities and systems of oppression (Grand Challenges for Social Work, 2016; Kaijser & Kronsell, 2013). Often, those least responsible for the causes of climate change disproportionately experience the brunt of its consequences, including relocation, financial burden, social isolation, and physical and mental health impacts (Besthorn & Meyer, 2010; Benevolenza & DeRigne, 2019; Hayes et al., 2019; Kaijser & Kronsell, 2013).

N. Mattocks (✉)
Department of Social and Behavioral Sciences, Colorado Mesa University, Grand Junction, CO, USA
e-mail: nmattocks@coloradomesa.edu

M. I. Singh
Suzanne Dworak-Peck School of Social Work, University of Southern California, Los Angeles, CA, USA

L. Smith
University of Maryland School of Social Work, Baltimore, MD, USA

G. L. Dolan-Reilly
School of Social Work, Sacred Heart University, Fairfield, CT, USA

M. C. F. Powers
Department of Social Work at UNC Greensboro, Greensboro, NC, USA

K. Smith
Columbia School of Social Work and Adelphi University School of Social Work, Garden City, NY, USA

C. Forde et al. (eds.), *Teaching and Learning in Ecosocial Work*,
https://doi.org/10.1007/978-3-031-58708-5_6

Such environmental and climate injustices to ecosystems and all species, including the human species, are considered wicked problems. "Wicked problems" (Koloko, 2012) result in the convergence of chronic and complex injustices, including colonialism, ableism, oppression, and violence based on race, gender, and sexuality, structural inequality, climate injustice, the COVID-19 pandemic, and economic fallout. bell hooks succinctly phrased the roots of such wicked problems as *imperialist, white supremacist, capitalist heteropatriarchy* (hooks, 2013) which results from anthropocentric worldviews, the flawed growth ideology, false dichotomies (e.g., people vs. nature), over-consumerism and materialism, oppression, exploitation, racism, and insufficient measures of well-being (Powers et al., 2021). The consequences of wicked problems include mental and physical health challenges, such as depleted "surge capacity," a type of energy we emotionally and physically rally that allows us to meet the needs of an acute crisis or special occasion. This has become overtaxed in most people, as multiple, ongoing crises continue to flood our current context (Haelle, 2020), leading to mental paralysis, ongoing and anticipatory grief, ambiguous loss, and physical distress (Engstrom & Powers, 2021).

In order to address wicked problems, we must see the connections between people and their environments, which often requires reevaluating our worldviews and embracing ecosocial worldviews. Ecosocial worldviews acknowledge a sense of interconnectedness through a process of personal and collective critical reflection that questions our current unjust systems, beliefs, and practices which value resources for some rather than wellness for all, including the planet and all of its inhabitants (Rambaree et al., 2019). When applied to the field of social work:

> Ecosocial Work is social work, with all its depth and breadth, but it approaches the analysis of social problems, issues, and concerns with an ecosocial paradigm or lens, rather than an anthropocentric lens (Matthies & Närhi, 2016). Thus, ecosocial work is not a specialty within social work, rather all social work can, and we argue should, be ecosocial work. (Rambaree et al., 2019, p. 205)

Ecosocial approaches are necessary for social workers to learn about environmental issues and take action using sustainable development principles (Närhi & Matthies, 2018). Boetto (2017) created an ecosocial model that is "congruent across the ontological (being), epistemological (thinking), and methodological (doing) dimensions" (p. 1) of social work practice. In this transformative ecosocial model, a connection to nature is the center of identity, which is the being. Thinking is connected to professional values and ethics, and professional knowledge, and doing incorporates holistic well-being, cultural sensitivity, sustainability, organizational change, and social action (Boetto, 2017). Ecosocial work practice should include educating communities on environmental issues to engage in environmental decision-making (Kang et al., 2019) and promoting ecological justice by disrupting systems of injustice and oppression (Power et al., 2019b).

Social workers have been engaging in ecosocial work since the profession was first established (Alston & Besthorn, 2012; Bowles et al., 2018; Dewane, 2011; Dominelli, 2012; Gray et al., 2008; IFSW et al., 2012; Kemp, 2011; Närhi & Matthies, 2001; Teixeira & Krings, 2015; Powers et al., 2019b; Zapf, 2009). As early as 1981, we find the first literature in the USA calling social workers to action to embrace ecosocial worldviews and practices (Berger & Kelly, 1993; Besthorn, 1997; Hoff & McNutt, 1994; Hoff & Rogge, 1996; Park, 1996; Rogge, 1993; Soien, 1987; Weick, 1981). Through the International Federation of Social Workers (IFSW) Climate Justice program, the profession at large has begun to acknowledge and emphasize the intersection of environmental crises and social justice and the profession's role in this work (International Federation of Social Workers et al., 2012). *The People's Charter*, a global interdisciplinary initiative, grew out of the collaborative efforts of a recent conference, "Co-Building a New Eco-social World: Leaving No one Behind' People's Summit," initiated and cohosted by the IFSW in June 2022 (The People's Summit, 2022).

However, inclusion of the environment within social work codes of ethics remains primarily focused on the social environment (Liu & Flynn, 2021). In their analysis of codes of ethics from IFSW member nations, Liu and Flynn found that mention of the physical environment was largely more theoretical and lacked guidance from code to practice, which limits the scope of understanding around how to integrate environmental issues into educational curricula. Further, the recommendations for practice are often reactive (i.e., natural disaster response) rather than preventative, hindering the profession from developing evidence-based interventions that could prevent or reduce environmental destruction and resulting physical and emotional harm.

While the social work profession has long embodied the person-in-environment approach, it has been slow to mainstream the ecosocial worldview within education curricula (Androff et al., 2017; Jeffery, 2014). Anecdotal and early empirical evidence suggests minimal explicit focus on ecosocial work and related topics (e.g., environmental justice) in social work curricula. Recent studies of social work curricula in Australia (Harris & Boddy, 2017), Greece, Switzerland, the UK, and Germany (Papadimitriou, 2022) reveal wide variability in the number of ecosocial work courses available, demonstrating inconsistent prioritization of ecosocial work education across the Global North. To date, no such study has been conducted in the USA. However, several studies have explored how ecosocial work content is infused into existing courses and teaching practices (Androff et al., 2017; Beltrán et al., 2016; Boddy et al., 2018; Harris & Boddy, 2017; Hayward et al., 2013; Hudson, 2019). Alternatively, other scholars have explored experiential pedagogical opportunities, from the integration of ecosocial work concepts into field practice (Crawford et al., 2015; Hayward et al., 2013) to non-internship immersive experiences with community members and policymakers (Magruder et al., 2022; Powers et al., 2019a). These studies highlighted the roles that activism, classroom, and field experience play in shaping student professional identity and values as an ecosocial worker (Dodd & Mizrahi, 2017; Hayward et al., 2013).

In response to the escalating climate crisis and inadequate focus on the human-environment connection in US social work education curricula, in 2015, the US Council on Social Work Education (CSWE) made a call to educate social workers on environmental justice, mandated by an update to their Education and Policy Accreditation Standards (or EPAS)—"Competency 3: Advance Human Rights and Social, Economic, and Environmental Justice" (Council on Social Work Education, 2015, p. 7). Similarly, Australia updated their educational standards to include recognition of the interconnectedness of human health and the environment (Australian Association of Social Workers, 2010). Nevertheless, while educational accrediting bodies mandate such changes to their curricula, this does not guarantee that all programs can and will implement them promptly. Many universities in the USA are facing declining enrollment (Binkley & The Associated Press, 2023), restricting resources and limiting their ability to hire new faculty with expertise, train faculty in emerging topics (e.g., climate change), or develop new courses.

There is currently no documented study in the USA reporting the extent to which ecosocial work content is being taught in MSW program curricula since the 2015 CSWE EPAS update. Additionally, there remains a lack of clarity regarding the best practices for when and how ecosocial work should be integrated into social work curricula (Beltrán et al., 2016). Some scholars suggest integration of ecosocial work content into select courses, while others argue this approach falls short and recommend an infusion of ecosocial work content into numerous courses, field experiences, and school culture due to its interconnectedness with all aspects of practice (Androff et al., 2017; Beltrán et al., 2016; Powers et al., 2019a). Hence, this chapter will fill a gap in the literature by examining if and how US social work programs meet this call to educate future social workers with an ecosocial worldview.

Objectives and Study Methods

We would like to recognize our ecofeminist collaborative work style (Mountz et al., 2015) that allowed us, as coauthors, to flourish in the development of this chapter. We come from various backgrounds and geographical and cultural locations, and we strive to promote ecosocial work through our teaching, scholarship, and service to the profession. This chapter has been an effort of collegiality and support amid multiple pandemics and the ongoing global climate crisis and in the context of women in academia. This ecofeminist collaborative style is unfortunately rare in academia and yet desperately needed if we are to expedite our impact in these critical times. We believe social workers play many critical roles in society which facilitate ecosocial well-being, and we seek to promote a professional mindset that ecosocial work is "everybody's thing."

This chapter seeks to highlight areas of strength and opportunities for improving MSW-level education on environmental justice and ecosocial work. The purpose of this chapter was to examine current US MSW course offerings to identify (1) "ecosocial work" courses offered in MSW programs, (2) patterns of course content based on program websites and a syllabi review, and (3) patterns and gaps in geographical distribution of programs. These findings are then applied to the global social work education context, and several creative solutions are offered to align social work education with the current climate crisis.

We used two data sources to achieve our objectives: (1) all CSWE-accredited MSW program websites as of 2022 ($N = 309$) and (2) syllabi of "ecosocial work" courses taught in MSW programs ($N = 10$) gathered from the Ecological Justice

Social Work Syllabi Clearing House, an online, publicly shared folder which includes syllabi submitted by educators and students. This study was deemed Non-Human Subjects Research by the University of Maryland, Baltimore Institutional Review Board.

MSW Program Website Review Procedures

Two research team members used a hybrid inductive-deductive approach to determine our sample and analyze the data for the website review. First, we visited the CSWE Directory of Accredited Programs (CSWE, 2022) to develop a list of all accredited MSW programs to date. Then, we divided the total number of CSWE-accredited MSW programs (N = 309) in half, with 1 reviewer assigned 155 programs and the other assigned 154 websites. Each researcher visited their assigned websites and read course catalogs, curriculum guides, MSW student handbooks, and class lists to identify all course titles and descriptions that included terms identified as having a potential focus on ecosocial work or an environment-related topic. For the sake of clarity, the term "course" refers to a semester-long class, also known as a "module" in many countries outside of the USA. A priori keywords and phrases included environment*, natur*, ecolog*, sustainab*, climate, outdoor, and wilderness. These words were selected because we wanted to start with a broad filter to ensure a comprehensive list of terms related to the natural environment. After reviewing 30 program websites, we observed additional terms that reflected our intent and expanded our keywords and phrases to include physical environment, adventure, food, global, displace*, health, animal, place*, and conservation. Courses with any of the above terms found in the title or course description were included in our sample for further review and analysis. Many of the terms selected during our inductive process reflect the deep ecology paradigm, which treats humans and nature (including all species) as equals and views humans as part of nature, not outside or against it (Besthorn, 1997; Devall, 1980).

The two website reviewers collected all data on an Excel spreadsheet. Content analysis was conducted by reviewing course descriptions and syllabi (if available) to determine if the course had a primary, secondary, or partial focus on ecosocial work content. Courses were categorized using the following codes: 0 = "Not an ESW-focused course"; 1 = "Maybe, not enough information available, or includes ESW content but not as a primary focus"; or 2 = "Yes, primary focus on ESW topics." For each course in the sample, a determination was made by the primary reviewer, and the code was entered into the shared spreadsheet. All courses coded "1" or "2" were reviewed by both researchers to verify the determination. Once an ESW-focused course ("2") was identified, other information was gathered about the course, including elective or mandatory, the topic(s) of focus, and if part of a certificate or dual-degree program. Additional programmatic information was gathered, such as the total number of ESW courses offered and whether an ESW concentration, specialization, or certificate program was available. Courses retained a code of "1" if the two reviewers agreed that the course (1) did not meet the eligibility criteria for a course with a primary focus on ESW or environment-related topics and (2) appeared to include some content focused on environment-related issues or (3) key terms were used in the course title or course description, but there was not enough information available to determine the extent of the focus.

Descriptive analyses (e.g., means, frequencies, percentages) were conducted using Microsoft Excel and Stata SE version 16.

Ecological Justice Social Work Syllabi Clearing House Review Procedures

One of the biggest hurdles to developing ecosocial work courses is not knowing where to start. Around the time this project began, there was no central location to see examples of ecosocial work course syllabi. To address this identified gap in 2020, two authors of this chapter created the Ecological Justice Social Work Syllabi Clearing House, which is open for public viewing to expand the accessibility of ecological course content. The Clearing House was developed through the International Green/EcoSocial Work Collaborative Network, a professional network for sharing ideas and resources and building solidarity among those concerned about ecological issues. It utilizes platforms such as an email listserv, Twitter, and Facebook, all of which have an engaged international audience. In 2020, members of the Collaborative Network who had previously offered syllabi to the listserv were invited to include their syllabi in the Clearing House. Next, an invitation was sent to the listserv and corresponding social media groups to submit ecosocial work syllabi. A second invitation was sent in early 2021, and a third in mid-2022. In keeping with anti-oppressive educational practices, submissions were not limited to educators' syllabi; student-developed independent study and continuing education courses were also encouraged.

Using a similar hybrid inductive-deductive approach as with the website review, three researchers from this team coded the syllabi in the Clearing House for the following information: school, year and level taught, course modality, textbooks, other resources, learning objectives, learning activities, subpopulations studied, intersections with other justice components (e.g., racial justice), and the terminology used (i.e., green social work, environmental social work, and ecosocial work).

By reviewing these syllabi separately from the website review, we systematically triangulated the reach and themes of this topic (Patton, 1999). The true nature of a phenomenon or practice is best uncovered by looking at multiple data sources, which allows for a deeper analysis of the problem and broader application of implications across cultures (Patton, 1999).

Limitations to Our Approach

The website review presented several limitations to the study team. First, some program websites offered minimal course information, such as only listing course titles without course descriptions or syllabi. As a result, courses that may have met the content criteria for sample selection but did not include any of our keywords in the course titles could not be included. Therefore, it was difficult to determine the precise number of current ecosocial work courses. In some cases, we were unable to determine the most current course offerings; therefore, some courses that are listed but no longer offered may have been included in our findings. These limitations suggest a general lack of transparency among social work programs or a lack of capacity to keep websites updated with the most current information about course offerings. This presents an obstacle not only for this study but also for prospective students.

The website review was also limited by the depth of information provided within the course descriptions. Numerous courses had an unambiguous primary focus on ESW, and these were the courses coded as "2." However, many other courses appeared to have at least some emphasis on ecosocial work topics, but the depth and breadth of this focus were unclear after reviewing a one-paragraph course description.

The syllabi review was limited by self-selection, as all syllabi were voluntarily supplied to the Ecological Justice Social Work Syllabi Clearing House. While an attempt was made to collect syllabi from all courses found through the website review, there was a lack of access because most syllabi were unavailable on program websites; in some cases, course descriptions were assessed. Additionally, the syllabus provides merely a glimpse into the content covered in a course, as faculty and students may bring resources and examples of ecosocial work to the class. This material is valuable for our study purposes and would not be on a syllabus.

Findings from the Website and Syllabi Reviews

MSW Program Website Review

Fifty-four MSW programs (17%) in the USA currently offer a course with either a primary, secondary, or partial focus on ecosocial work or environment-related content. These programs are relatively well distributed across the four US Census-based regions (United States Census Bureau, 2021); however, a small majority are concentrated in the West and Northeast regions of the country. Among the 54 programs, 15 (28%) offer more than 1 ESW course with a primary or secondary focus, or both. A more detailed discussion of results is organized by primary and secondary ESW courses.

Courses with a Primary ESW Focus

Findings from the website review indicate that only 24 out of 309 (8%) US MSW programs offer courses primarily focused on ecosocial work or related topics; 35 total courses met our criteria for a primary ESW course. Among these 24 programs, the majority (n = 18) only offer 1 ESW course, and the remainder (n = 6) offer between 2 and 7 courses. The University of Denver offers the most courses (n = 7), as well as an Ecological Justice Concentration and a Certificate in Human-Animal-Environment Interactions in Social Work; this is the only ESW-related concentration offered in a US-based MSW program. Among the 35 total courses with a primary ESW focus, 21 are offered as general social work electives, and the remaining 14 satisfy requirements for a concentration, certificate, or dual-degree program.

Topics covered in these courses included sustainable development policy and practice, climate change and resilience, environmental justice, adventure- and wilderness-based practice, food insecurity and the food system, social and environmental impact assessments, conservation in a global context, natural environment and health, outdoor behavioral health, and environmental policy. To demonstrate an exemplar course meeting criteria for a primary ESW focus, below is an excerpt from a course description for *Environmental Justice and Social Work*:

> This course will highlight theories to enhance the students' understanding of "Person-in-Environment" to be inclusive of the physical environment. Students will develop a paradigm regarding the interconnection of social work and environmental justice with an understanding of the bi-directional relationship of people and nature. Students will examine social work history and create an emerging perspective of social workers' role with respect to the environment. (Rutgers School of Social Work, 2022, Generalist curriculum electives section)

Programs offering ESW-focused courses were relatively evenly distributed across the four Census-based regions of the country (U.S. Census Bureau, 2021): three programs were in the Midwest, five in the Northeast, six in the West, and eight in the South. However, the programs located in the West were more likely to offer multiple ESW courses. Therefore, there is a geographical concentration of courses offered in the West compared to other regions. For instance, out of the 35 primary ESW courses, nearly half (n = 15) were located in the West, indicating that social work students with an interest in ecosocial work will have more educational opportunities if living in the Western region of the country, which includes Montana, Wyoming, Colorado, New Mexico, and all states west. However, the West is only the third most populated out of the four US Census regions (U.S. Census Bureau, 2021), indicating a misalignment of ecosocial work course availability and population density.

Courses with a Secondary or Ambiguous ESW Focus

There were 45 courses coded as "1," indicating that either (1) the course may cover ESW-related content, but it did not appear to be the primary focus, or (2) there was not enough information available on the website to determine the primary focus of the course. An example of a course description that met these criteria comes from *Global and Local Inequalities: Critical Analyses of the Processes and Policies of Globalization*: "Discussion of the health of the planet, economic and cultural globalization, the enduring legacies of imperialism and colonialism in this global era, and their local impacts. Foci include international agreements, UN conventions, immigration, and refugee policies" (University of Washington, Office of the University Registrar, 2022, Detailed course offerings section). These courses were offered across 36 MSW programs, 5 of which were on the list of programs with a primary ESW course; this brings the total of MSW programs with a primary or secondary ESW focus to 54 out of 309 (17%).

In many cases ($n = 15$), there was a brief mention of "environmental justice" in the course description, listed along with other injustices, as required by Competency 3 of the CSWE Social Work Competencies (Council on Social Work Education, 2015). In these cases, environmental justice does not appear to be a primary or even secondary focus. Instead, it is listed as part of a package of injustices the program must cover. For example, below is an excerpt from a course description for *Advanced Integrated Practice with Organizations and Communities*:

> This course prepares social work practitioners who operate within and through frameworks of human rights, social, economic and environmental justice, and empowerment practices. Learners will be trained to engage in social change at all levels of society in order to become social workers who are committed to equity [in] all forms .... (Fordham University, 2022, Course listings section)

As with the primary ESW courses, programs with secondary ESW courses were relatively evenly distributed across the four Census regions; 9 programs were located in the Northeast, 5 in the South, 13 in the Midwest, and 8 in the West. Many programs ($n = 9$) offered more than one course that met these criteria, and they were evenly distributed across three of the four Census regions, with none located in the Northeast. When examining the availability of primary and secondary ESW courses combined, 26 are in the West, 19 in the Midwest, 19 in the South, and 17 in the Northeast.

Ecological Justice Social Work Syllabi Clearing House Review Findings

At the time of this study, the Ecological Justice Social Work Syllabi Clearing House was home to ten syllabi, including bachelor's level (BSW) courses ($n = 5$), master's level (MSW) courses ($n = 2$), self-study courses ($n = 1$), postgraduate continuing education courses ($n = 1$), and co-listed MSW and BSW courses ($n = 1$). Two of the three MSW courses were also found through the MSW Program Website Review. Interestingly, all syllabi came from US universities despite the international platforms used to invite participation, as described above. It is unknown why international participation was not present; however, the findings described below apply to a broad audience within and outside of the USA.

Due to the limited number of course syllabi at the MSW level in the Clearing House at the time of this study, the research team felt it would be most meaningful to include findings from all syllabi, regardless of their level. The syllabi review indicated courses focused on introductory-level environmental issues and emphasized ecosocial work's presence within the field by citing the NASW Code of Ethics and CSWE Competencies. The course objectives were achieved through classroom activities, policy analysis, service-learning projects, and reflective journaling. Several courses explored the individual versus systemic nature of environmental problems, employing activities such as "understanding your carbon footprint," a "green office audit," and analyzing environmental policies. Despite all syllabi originating from US universities, each course incorporated an international focus by highlighting the global nature and implications of climate change and environmental injustice, and several courses explicitly centered the voices and experiences of Indigenous, minoritized, and Global South communities.

Course Description Review Findings

Course descriptions indicated that most courses focused on vulnerable populations in the USA and the local community where the program was located. A few courses offered a service-learning component or a global immersion experience. Often, environmental injustices were coupled with social, racial, and economic injustices to understand the intersecting impact on clients. Multiple courses applied a macro focus, examining policy implementation and impact, resource management, and large-scale environmental degradation. Theoretical frameworks and models included adventure-based practice, green social work, community food security model, resilience-oriented prevention, and restorative justice. Key focal areas across all syllabi and descriptions included the impacts of neocapitalism, colonialism, and structural and environmental racism on the inequitable distribution of environmental protections and harms.

Putting Our Findings into Context

Findings from the website review suggest that an insignificant number ($n = 24$) and proportion (8%) of MSW programs in the USA offered courses with a primary ESW focus in 2022. When broadening the criteria to include courses with a possible secondary focus on ESW topics, this increased to include 17% of MSW programs ($n = 54$). However, many courses that fall into this secondary category do not appear to give much targeted attention to environmental issues but rather include the environment as one among many injustices faced by individuals and communities; this finding was reflected in the syllabi review as well. Arguably, any integration of environmental content into the MSW curriculum is beneficial; however, such a topic warrants greater emphasis across courses and its own course when possible. It is clear that US-based MSW programs have a long way to go in educating and preparing future social workers to address environmental concerns faced by the populations we serve.

Programs that offer ESW courses are relatively evenly distributed across the four US Census regions, which is encouraging; students from many parts of the country have access to an ESW course. However, there is a more significant concentration of courses in the West, as numerous programs in that region offer multiple ESW courses. There are currently only five MSW programs that offer two or more courses with a primary ESW focus, three of which are in the West, the others located in the Northeast. Therefore, a student looking for a more robust set of courses focused on ecosocial work will be limited in their options, and students living outside of the West or Northeast may have to relocate, which is not always feasible. This aligns with an earlier study of the spatial distribution of American environmentalism (Mazur & Welch, 1999), which revealed that the Northeastern and Western regions of the USA had the highest scores on an environmentalism indicator (measured by the number of advocacy organizations, state policies, legislative action, and public opinions pertinent to environmental protections). Generally, this pattern reflects the politicization and polarization of environmental issues in the USA whereby conservative voters generally deprioritize environmental issues and climate change policy, and liberals report more significant concern with environmental issues, comparatively (Funk & Hefferon, 2019; Mildenberger et al., 2017; Tyson et al., 2023). The spatial patterns of political affiliation within the USA generally mirror our findings, such that more liberal regions have more ESW courses.

Comparing our findings to the international literature, Harris and Boddy (2017) conducted a similar review of social work curricula in Australia and found that 31 courses from 15 universities (out of 29) contained terms in the title or description related to the natural environment. Putting this in the context of our analytic framework, 51% of Australian universities with social work programs offered at least one course with ESW content. While our analysis was at the program level rather than school or department level, we found that only 17% of US MSW programs offer a course with an ESW focus, compared to 51% in Australia. Harris and Boddy (2017) also explained that 24 of these courses (77%) were part of the core curriculum (rather than electives); our findings suggest only 29 US courses (36%) were required and primarily for concentrations or dual degrees rather than the general curriculum.

Another recent study examining social work curricula in the UK, Germany, Switzerland, and Greece (Papadimitriou, 2022) found that 23 out of 2613 (or less than 1% of) courses included content that referred to the natural environment. Applying our analytic framework to their study, 75% of Greek universities, 38% of Swiss universities, .8% of German universities, and 2% of UK universities with social work programs offer a course that covers environmental issues, compared to 17% of US universities. Among the 23 courses in these 4 countries, 7 were required (30%) and 16 were optional; this proportion is comparable to that found in our study, with only 36% of ESW courses required. However, it is worth noting that the number of required and optional courses in Papadimitriou's study was aggregated across all four countries; thus, we cannot directly compare these figures between countries individually.

A cross-national comparison of the proportion of programs with ESW courses to total social work programs demonstrates that the USA ranks below Australian, Greek, and Swiss programs but higher than German and UK programs. Considering the proportion of ESW courses that are required (versus elective) suggests the majority of Australian ESW courses are required of all social work students, which contrasts our study findings as well as that of Papadimitriou (2022) in which the majority of ESW courses were electives. Overall, this comparison suggests Australian, Greek, and Swiss social work students are more likely to be exposed to environmental content during their studies than US students.

In triangulating the data collected from the website and syllabi reviews, we discovered several ESW courses at the MSW level that were not previously in our Clearing House. This discrepancy may indicate a lack of awareness about the International Green/EcoSocial Work Collaborative Network or Clearing House, an unwillingness to share syllabi, and ultimately, an obstacle to integrating ESW content into the curriculum more broadly. These concerns were echoed in Harris and Boddy's study (2017) of Australian social work education and may also be relevant to other countries. Despite this challenge, our review revealed relative consistency among the topics, theories, and approaches to ecosocial work education across courses, with the concepts of environmental justice and advocacy at the core.

Implications for Social Work Educators

This chapter suggests significant room for growth in offering ecosocial work courses across schools of social work in the USA. Currently, most programs in the USA do not offer any courses that include ESW content, and among those that do, the ecosocial focus is often secondary or tertiary. Among the courses with a primary ESW focus, most are electives or part of specializations and therefore not required for all students. An emphasis should be placed on developing ecosocial work courses and infusing this content into the core curriculum for all students (Powers, 2016; Powers et al., 2019a). While we recognize some hindrances may include the absence of faculty expertise, this can be overcome with faculty curriculum workshops and trainings (Mattocks et al., 2023; Schmitz et al., 2017), continuing education, and student experiences (Powers et al., 2019a; Powers, 2016).

Faculty should also rely on existing resources such as the International Federation of Social Worker's Environmental Social Work publications and resources, Australian Association of Social Worker's Green Social Work Practice Network, and content experts across the globe. For example, the CSWE Curricular Guide for Environmental Justice (2020) was developed in the USA to help educators map environmental justice and climate change topics to the educational competencies required of all accredited programs. The Guide offers lecture materials, reading lists, other media resources, and recommended activities that can be applied in most countries and communities. Other resources available to the international community include the Ecological Justice Social Work Syllabi Clearing House and the International Green/EcoSocial Work Collaborative Network. Materials shared in these venues are applicable and relevant to all educators regardless of culture or geography. In response to the findings of this chapter, effort has been made to expand the national and cultural representation of the Clearing House by inviting participation from international scholars.

Recognizing that many social work programs face challenges aligning with new educational policies and accrediting standards due to limited resources and declining enrolment, introducing a new core course into the curriculum is not always feasible. Some other options include developing a partial term or elective course an interdisciplinary course partnering with another discipline (e.g., public health, environmental science, or urban planning), an independent study, and material to integrate into an international course or field seminar.

In recognition of the anti-oppressive foundation of the profession and this topic, we also encourage student engagement in course development. Collaboration between students and faculty mirrors the critical pedagogy of Paulo Freire who advocated for an anti-oppressive approach to education in which educators and students cocreate knowledge (Freire, 2000). What follows is an example of a student-driven approach to ecosocial work education for a program that could not offer an ecosocial work course. One author of this chapter recently collaborated with an MSW student interested in ecosocial work practice to codevelop the first ecosocial work student club in their MSW program; the faculty member served as Faculty Advisor and the student as Club President. During the academic year, the club gained significant momentum, and the student members advocated for more environmental justice content in their program curriculum. With support from the Faculty Advisor, the students formed an Environmental Justice Curriculum Committee, which meets weekly to strategize and organize their recommendations for environmental justice content integration into selected courses. The students have facilitated meetings with two course developers and are collaborating on discussions around lecture materials, assignments, readings, guest speakers, and other resources to integrate into these courses. Such a student-driven approach to curriculum updates is particularly apt in this context, considering younger generations are generally more concerned with climate change and environmental injustice compared to older generations (ecoAmerica, 2014; Funk & Hefferon, 2019).

Implications for Researchers

This chapter serves as a foundation for additional investigation. Future research in the USA should expand this work by including BSW programs to determine if they follow similar trends as MSW programs. Additionally, we offered a starting point for a cross-national comparison of several countries within the Global North. Expanding the scope of this study to investigate social work offerings in additional countries, particularly those in the Global South, could allow for a broader understanding of ecosocial work education.

To gain a clearer picture of successes and obstacles to ESW course development, an in-depth exploration of existing ecosocial work courses, including how each course was conceived, authorized, implemented, and evaluated, would provide critical information for future ecosocial work curriculum planning. Additionally, course evaluations could provide a student perspective on how the course was received. In contrast, for programs without ecosocial work courses, a qualitative investigation of potential obstacles faced by faculty could indicate areas for growth and change within social work programs.

To conduct a more coordinated and systematic review of ecosocial work education within a country, we recommend research collaborations with professional associations or coalitions that have a vested interest in ecosocial work education. For example, the US Council on Social Work Education (CSWE) formed a Committee on Environmental Justice in 2015, which provides recommendations on how US social work education programs should integrate environmental justice into their curriculum (Council on Social Work Education, 2023). Such a committee would be an ideal entity to collaborate with for a nationwide investigation, and if such an entity does not exist, this may be an ideal place to start. Another avenue is to develop an "Ecosocial Work Report Card," following the lead of the Sustainable Endowments Institute's College Sustainability Report Card, an independent and comparative evaluation of sustainability practices across 300 universities and colleges in the USA and Canada. According to the Sustainable Endowments Institute (n.d.):

> The Report Card focused on policies and practices across nine categories: Administration, Climate Change & Energy, Food & Recycling, Green Building, Student Involvement, Transportation, Endowment Transparency, Investment Priorities, and Shareholder Engagement.

While the Sustainable Endowments Institute is not actively conducting an evaluation at this time, their work provides a foundation that can be applied to social work programs.

Adapting the "Report Card" concept to the focus of this chapter, an Ecosocial Work Report Card might include an evaluation of schools of social work based on ecosocial work content present within curricula, field placements, academic service, and research, along with some of the core areas noted by the Sustainable Endowments Institute, such as energy use, food waste, recycling, and student involvement. Such an effort would require system-wide collaboration across social work programs but has the potential to hold programs accountable to a global commitment to combat environmental and climate injustices.

Conclusion

This chapter highlights significant shortcomings in US MSW program curricula for preparing future social workers to engage in environmental justice and ecosocial work practice. A cross-national comparison with studies conducted in five countries of the Global North (Harris & Boddy, 2017; Papadimitriou, 2022) revealed that the USA falls somewhere in the middle in terms of ecosocial work education offerings but lags significantly behind Australia, the country among these most similar in size and scale to the USA. The geographical distribution of ecosocial work courses in the USA reflects a national pattern correlating environmentalism with political affiliation, such that more liberal parts of the country have higher concentrations of ecosocial work course availability.

Implications for educators include a clear call to action to develop ecosocial work courses emphasizing the critical role that social workers should and do play in addressing environmental injustices. For under-resourced programs or those without instructors who possess ecosocial work expertise, educators should consider alternatives such as partnering with environmentally focused programs (e.g., public health, environmental science) to develop interdisciplinary courses or create electives or independent study course options. A student-driven approach, such as a curriculum committee advising on content integration into existing courses, is another option for programs with expressed student interest but lacking faculty expertise. Social work educators in more conservative regions of the USA, in particular, have a significant gap to fill as students have considerably less access to ecosocial work content. Once students have transitioned from the classroom to the field, international social work codes of ethics must shift to incorporate the physical environment into the profession's conception of the person-in-environment approach (Liu & Flynn, 2021).

Implications for researchers and administrators include a coordinated effort to develop an ESW Report Card to evaluate all social work programs within a country, taking inventory of how and to what extent each program incorporates environmental content into the curricula. It is also imperative to investigate the obstacles faculty face in integrating ecosocial work content into the curricula, to thwart such barriers and make this content more widely available across programs.

Findings from this chapter demonstrate that we have a long way to go in preparing professionals for ecosocial work in the USA. Ecosocial work should not be a niche role relegated to environmental scientists or certain ideological groups. Rather, it is time for all social workers to embrace a professional mindset that "ecosocial work is everybody's thing," to address environmental injustices in practice, policy, research and education and provide the next generation of social workers with the tools to combat the ongoing climate crisis.

Acknowledgments We would like to acknowledge Indigenous and First Nations peoples, worldwide, whose land, winds, and waters we now all share. We pay respect to their beliefs and values and their continuing and enduring cultures which deepen and enrich the life of our global community and our specific educational communities throughout Turtle Island (the Indigenous name for the North American continent).

References

Alston, M., & Besthorn, F. (2012). Environment and sustainability. In K. Lyons, T. Hokenstad, M. Pawar, N. Huegler, & N. Hall (Eds.), *The SAGE handbook of international social work* (pp. 56–69). SAGE.

Androff, D., Fike, C., & Rorke, J. (2017). Greening social work education: Teaching environmental rights and sustainability in community practice. *Journal of Social Work Education, 53*(3), 399–413. https://doi.org/10.1080/10437797.2016.1266976

Australian Association of Social Workers. (2010). *Australian Social Work Education and Accreditation Standards*.https://www.academia.edu/32051559/CODE_OF_ETHICS_Australian_Association_of_Social_Workers.

Beltrán, R., Hacker, A., & Begun, S. (2016). Environmental justice is a social justice issue: Incorporating environmental justice into Social Work practice curricula. *Journal of Social Work Education, 52*(4), 493–502. https://doi.org/10.1080/10437797.2016.1215277

Benevolenza, M. A., & DeRigne, L. (2019). The impact of climate change and natural disasters on vulnerable populations: A systematic review of literature. *Journal of Human Behavior in the Social Environment, 29*(2), 266–281.

Berger, R. M., & Kelly, J. J. (1993). Social work in the ecological crisis. *Social Work, 38*(5), 521–526.

Besthorn, F. H. (1997). *Reconceptualizing social work's person-in-environment perspective: Explorations in radical environmental thought* (Doctoral dissertation). ProQuest Dissertation Publishing.

Besthorn, F. H., & Meyer, E. E. (2010). Environmentally displaced persons: Broadening social work's helping imperative. *Critical Social Work, 11*(3), 123–138.

Binkley, C. & The Associated Press. (2023, March 9). The labor shortage is pushing American colleges into crisis, with the plunge in enrollment the worst ever recorded. *Fortune*. https://fortune.com/2023/03/09/american-skipping-college-huge-numbers-pandemic-turned-them-off-education/

Boddy, J., Macfarlane, S., & Greenslade, L. (2018). Social work and the natural environment: Embedded content across curricula. *Australian Social Work, 71*(3), 367–375.

Boetto, H. (2017). A transformative eco-social model: Challenging modernist assumptions in social work. *British Journal of Social Work, 47*(1), 48–67. https://doi.org/10.1093/bjsw/bcw149

Bowles, W., Boetto, H., Jones, P., & McKinnon, J. (2018). Is social work really greening? Exploring the place of sustainability and environment in social work codes of ethics. *International Social Work, 61*(4), 503–517. https://doi.org/10.1177/0020872816651695

Bullard, R. D., Mohai, P., Saha, R., & Wright, B. (2008). Toxic waste and race at twenty: Why race still matters after all of these years. *Environmental Law, 38*(2), 371–411.

Council on Social Work Education (CSWE). (2015). *2015 Educational policy and accreditation standards*. https://www.cswe.org/getattachment/Accreditation/Standards-and-Policies/2015-EPAS/2015EPASandGlossary.pdf

Council on Social Work Education (CSWE). (2020). *Curricular guide for environmental justice*. https://www.google.com/url?sa=t&rct=j&q=&esrc=s&source=web&cd=&ved=2ahUKEwjlm_rStqH6AhVOtoQIHTfuBooQFnoECBEQAQ&url=https%3A%2F%2Fwww.cswe.org%2Feducation-resources%2F2015-epas-curricular-guides%2F2015-environmental-justice-guide-web-version%2F&usg=AOvVaw1u_FqgigutNe5LJwR5bRRL

Council on Social Work Education (CSWE). (2022). *Directory of Accredited Programs*. https://www.cswe.org/accreditation/about/directory/?

Council on Social Work Education (CSWE). (2023). *Committee on Environmental Justice*. https://www.cswe.org/about-cswe/governance/governance-groups/commission-on-global-social-work-education/committee-on-environmental-justice/

Crawford, F., Agustine, S. S., Earle, A., Kuyini-Abubakar, A. B., Luxford, Y., & Babacan, H. (2015). Environmental sustainability and social work: A rural Australian evaluation of incorporating

eco-social work in field education. *Social Work Education, 34*(5), 586–599. https://doi.org/10.1080/02615479.2015.1074673
Devall, B. (1980). The deep ecology movement. *Natural Resources Journal, 20*(2), 299–322.
Dewane, C. J. (2011). Environmentalism & social work: The ultimate social justice issue. *Social Work Today, 11*(5), 20–24.
Dodd, S. J., & Mizrahi, T. (2017). Activism before and after graduate education: Perspectives from three cohorts of MSW students. *Journal of Social Work Education, 53*(3), 503–519. https://doi.org/10.1080/10437797.2016.1272514
Dominelli, L. (2012). *Green social work: From environmental crisis to environmental justice*. Polity Press.
EcoAmerica. (2014). *American Climate Values 2014: Psychographic and demographic insights*. https://ecoamerica.org/wp-content/uploads/2017/03/ea_american_climate_values_2014_psychographic_and_demographic_insights.pdf
Engstrom, S., & Powers, M. C. F. (2021). Embracing an ecosocial worldview for climate justice and collective healing. *Journal of Transdisciplinary Peace Praxis, 3*(1), 120–144. https://jtpp.uk/
Fordham University. (2022). *Graduate School of Social Service Bulletin, Course Listings*. https://bulletin.fordham.edu/gss/courses/#text
Freire, P. (2000). *Pedagogy of the oppressed* (30th anniversary ed.). Continuum.
Funk, C., & Hefferon, M. (2019, Nov 25). *U.S. public views on climate and energy*. Pew Research Center. https://www.pewresearch.org/science/2019/11/25/u-s-public-views-on-climate-and-energy/
Grand Challenges for Social Work. (2016). *Creating social responses to a changing environment*. https://grandchallengesforsocialwork.org/create-social-responses-to-achanging-environment/. https://doi.org/10.1080/0312407X2016.1263352.
Gray, M., Coates, J., & Yellow Bird, M. (Eds.). (2008). *Indigenous social work around the world: Towards culturally relevant education and practice*. Ashgate Publishing.
Haelle, T. (2020). Your 'surge capacity' is depleted – it's why you feel awful. *Elemental*. https://elemental.medium.com/your-surge-capacity-is-depleted-it-s-why-you-feel awful-dc285d542f4c
Harris, C., & Boddy, J. (2017). The natural environment in social work education: A content analysis of Australian social work courses. *Australian Social Work, 70*(3), 337–349.
Hayes, K., Blashki, G., Wiseman, J., Burke, S., & Reifels, L. (2019). Climate change and mental health risks, impacts, and priority actions. *International Journal of Mental Health Systems, 12*, 28.
Hayward, R. A., Miller, S. E., & Shaw, T. V. (2013). Social work education on the environment in contemporary curricula in the USA. In M. Gray, J. Coates, & T. Hetherington (Eds.), *Environmental social work* (pp. 246–259). Routledge.
Hoff, M. D., & McNutt, J. G. (1994). *The global environmental crisis: Implications for social welfare and social work*. Ashgate.
Hoff, M. D., & Rogge, M. E. (1996). Everything that rises must converge: Developing a social work response to environmental injustice. *Journal of Progressive Human Services, 7*, 41–57. https://doi.org/10.1300/j059v07n01_04
hooks, b. (2013). *Writing beyond race living theory and practice*. Routledge.
Hudson, J. (2019). Nature and social work pedagogy: How U.S. social work educators are integrating issues of the natural environment into their teaching. *Journal of Community Practice, 27*(3-4), 487–502. https://doi.org/10.1080/10705422.2019.1660750
International Federation of Social Workers (IFSW), International Association of Schools of Social Work (IASSW), & International Council on Social Welfare (ICSW). (2012). The global agenda for social work and social development: Commitment to action. *Journal of Social Work Education, 48*(4), 837–843. https://doi.org/10.1080/10437797.2012.10662225
IPCC. (2021). Summary for policymakers. In V. Masson Delmotte, P. Zhai, A. Pirani, S. L. Connors, C. Péan, S. Berger, N. Caud, Y. Chen, L. Goldfarb, M. I. Gomis, M. Huang, K. Leitzell, E. Lonnoy, J. B. R. Matthews, T. K. Maycock, T. Waterfield, O. Yelekçi, R. Yu, & B. Zhou (Eds.), *Climate change 2021: The physical science basis. Contribution of Working Group I*

to the sixth assessment report of the intergovernmental panel on climate change. Cambridge University Press. In Press. https://www.ipcc.ch/report/ar6/wg1/#SPM

Jeffery, D. (2014). Environmentalism in social work: What shall we teach? *Affilia, 29*, 492–498. https://doi.org/10.1177/0886109914533697

Kaijser, A., & Kronsell, A. (2013). Climate change through the lens of intersectionality. *Environmental Politics, 23*(3), 417–433.

Kang, J., Fabbre, V. D., & Ekenga, C. C. (2019). "Let's talk about the real issue": Localized perceptions of environment and implications for ecosocial work practice. *Journal of Community Practice, 27*(3-4), 317–333.

Kemp, S. P. (2011). Recentering environment in social work practice: Necessity, opportunity, challenge. *British Journal of Social Work, 41*, 1198–1210.

Koloko, J. (2012). *Wicked problems: Problems worth solving*. Self Published.

Liu, K., & Flynn, C. A. (2021). How the environment is conceptualised and represented globally in social work codes of ethics. *International Social Work, 66*(3), 868–888.

Magruder, K. J., Mitschke, D. B., & Slater, H. M. (2022). Ecological devastation bus tour: A new pedagogy for environmental social work? *Journal of Social Work Education*. https://doi.org/10.1080/10437797.2022.2062506

Matthies, A. L., & Närhi, K. (Eds.). (2016). *The Ecosocial Transition of Societies: The contribution of social work and social policy*. Routledge.

Mattocks, N., Singh, M. I., Smith, L., Dolan-Reilly, G., Powers, M. C. F., Smith, K., & Zhao, L. (2023, Jan). *Ecosocial work is everybody's thing*. Poster presentation. Society for Social Work Research (SSWR) annual conference, 2023.

Mazur, A., & Welch, E. W. (1999). The geography of American environmentalism. *Environmental Science & Policy, 2*, 389–396.

Mildenberger, M., Marlon, J. R., Howe, P. D., & Leiserowitz, A. (2017). The spatial distribution of Republican and Democratic climate opinions at state and local scales. *Climatic Change, 145*, 539–548.

Mountz, A., Bonds, A., Mansfield, B., Loyd, J., Hyndman, J., Walton-Roberts, M., Basu, R., Whitson, R., Hawkins, R., Hamilton, T., & Curran, W. (2015). For slow scholarship: A feminist politics of resistance through collective action in the Neoliberal University. *ACME: An International Journal for Critical Geographies, 14*(4), 1235–1259. Retrieved from https://www.acme-journal.org/index.php/acme/article/view/1058

Närhi, K., & Matthies, A. L. (2001). What is the ecological (self)consciousness of social work? Perspectives on the relationship between social work and ecology. In A. L. Matthies, K. Närhi, & D. Ward (Eds.), *Eco-social approach in social work* (pp. 16–53). Sophi.

Närhi, K., & Matthies, A. L. (2018). The ecosocial approach in social work as a framework for structural social work. *International Social Work, 61*(4), 490–502.

Papadimitriou, E. (2022). *Coverage of environmental issues in undergraduate curricula in social work in four European countries: The UK, Switzerland, Germany and Greece*. Social Work Education. https://doi.org/10.1080/02615479.2022.2028763

Park, K. (1996). The personal is ecological: Environmentalism of social work. *Social Work, 41*, 320–323.

Patton, M. Q. (1999). Enhancing the quality and credibility of qualitative analysis. *Health Services Research, 34*(5 Pt 2), 1189–1208. https://pubmed.ncbi.nlm.nih.gov/10591279/

Powers, M. C. F. (2016). Transforming the profession: Social workers' expanding response to the environmental crisis. In A.-L. Matthies & K. Narhi (Eds.), *Ecosocial transition of societies: Contribution of social work and social policy*. Ashgate.

Powers, M., Schmitz, C., & Moritz, M. B. (2019a). Preparing social workers for ecosocial work practice and community building. *Journal of Community Practice, 27*(3-4), 446–459. https://doi.org/10.1080/10705422.2019.1657217

Powers, M. C., Rambaree, K., & Peeters, J. (2019b). Degrowth for transformational alternatives as radical social work practice. *Critical and Radical Social Work, 7*(3), 417–433.

Powers, M. C. F., Rinkel, M., & Kumar, P. (2021). *Co-creating a "new normal" for our profession and beyond: Embracing an ecosocial worldview*. MDPI: Sustainability.
Rambaree, K., Powers, M. C. F., & Smith, R. J. (2019). Ecosocial work and social change in community practice. *The Journal of Community Practice, 27*, 3–4. https://doi.org/10.1080/10705422.2019.1660516
Rogge, M. E. (1993). Social work, disenfranchised communities, and the natural environment: Field education opportunities. *Journal of Social Work Education, 29*, 111–120. https://doi.org/10.1080/10437797.1993.10778803
Rutgers School of Social Work. (2022). *Generalist curriculum electives*. https://socialwork.rutgers.edu/academics/master-social-work-msw/msw-syllabi
Schmitz, C. L., Powers, M. C. F., Nesmith, A., & Forbes, R. (2017, Oct). *Environmental justice across the curriculum: 2015 EPAS and beyond*. Grand Challenges Teaching Institute, Council on Social Work Education Annual Program Meeting.
Soine, L. (1987). Expanding the environment in social work: The case for including environmental hazards content. *Journal of Social Work Education, 23*(2), 40–46.
Sustainable Endowments Institute. (n.d.) *Initiatives*. https://www.endowmentinstitute.org/our-initiatives/
Teixeira, S., & Krings, A. (2015). Sustainable social work: An environmental justice framework for social work education. *The International Journal on Social Work Education, 35*(5), 513–527.
The People's Summit. (2022). *The people's charter for an eco-social world*. https://newecosocialworld.com/wp-content/uploads/2022/09/Peoples-Charter-EN.pdf
Tyson, A., Funk, C., & Kennedy, B. (2023, April 18). *What the data says about Americans' views of climate change*. Pew Research Center. https://www.pewresearch.org/short-reads/2023/04/18/for-earth-day-key-facts-about-americans-views-of-climate-change-and-renewable-energy/
United States Census Bureau. (2021). *2010 census regions and divisions of the United States*. https://www2.census.gov/geo/pdfs/maps-data/maps/reference/us_regdiv.pdf
University of Washington, Office of the University Registrar. (2022, Sept 17). *School of Social Work (MSW) detailed course offerings*. https://www.washington.edu/students/crscat/socwk.html
Weick, A. (1981). Reframing the person-in-environment perspective. *Social Work, 26*(2), 140–143.
Zapf, M. K. (2009). *Social work and the environment: Understanding people and place*. Canadian Scholars' Press Inc.

Ecological Justice Social Work Syllabi Clearing House

Ecological Justice Syllabi

Recommended Resources

Radical Self-Care for Social Workers in the Global Climate Crisis. https://www.semanticscholar.org/paper/Radical-Self-Care-for-Social-Workers-in-the-Global-Powers-Engstrom/05e067c02c8d18b2c5a9bbc0dcb0ab7b1749819b
Braiding Sweetgrass: Indigenous Wisdom, Scientific Knowledge, and the Teachings of Plants (See: An Interview with Dr. Robin Wall Kimmerer. https://sustainabilitystudies.nd.edu/resources/take-10-for-the-planet/video-an-interview-with-dr-robin-wall-kimmerer/
"Kincentric Ecology: Indigenous Perceptions of the Human-Nature Relationship." By Enrique Solomon. https://esajournals.onlinelibrary.wiley.com/doi/abs/10.1890/1051-0761(2000)010%5B1327:KEIPOT%5D2.0.CO%3B2
Taking Root: The Vision of Wangari Maathai. https://www.greenbeltmovement.org/wangari-maathai/taking-root-documentary
Music of Wade Fernandez. https://wadefernandezmusic.com/
Sustaining Spirit by Naomi Ortiz

IFSW Workbook Series: Social Work Promoting Community and Environmental Sustainability (Vol 1: https://www.ifsw.org/product/books/social-work-promoting-community-and-environmental-sustainability/; Vol 2: https://www.ifsw.org/product/books/social-work-promoting-community-and-environmental-sustainability-volume-2/; Vol 3: https://www.ifsw.org/product/books/social-work-promoting-community-and-environmental-sustainability-volume-3/)

IFSW Climate Justice Program: Statement on IPCC climate report "Encouragement as we collectively continue co-building a new ecosocial world with a sustainable future for all". https://www.ifsw.org/ifsw-climate-justice-program-statement-on-ipcc-climate-report-encouragement-as-we-collectively-continue-co-building-a-new-ecosocial-world-with-a-sustainable-future-for-all/

IFSW Climate Justice Program Statement on COP26: A commitment to ongoing action and accountability. https://www.ifsw.org/ifsw-climate-justice-program-statement-on-cop26-a-commitment-to-ongoing-action-and-accountability/

The people's charter for an eco-social world. https://newecosocialworld.com/the-peoples-charter-for-an-eco-social-world/

Case Study 2: WiseWater Wellness: In a Disconnected World—Water Connects

Katie McShane and Elaine Moynihan

WiseWater
Academy

In a world dominated by capitalist expansion, the gaps between contemporary society and environmental education are becoming wider. While young people learn about the environment and climate change within formal education, the current school curriculum fundamentally lacks an ecocentric approach. To achieve a sustainable future, it is crystal clear that human behaviours, values, and attitudes need to change. Ecocentric education is an important tool in reshaping the unsustainable worldviews that have taken this planet to tipping point. As two youth and community workers and passionate water activists we have worked with young people and communities in various water projects throughout Ireland. Our work revealed that people have a thirst for dialogue and education that is relational and which inspires purposeful action in these critical times. Furthermore, with the reality of the multitude of environmental crisis, we experienced many people suffering from eco anxiety, despondency, and a high prevalence of nature deficiency. We knew we had a responsibility to restore hope, and so we developed the WiseWater Wellness (WWW) educational programme. This is a uniquely designed ecocentric wellness education programme that supports reconnecting young people to themselves, their communities, and the living world. WiseWater has many tributaries of ecosocial work all flowing from a source of connection, awareness, and appreciation of our interrelationship with the environment.

The environmental crisis can cause fear, overwhelment, and despair, yet WiseWater recognises the importance of hope. Through celebrating our

K. McShane (✉) · E. Moynihan
University College Cork, Cork, Ireland

C. Forde et al. (eds.), *Teaching and Learning in Ecosocial Work*,
https://doi.org/10.1007/978-3-031-58708-5_7

environment and focusing on positive change, active hope ripples throughout WiseWater with the message that 'every drop counts'. Two major challenges facing society today are the need to protect and preserve the environment and to support young people's wellbeing. WiseWater Wellness uniquely addresses both these issues through the interconnected relationship of environmental health and human health. The value of water is normally framed as a commodity or resource to be managed. This view has initiated increasing action and environmental programmes to conserve it; however, we feel it is far more valuable and sustainable to connect with water's intrinsic worth. By exploring water's mysterious properties and behaviour, we provide the students with opportunities to develop a new relationship with water. To recognise themselves as internal bodies of water supports connection with external water bodies in the natural environment. Therefore, while we are the facilitators of the programme, water is the teacher. Through various activities and experiments, the programme creatively brings the magic and power of water into the classroom. For instance, water is receptive, supporting our ability to take in sound and information. We demonstrate this through sound bowls filled with water, which when struck reveal unique patterns. Following this we create a human sound bowl with students standing in a circle using their voices to create energising or calming sounds. Exploring how sound influences us utilises another quality of water, its ability to reflect. Reflection is an important tool for self-awareness, which we encourage throughout the programme with art, discussion, and creative writing. Activities such as enacting the movement of water molecules in an ocean wave demonstrate how water is responsive and its ability to bond makes it resilient, teaching the importance of social connection. Water teaches us the power of unity and noncompetitiveness, each individual has their own strengths, and more can be achieved when we learn and work together. Water is both peaceful and powerful, which we explore in our ability to calm the stormy waters by understanding our thoughts and emotions or rise up and support action to create waves of change. Water is also vulnerable and needs to be minded, as is humanity and the planet in these times of accelerated change.

The correlation of keeping the water environment well and maintaining personal wellbeing is woven throughout all the activities and experiments. Therefore, this programme brings not only awareness to the water environment but also self-awareness for the people involved. Through the journey of learning about water, students are given different tools of how to care and protect themselves, each other, and the water. By exploring how ecosystems and biodiversity work together, young people can recognise the importance of diversity and cooperation within community. Additionally, the young people learn about the concept of 'one well' which portrays that beyond our community and our country, we are a part of 'one well' of water; therefore our actions impact the global water crisis. With the message 'every drop counts', the young people are empowered that every action makes a positive difference. Through building a reciprocal relationship with water, an empathic connection develops.

One activity that has an extraordinary impact is the recreation of the water cycle. As part of the geography curriculum young people learn about the different stages of the water cycle, yet it continues to be taught as if free from human interference. In a recent analysis of over 450 water cycle diagrams, 85% omitted human interaction (Abbott et al., 2019). However, it is crucial that young people along with all society recognise and take responsibility for the effect humans have on the water cycle. The WiseWater activity brings the anthropocentric water cycle to life. Through a multisensory and immersive lesson, we engage the students by recreating the water cycle with the important emphasis of putting human activity in the picture. Students work together in small groups to make a physical model of a river catchment area; this includes a home, farm, town, and forest section in which they creatively enact the activities that occur. This demonstrates, for example, that what we put down the drain eventually falls down as rain. We encourage discussion and ideas for change in how society cares for water. The aim is to raise awareness of our human involvement in the water cycle bringing a sense of shared responsibility to protect and care for the 'one well'.

One drawback to how we currently deliver the programme is that it is taught within a classroom environment. Currently only one session provides contact with a local water source; however, having more interaction with the natural environment creates a more direct relationship with nature. Yet within the limits and bounds of the classroom, we explore the extraordinary within the ordinary. Throughout the sessions we explore different facets of water's life-giving nature, in preparation for a citizen science activity with the wider community at a local water body. Depending on the location of the school, this activity can vary from river testing to indicate water quality, a beach clean-up, or visiting a local well to help it run clear. WiseWater recognises the power education has to enable change in the way we treat and interact with the natural environment. By providing lifelong learning through an appreciation of nature, WiseWater aims to encourage young people to become active citizens in their own homes, schools, and communities. On completion of the programme, young people have expressed a greater sense of awareness and appreciation for water. Some feedback on the positive changes young people have made include saving water, educating family members, participating in beach clean ups, and feeling inspired to help save the planet. Each individual who makes an action for change becomes a drop of hope that ripples throughout the ocean of sustainability.

The WWW programme has been designed as we believe the curriculum fails to adequately address environmental education and the wellbeing needs of young people.

Recently WiseWater conducted research and interviewed eight primary school teachers regarding the necessity of the WWW programme in Irish national schools. This research conveyed that teachers are overwhelmed in their workload and are greatly concerned by the level of nature deficiency in their students. Despite Education for Sustainable Development (ESD) being part of Irish educational policy since 2014, surprisingly none of the teachers interviewed had heard of ESD. This highlighted that while ESD informs policy, it is yet to be successfully implemented in schools. Our programme aligns itself with the curriculum and implements ESD practices by blending subject learning within the classroom. This consequently supports teachers and students to experience how ecocentric education can bring topics such as geography, science, history, art, and social, personal, and health education strands into relative and creative learning about the environment. This benefits those involved by bringing a greater sense of awareness of our interdependency and inspires stewardship of the natural world.

Central to WiseWater is an ethics of care which is vital to provide education that empowers young people to enact change. Through innovative and playful techniques, the programme addresses important issues, thereby empowering young people to become active drivers of change. Carl Rogers's (2003) person-centred approach provides insight for valuing each young person's unique perspective. Dialogue, understanding and raising the critical consciousness of young people, as outlined by Paolo Freire (1972), supports the delivery of the programme. Recognising that there are many types of learning, the programme is delivered in a multiplicity of ways to encourage engagement with all students. The WiseWater programme consciously uses pedagogies that engage the cognitive domain (head), psychomotor domain (hands), and affective domain (hearts). As facilitators of the programme, we emphasise that the young people are equal within the learner and educator relationship. We are all learning and working together towards the same goal of finding ways to care for ourselves, each other, and the water environment. At the heart of the programme is the source of wellness and education that nature provides us about ourselves and our place in the world.

Alongside the schools' programme, WiseWater creatively engages in community work through citizen science initiatives and water heritage projects. This includes regular beach clean ups, community well clearing, gathering stories of the cultural heritage of water bodies in Ireland, community pilgrimages to river sources, and providing kick sampling data to the local authority. Kick sampling is a technique used to investigate the invertebrates living in a river or a stream to determine the health of the water. Again, through our message of 'every drop counts', whenever possible and however small, WiseWater takes action to help protect and care for our water sources. These blue spaces provide opportunities where connection is easily established between people, inspiring community work core values of empowerment, anti-discrimination, and sustainable development. WiseWater supports the implementation of ESD through direct action by working with community and education groups to provide opportunities for intergenerational learning and nature-based education. WiseWater aims to bring a greater understanding of how important it is to know that we need water and water needs us. While we cannot predict what is on the horizon, WiseWater supports young people and communities to stay connected through the tides of change.

Bibliography

Abbott, B. W., Bishop, K., Zarnetske, J. P., Hannah, D. M., Frei, R. J., Minaudo, C., Chapin, F. S., Krause, S., Conner, L., Ellison, D., Godsey, S. E., Plont, S., Marçais, J., Kolbe, T., Huebner, A., Hampton, T., Gu, S., Buhman, M., Sayedi, S. S., Ursache, O., Chapin, M., Henderson, K. D., & Pinay, G. (2019). A water cycle for the Anthropocene. *Hydrological Processes, 33*, 3046–3052. https://doi.org/10.1002/hyp.13544

Friere, P. (1972). *Pedagogy of the oppressed*. Penguin Education.

Rogers, C. (2003). *Client centered therapy*. Constable and Robinson Ltd.

Chapter 5: Eco-Activism and Greening the Social Professions Curriculum: Teaching and Learning Ecosocial Work in an Irish University

Mary Hurley, Catherine Forde, and Fiachra Ó Suilleabháin

Introduction

> When we fall in love again with the earth, we will know what to do and we need a collective awakening that individuals do not awaken in isolation, but we need to awaken together in community. (Thich Nhat Hanh, 2006, n.p.)

These words by the Buddhist monk Thich Nhat Hanh resonate with the profound concern felt by many of us engaged in professional education who are alarmed by the existential crisis that is the climate emergency, pollution, and biodiversity loss. As educators, we have been deliberating and reflecting on how to engage and educate our students to respond to this crisis in their professional and personal lives. In a period where the effects of advancing climate chaos, critical biodiversity loss, and species extinction are no longer remote or abstract possibilities (Intergovernmental Panel on Climate Change, 2022), Hanh's emphasis on the "collective nature of awakening together" has never been more compelling and urgent. This collective element connects well with the social work value of the importance of relationships and the principle of collective responsibility (IFSW, 2024). While recognition of the importance of collective responsibility and collective responses is not new, especially in critical and radical social work approaches (Mullaly, 2007; Payne, 2005; Lavalette, 2011), as educators we argue for the need to expand the meaning of a collective response to include an obligation to preserve and protect the natural world for current and future generations. Our position aligns somewhat with ecological social work (Jones, 2010, 2017; McKinnon & Bay, 2013), deep ecology (Ungar, 2002) and ecofeminist social work (Besthorn & McMillen, 2002; Klemmer & McNamara, 2020). Our stance requires a pivot from an anthropocentric worldview

M. Hurley (✉) · C. Forde · F. Ó. Suilleabháin
School of Applied Social Studies, University College Cork, Cork, Ireland
e-mail: M.Hurley@ucc.ie

C. Forde et al. (eds.), *Teaching and Learning in Ecosocial Work*,
https://doi.org/10.1007/978-3-031-58708-5_8

that positions humanity at the center of the Earth with nature and all other species in service to it. We advocate for an ontological shift that recognizes humanity as one of many species that share the Earth's resources with all others on the planet and that acknowledges the finite nature of the world and its resources.

As educators in the allied social professions of social work and youth and community work, we have been concerned for some time about how to incorporate teaching about the environment into our professional degree programs. In 2018, our discussions led to the development of an interdisciplinary environmental module for students on a professional postgraduate Masters in Social Work (MSW) program in University College Cork. This elective module is now in its third year, and a sister module subsequently rolled out to students of the qualifying BSocSc (Youth and Community Work) degree. Our longer-term goal is to teach these modules as an interdisciplinary offering to students of the social and helping professions across the university. We hope that this interdisciplinary approach to teaching will foster strong relationships between students and encourage knowledge exchange and collaborative multidisciplinary practice. This chapter explores the development and content of the sister modules and discusses some of the challenges and opportunities of introducing ecosocial work to the curriculum of the social professions in the university.

Climate Change and Ireland

While it is true that Ireland and its people have not yet had to deal with the extreme challenges of climate change experienced by other regions of the world, our weather patterns are changing and include rising temperatures and increased rainfall (Met Eireann, 2020). As an island nation, Ireland's climate is heavily influenced by the North Atlantic Ocean and oceanic changes including rising sea levels, sea warming, carbonization and storm surges, with the probability of stronger impacts on sea life, coastal areas, and communities. These changing human-induced patterns have recently been punctuated by extreme weather events that have caused major disruption. In October 2017, Storm Ophelia resulted in widespread environmental and property damage, and three people lost their lives in the storm (Towey, 2018; Met Eireann, 2018). Fallen electricity poles and electric cables left 300,000 households and 385,000 premises without power and impacted on other infrastructure and services including water treatment and wastewater systems (ibid). In 2018, the "Beast from the East" snowstorm led to extreme hardship across the country with major disruption to transport services and business and school closures. In October 2023, Storm Babet caused devastating flooding in several towns in the south of the country (Met Eireann, 2024). These and other weather events have triggered widespread disruption to services including schools, health services, businesses, and food production at a time when Ireland was still recovering from the impacts of the COVID-19 pandemic and its considerable effects on public services including health and education (Crowley & Hughes, 2021).

Ireland has been slow to face up to its climate challenges and responsibilities. Most recent figures show that Ireland was one of the just six EU countries that failed to achieve their own 2020 climate targets (Climate Change Advisory Council, 2021) and Ireland's current climate measures fall short of the emission reductions required to meet its 2030 and 2050 domestic climate obligations (Climate Change Advisory Council, 2022). Major issues that remain to be adequately addressed include the widespread pollution of Irish waterways (Environmental Protection Agency, 2022), high greenhouse gas levels and global carbon footprint (De Bruin & Yakut, 2022), and increased risk of storms, flooding, and drought (Toreti et al., 2023) with implications for communities and the environment across the country.

Inciting Eco-Activism and Resistance in Social Work

As an applied practice, the social work profession is tasked with the performance of social policies (Pawar, 2019). This definition from the International Federation of Social Workers (IFSW) and International Association of Social Work (IASW) places a strong emphasis on social action, social justice, empowerment, human rights, and solidarity:

> Principles of social justice, human rights, collective responsibility, and respect for diversities are central to social work. Underpinned by theories of social work, social sciences, humanities and Indigenous knowledge, social work engages people and structures to address life challenges and enhance wellbeing. (IFSW, IASW, 2020, n.p.)

The emancipatory, principled nature of this definition appears somewhat aspirational in the increasingly neoliberal global context. The values underpinning social work are under attack in a world that honors individualized self-interest and self-sufficiency, consumerism and self-gain (Weinberg & Banks, 2019). The current political landscape complicates and frustrates the social work *mission* and highlights how "the commodification of the helping relationship, with its narrow focus on recipients of care as consumers, contradicts the idea of social work as relationship-based and not easily reduced to concrete, short-term and measurable indicators" (ibid, p.3). The entrepreneurial and consumerist values of neoliberalism have resulted in a change in the relationship between human beings from one of care and concern to a transactional one of cash and contracts (Tsui & Cheung, 2004). Mutuality and collectivism have been replaced with individual self-interest. In this environment, the desire for greater social justice and equality is null and void, and welfare provision is not a priority. Pascoe et al. (2023) skewer the destructive consequences of these values which demonstrate the:

> continued dominance of bureaucratic structures and managerialist policies, confirming the predicted sense of alienation and frustration in social work practice. Once a 'bedrock of successful support and treatment' (Howe, 1996, p. 93) … managerialism and bureaucracy devalues relationships by reducing time spent with service users, prioritising outputs, increasing paperwork and diminishing the sense of trust in the profession. (p. 528)

As educators, we believe the role of social work education demands a pedagogical commitment to develop social responsibility and awareness among our students and proactively resist the "neoliberalization" of social work and to evidence an alternative vision of social work which authentically coheres with social work's ethical commitment to social justice. Reisch (2013, cited in Jones et al., 2017) describes what is needed:

> A shift towards a global way of thinking can help propel social work education beyond the relatively narrow focus programs have adopted in response to dominant neo-liberal forces that tend to privilege professionalisation, market-based solutions, individual concerns, and clinically oriented work. (p. 2)

This vision provided us with a clear rationale to reassess our teaching portfolios, to reflect on our teaching philosophies and value-base, and to publicly demonstrate our commitment to act "as stewards of [the] discipline, both for the sake of the discipline and its integrity, and for the ways in which the discipline gets used in the rest of the world" (Shulman, 2007, p. 14). We felt compelled to foster, mentor, and demonstrate "ethical resistance" (Weinberg & Banks, 2019) in our classrooms. Weinberg and Banks (2019) describe resistance as opposing or withstanding something, while Couzens Hoy (2005) sees it as both an activity and an attitude: "It is the activity of refusal. It is also an attitude that refuses to give in to resignation" (p. 5). Weinberg and Banks (2019) say, "Resistance, unlike resignation, can lead to hope that a better world is possible, even if we do not necessarily know what this might look like" (p.4). This sentiment of resistance encapsulates our pedagogical journey in developing a module on ecological justice and sustainability in social work. We wanted to resist the consumerist and market values of neoliberalism and sought to convey hope to students during their professional and academic formation. We also wanted to radically stand against human consumption of the living planet, however, we were unsure how to embed green social work into already packed social professions curricula.

Coloring the Curriculum: Putting Green on the Social Work Agenda

There is an urgent need for placing the environment and sustainability on the social work curriculum. Teixeira and Krings (2015) emphasize that:

> environmental degradation is a pressing, global problem and is concentrated in oppressed populations and oppressed geographical regions. This matters to our [social work] clients. The human impacts of environmental challenges fall most heavily on those to whom social workers are most accountable. (p. 524)

This powerful statement resonates with the humanistic and emancipatory values of social work with roots in social justice (Howe, 2009). However, it is this very emphasis on social justice, with its anthropocentric and speciesism focus, which social work needs to address if the profession is going to play its part in helping service users and communities to deal with the worst impacts of climate change. While post-anthropocentric approaches have received growing attention since the start of the twenty-first century (Ranta-Tyrkkö & Närhi, 2021), in Ireland, ecosocial work, environmental justice, green social work, and ecological justice have been largely absent from the social work education curriculum and from the curricula of the other social professions, including youth and community work. In response to a growing awareness of the need to engage students about the impact of the climate change—the most pressing existential issue facing humanity—we have sought to make explicit the connections between core social work values and approaches, ecological justice, and the "real-world" implications of climate change for social work service users. Clement Atlee (1920) argued that "Every social worker is almost certain to be also an agitator" (p. 237). As social work educators, we want to foster agitation among future social work practitioners about the centrality of the environment in the social work mission.

Developing a brand-new module is challenging within a crowded social work curriculum. With the support of the core MSW program team and colleagues within our school, we initially designed a workshop to discuss activism and the environment with one cohort of first-year postgraduate social work students. The 3-hour workshop was held during a short 2-week third term that is taught in the second semester of the academic year following the students' initial placement practicum. As a stand-alone workshop, we (Mary and Fiachra) developed "Activism and Environmental Campaigning in Social Work" as the generative topic for the workshop with the following learning outcomes:

By the end of the workshop, students should understand the following:

1. Their own working definition of social work activism.
2. The relationship between social activism and social work.
3. The inextricable links between social justice and equality principles and activism: The "personal is political."
4. The internal (personal) barriers and benefits of engaging in activism.
5. The external (systemic) barriers and benefits of engaging in activism.

We were inspired by the work of Wehbi and Straka (2011) who have engaged in participatory, reflective social work education. They argue that "a very important goal of an anti-oppressive social work educator is to help students 'catch' a social justice vision to motivate their practice" (Wehbi & Straka, 2011, p. 46). We used ourselves as "case studies" in the classroom in order "foster critical thinking and reflection so that students learn how to learn on their own" (McDade, 1995, p. 9). For instance, one of the authors (Mary Hurley) is heavily involved in a long-standing movement to challenge the installation of an incinerator in the Cork Harbour area in the south of Ireland. McDade (1995, p.10) summarizes 11 powerful arguments for the use of a case study/discussion method pedagogy. Four of these arguments particularly underpinned the curricular design and teaching enactment of this workshop:

- It provides a "laboratory" for practicing and advancing critical thinking.
- It is contextually based and therefore helps students to appreciate contextual nuances.

- It challenges students' assumptions and beliefs.
- It encourages students to imagine alternatives.

The workshop was discursive, reflective, and based on the principle of active partnership in the learning process. Its starting point was where the students were at, recognizing that they come with prior conceptions and a vision of the world. In keeping with Wehbi and Straka's (2011) argument that the "living classroom" should become a site where social justice is brought to life, we came to the learning space as co-learners with our students. We rejected the "banking method of teaching" (Freire, 2014) and consciously chose to engage in "dialogic action" with our students in the co-construction of knowledge and understanding about the place of social activism, social justice, and ecological justice in social work education and practice. Rather than seeing ourselves as the instructional "experts" in the classroom, we invited students to perform their own understandings of activism. We engaged students in a performance of their understanding of activism by getting them to define it. This was in keeping with Perkins's (1998) description of understanding as "the ability to think and act flexibly" (p. 42); the students were engaged in their learning and invited to work things out both *for* themselves and *with* us. Students identified a broad range of methods of social activism from donations and fundraising to democratic participation (voting) all the way to direct action and protest. The students also discussed their views about the reasons for activism in social work, which varied from their own personal microlevel self-interest up to macro-level reasons, which focused on responding to structural oppression.

Absence of Green Social Work Activism Among Students

The climate crisis, sustainability, and ecological justice were not identified by any of the 31 master students who attended the workshop as subjects worthy of, or of interest to, social work activism. As social work educators with a deep concern for the living planet, we believe that the role of social work education demands a pedagogical commitment to develop social responsibility and awareness of humanity's inextricable links to nature/environment in our students. We agree with Dominelli's (2011) argument that:

> Climate change debates can benefit from social workers' involvement. They have important roles to play in reducing carbon emissions, promoting clean energy consumption, and protecting vulnerable populations from the deleterious impact of climate change by enhancing individual and community resilience and helping residents access green technologies. (p. 437)

Despite being infused with the language of ecological systems theory (Bronfenbrenner, 1979), social work education's focus to date has been anthropocentric and anthropogenic (Boetto, 2017), emphasizing social ecology. In other words, when the social work profession talks about taking an ecological perspective, the environment is discussed only in the context of how we as humans use and benefit from it—the environment in service to people. We tend to view our ecological systems as a system of our human interaction as if we only interact with humans in our daily lives and live separately from the ecologies around us. In promoting this focus, the profession perpetuates consumerist hegemony, which conceptualizes humans as benign consumers of the environment, and focuses on "the person in the social environment, more than the person in the social *and* natural environment" (Bretzlaff-Holstein, 2018, p. 926; own emphasis).

The evidence from the workshop of our students' learning highlights a basic gap in their critical awareness of globalization, sustainability, and the climate crisis and its impacts on social work at local and international levels. This deficit needed to be addressed as part of our commitment to integrating social responsibility into our teaching on a social work program. On foot of this workshop and further discussion with colleagues on the program and across the island of Ireland, we codesigned and co-delivered an elective module for second year postgraduate social work students called *Sustainability, Environmental and Social Justice Issues in Professional Practice*. Designed with interdisciplinarity in mind, the module is co-taught with its sister module *Sustainability, Environmental and Social Justice Issues in Youth and Community Work* to these social work students and final year undergraduate youth work and community development students. While the youth and community work module is a core/compulsory module, the social work module is currently an elective. Our preference was for both modules to be compulsory, but there were few opportunities to embed an additional core module in the crowded social work timetable.

Drawing Inspiration from Environmental Practitioners and Educators

Key texts on the environment (Zapf, 2009), ecological justice (Philip & Reisch, 2015), environmental social work (Ramsey & Boddy, 2017), green social work (Dominelli, 2012; Noble, 2016), humane education and social work (Bretzlaff-Holstein, 2018), environmental justice and community development (Harley & Scandrett, 2019), and social justice (Mehrotra et al., 2019; Weinberg & Banks, 2019) have helped to develop the instructional and conceptual content of the modules. We have also drawn on specific literature which examines approaches and strategies for learning and teaching about sustainability and the environment in social work education, including the work of Jones (2010), Jones et al. (2017), Dominelli (2011), Drolet et al. (2015), Melekis and Woodhouse (2015), and Teixeira and Krings (2015). These readings synchronously align with Teaching for Understanding (TfU) and Universal Design for Learning (UDL) principles because the authors reiterate the need for curricula that enable students to engage with the generative topic and perform multiple ways of knowing and understanding throughout the module. They offer helpful suggestions for engaging students in multiple entry points of learning including the following:

- Use of national and global case studies and vignettes (stories, *YouTube* clips, artwork, and digital humanities).
- Community walks and environmental site visits.
- Mind-mapping.
- Guest speakers (including, e.g., indigenous ways of knowing).
- Animals in social and community work practice (companion animals, therapeutic work with animals, e.g., support pets in courtrooms). The University College Cork has two therapy dogs who provide support to students and staff.

These strategies for engaging students cohere with Gardner's (2006) multiple intelligences and Wiske's (1998) teaching for understanding frameworks and Kornhaber, Fierros and Veenema's (2004) work on multiple entry points of learning that enable students to gain different perspectives on the same topic.

Interdisciplinarity has been central to the development and delivery of the modules. Working on an interdisciplinary basis helps us to acknowledge the deep-rooted connections between social work and community development and youth work, the significance of community in ecosocial practice, the need for holistic approaches that are not siloed, and the recognition of the importance of "community-led efforts for building community resilience and strategies to respond to environmental vulnerabilities" (Rambaree et al., 2019, p. 207). Without interdisciplinarity and cross-disciplinarity within and outside the academy, it is arguable that little can be achieved.

Key learning outcomes have been developed for both modules. In the social work module, there is an expectation that by the end of the module, students should be able

1. To define sustainable social development and understand the multiple dimensions of sustainability
2. To have an informed conversation about the impacts of the climate emergency on the natural environment and how this affects social work service users
3. To examine and evaluate the concepts of "green social work," "humane social work," "environmentalism," "ecological justice," "eco-spirituality," and "activism"
4. To critically discuss and evaluate the role of social work in promoting sustainable development and identify sites of practice at local and global levels
5. To identify issues surrounding vulnerability, resilience, activism, and indigenous knowledge around sustainability and ecological justice

The module design incorporates Universal Design for Learning (UDL) principles and includes the range of Teaching for Understanding (TfU) performances of learning. The framework enables narrational, logical, numerical, foundational, esthetic, experiential, and interpersonal entry points to learning across the six classes. In addition, at least one of the classes takes place off-campus. Jones (2010) has commented that "taking students out of the classroom often opens them up to new ways of looking at issues" (p.82). The idea of a community walk was inspired by Jones (2010) but adapted due to the global COVID-19 pandemic to become a "My World, My Ecology" walk, where students were asked to take a walk around their home community and to assess their ecology by responding to some guided questions. In 2023, the youth and community work students visited a community garden in Cork City and spoke to the community development workers and volunteers who oversee the running of the garden and use it to facilitate a range of community health and well-being activities, including groups for men and women, fathers and sons, older people and children and young people.

Guest speakers including ecosocial practitioners and international academics have facilitated students to understand the "real-world" implications of the climate crisis and the critically important roles that social work, youth work, and community development can play in sustainable development. We agree with Drolet et al. (2015) that guest speakers share "their practice experience and demonstrate(d) how the concepts of sustainable development and social development are integrated in the field" (p. 538). It is our intention to draw more heavily on the experience and expertise of social work and youth and community work graduates who are engaging in ecosocial work practice. Flipped classroom techniques and case-based discussions enable students to demonstrate their understanding and learning throughout the modules. Drolet et al. (2015) argue that "with case-based teaching, students develop skills in analytical thinking and reflective judgment by reading and discussing complex, real-life scenarios" (p. 537). Table 1 offers a tabular outline of the module classes and flipped classroom elements.

An ePortfolio (total word count 2000 words) is used as the formal mode of assessing students' learning on both modules. The ePortfolio is comprised of entries which connect to each week's class-based learning across the module. This enables students to compose their ePortfolios in a timely and relevant fashion, inviting them to critically reflect on core themes as the module unfolds. The elements of the social work ePortfolio assignment are as follows:

(a) 1 × a short introduction setting out the student's thoughts about the connections between social work and ecological issues (200 words)
(b) 1 × critical discussion about the facts and figures of the climate crisis and climate justice and the response of social work as a discipline to the global climate emergency (800 words)
(c) 1 × My World, My Ecology Walk Record Sheet 1 × a reflective journal entry about the student's experience of completing the "My World, My Ecology" exercise based on class discussions (300 words)
(d) 1 × reflective journal entry about two of the examples of ecosocial practice, appraising the mode of service delivery (tips: how could such services be extended to other areas of practice/with other service user cohorts? Do they relate at any of your placements or other social work-related practice? How do international examples help you to understand ecological social work as a global issue? How do you relate to ecofeminism and other social work approaches?) (400 words)
(e) 1 × artifact (photograph/artwork image/poem/song) to represent the student's learning from the module—with short narrative explanation (100 words)

Table 1 Module classes year 1

Week	Asynchronous element	Live class	What pre-class work?	Theme
1	Yes	Yes	Reading Green Social Work catches on. (2019, Apr 07). *The Sunday Observer* https://www.proquest.com/newspapers/green-social-work-catches-on/docview/2204065833/se-2 Viewing TEDx Talk Environmental Justice Peggy Shepard at TEDx Harlem	Introduction
2	Yes	Yes	Watch YouTube video (https://www.youtube.com/watch?v=dREtXUij6_c) Reading Powers, M., and Rinkel, M. (Eds.). (2018). Social work promoting community and environmental sustainability: A workbook for global social workers and educators (Volume 2). Switzerland: International Federation of Social Work (IFSW). Jones, P. (2010). Responding to the ecological crisis: Transformative pathways for social work education, *Journal of Social Work Education, 46*(1), 67–84. https://doi.org/10.5175/JSWE.2010.200800073	Climate Justice for Nonlegal Social Professions
3	Yes	Yes	Take a sensory reflective walk Complete reflective *My World, My Ecology* Exercise	My World, My Ecology
4	Yes	No	Review examples of ecosocial practice Mindfulness, beekeeping Adventure therapy Social farming Community gardening Forest bathing Green social work in Zimbabwe Green social work in China Discussion forum	Ecosocial Practice
5	Yes	Yes	Reading Alston, M. (2015). Social work, climate change and global cooperation. *International Social Work, 58*(3), 355–363. https://doi.org/10.1177/0020872814556824 Dominelli, L. (2014). Promoting environmental justice through green social work practice: A key challenge for practitioners and educators. *International Social Work, 57*(4), 338–345. https://doi.org/10.1177/0020872814524968 Reviewing website	The Climate Emergency Facing Our Living Planet
6	No	Yes	Professor Margaret Alston, Keynote: "Adapting Social Work and Health Practice following Covid 19 and other Disasters," 23rd February 2022	Environmental Education Masterclass

(f) 1 × narrative conclusion at the end of the module about the student's personal/professional response to the climate crisis (200 words)
(g) Bibliography

Assessment of the ePortfolio
Students are invited to complete the different elements of the portfolio alongside their attendance of the module. There is no written exam. The assessment rubric is divided under the following criteria:

1. Knowledge and understanding of climate justice and the response of social work to the global climate emergency
2. A critical reflective appraisal of ecosocial practices in Irish contexts
3. A critical reflection including the student's reflections about the climate, their world, facts and figures of climate change, their learning throughout the module, and their artifact
4. Structure and coherence of the portfolio
5. Use of literature: reading and references
6. Use and understanding of ecosocial language.

A maximum of 5 points is given for each element (total 30 points).

Student engagement and feedback
Students of both programs are given the opportunity to evaluate their experiences of the new modules after their completion. Their responses have been very positive, and they have commented on how the module has enabled them to think critically about the immediacy of climate change and the role of the social professions in engaging with its causes and effects:

> Social work has an important role going forward as we face climate change. The importance of just transition. The importance of seeing this as a current issue, not something far away or in the future. (Social work student)

> Where climate change is concerned we have to act right now. My ecological carbon footprint is too high. (Youth and community work student)

> Most importantly the hope that came from prioritising the planet in our work for society, that this is a matter of urgency and is not separate to the wellbeing of the people we work with and we need to focus on it together. (Youth and community work student)

Student feedback about the ePortfolio mode of assessment indicates that they enjoy engaging in critical discussion about sustainability and ecological justice:

> Completing the assignment made me realise how more advocates from social work are needed in the climate crisis. (Social work student)

This student indicates a motivation to pursue a career in green social work:

> Green Social Work, I feel this really makes sense for me and fits in with how I would like to pursue my career. I felt it was such a positive lens to tackle very difficult issues. In my opinion Green Social Work is inclusive and pro-active. I Love It. (Social work student)

Ecosocial Work Teaching and Learning in the University: Challenges and Opportunities

In an era where neoliberal economics and policies dominate so many aspects of our lives including education in the social professions, creating teaching spaces in the timetables of busy qualifying programs is challenging. To date, it has only been possible to teach these modules as an elective in second semester of the MSW final year and a core course in the final year of the BSocSc (Youth and Community Work) degree, respectively. Our goal is to embed teaching and learning on sustainability and the environment from the start of these programs and as core modules for professional formation to enable students to recognize the critical and urgent nature of the climate emergency and its relevance for professional practice. A less comprehensive approach risks tokenism or even green washing. In this goal, we are motivated by the recognition that planetary boundaries (Angus, 2018) should take precedence over all other needs and should be the foundation of all thinking, planning, and action, including in the social professions. Arevalo (2017) asks:

> Who better to confront the environmental/ecological injustices than a social worker; our very profession exists for the betterment of humanity. Some may think it is taking a leap if social workers move towards a more 'structural' melee that includes the environment, but indeed, it should come quite naturally to our profession considering our history and our values, skills, and roles.

We are attempting to embed environmental thinking in a number of other modules on our programs, including research methods modules which are a core element of all qualifying programs in the social professions. In research modules, we introduce students to ecological paradigms including posthumanism, thereby encouraging them to think beyond "conventional paradigms" (Boetto, 2017, p. 49) and consider holistic ecocentric and post-anthropocentric perspectives. We also try to model environmental thinking through our collective membership and leadership in research and action groups including the Research for Civil Society, Environment and Social Action (REACT) cluster at University College Cork and the Environment Special Interest Group (SIG) of the European Association of Schools of Social Work (EASSW) (Catherine Forde). Since 2020, the REACT Cluster has organized four high-profile online events on the themes of environmental education, environmental policy, and the university, community, and environment. Meanwhile, environmental thinking is becoming more influential in our school. We will contribute to a new Masters in Social Policy and Social Justice module on *Sustainability, Climate Justice and Eco-Social Policies* to be delivered from 2025. While this module will also start as an elective, we believe that interdisciplinary collaboration is important to increase the profile of the environment and sustainability among other students and colleagues across all programs and disciplines in the school.

One of the most immediate challenges for us and our students is the lack of environmental and sustainability awareness across placement settings in the social professions. Social work and youth and community work agencies including the larger service providers have yet to meaningfully recognize or acknowledge the significance of climate emergency for service users and communities in the planning and delivery of services. While concepts like sustainability, climate crisis, and ecological and climate justice appear in policy documents, they are less visible in the work of agencies or in practice on the ground. It is unsurprising that this lack of awareness is evident in student portfolios and discussions on return from fieldwork education (placement); the focus of their writing is exclusively on social justice and social ecologies. As educators, we need to improve on supporting students to reflect and write about the climate emergency and its effects on the lives of service users when placement agencies are silent on these issues. When we speak to social work students on their return from placement about connections between the natural and social environments, they say that it was never a topic of discussion with practice teachers or social work colleagues. Yet we know social workers and youth and community workers who engage in green and ecosocial work in their practice. Elsewhere in this volume, two graduates of the BSocSc (Youth and Community Work) degree (Elaine Moynihan and Katie McShane) describe their development of Wise Water, a unique educational program aimed at developing children's awareness of the importance of water in our lives and ecosystems. Social work practitioners use social prescribing to refer clients to social farming settings and outdoor adventure therapies and use green spaces for meetings with service users, while community gardens and other green initiatives have become significant features of community work in Ireland (Doyle, 2022). However, these practitioners do not necessarily identify as ecosocial or green social workers or youth and community workers. Despite a small but growing body of Irish literature on eco-youth work and community work (Gorman, 2016; Nolan & Gorman, 2018; Forde et al., 2021), there is very little Irish research on eco- or green social work. These are still emerging areas of practice, and many practitioners may not be familiar with the growing body of theory and international practice.

In a recent small-scale qualitative comparative study of Australian and Finnish ecosocial practitioners, researchers developed a series of workshops to enhance ecosocial work interventions. The findings suggest that their involvement with colleagues in the workshops helped to increase workers' personal and professional awareness of ecosocial practice. In addition, being involved in "communities of practice" created through participation in the workshops improved their ability to use ecosocial work practices (Boetto et al., 2022). The study also found that while positive changes were observed in the practitioners' organizations, organizations could also present a range of barriers to ecosocial work practice (Boetto et al., 2022).

Critical to building the capacity of social workers and other staff is the role and commitment of organizations to incorporate the Sustainable Development Goals (SDGs) at a systemic and structural level. University College Cork is ranked first in Ireland for the advancement of sustainability in the 2023 Quacquarelli Symonds (QS) rankings and is among the top-performing universities globally in the areas of sustainability, with a global ranking of 6th in the 2023 UI Greenmetric World University Rankings. University College Cork recognizes the centrality of the environment in its *Strategic Plan 2023–2028* (2023a) and *Sustainability and Climate Action Plan 2023–2028* (2023b), which sets out nine key goals in areas including climate action, sustainability citizenship, and waste reduction and circular economy. Just Transition is a key plank of University College Cork's sustainability and climate action plan which seeks to embed sustainability goals and targets across all aspects of the University's educational, research, and ancillary operations, infrastructural developments, and interactions with the community. Among these aims, UCC is endeavoring to become a zero-waste campus by 2030 and a carbon neutral campus by 2040. For example, the university went plastic-free on campus in 2023, thereby eliminating the waste generated by one million paper cups.

Other large Irish public organizations are moving toward recognition of the significance of climate change in the context of public and social service provision. In its *Climate Action Strategy 2023–2050*, the Health Service Executive (HSE) identifies climate change is one of the greatest threats to public health. A recent statement by the CEO of the HSE announced a commitment to training its 100,000 employees, including social workers and community workers, as "climate activists" (HSE, 2023, p. 36). While this recognition of the interconnectedness of climate change and health is welcome, it needs to be translated into meaningful and resourced work practices and not simply left to the goodwill of individual employees.

In the professional context, in its sixth *Strategic Plan 2022–2024* (2022), the Irish Association of Social Workers (IASW) identifies the prioritization of Sustainable Development Goals including climate and environmental justice as a strategic objective. In its values statement, the IASW also highlights the need to stand in solidarity with communities and individuals affected by climate change. CORU, the body that regulates social work education and professional practice in Ireland, does not however include any reference to the advancement of environmental justice or the promotion of SDGs in any of the core competencies for social workers. The absence of such a competency is a critical weakness if the profession is to meet and respond to the challenges of climate change. As educators, we must strongly advocate for its inclusion as a professional requirement for social work education and practice. To contribute to this process, in 2023 one of the authors (Mary Hurley) jointly facilitated an ecosocial work online event for social work practitioners through the IASW. This continuing professional development (CPD) event was jointly offered by the IASW and UCC to examine the current state of ecosocial work in Ireland with a group of practitioners, and the occasion offered an opportunity to create and foster "communities of practice" amongst social workers (Boetto et al., 2022). The impetus for the event emerged from conversations between Mary and the IASW CPD officer who has a keen interest in expanding social work beyond its current anthropocentric focus. The webinar included a contribution by a social work colleague who uses ecosocial practices in his work in adult mental health services. It is hoped that the webinar will foster ongoing opportunities for connections between like-minded colleagues in practice and professional education and their professional bodies to meet and discuss ideas about incorporating ecosocial work into mainstream professional practice and professional education.

The All Ireland Community Work Educators Board (AIEB) which has a role in regulating community work education in Ireland recognizes social justice and sustainability as a core value of community work. In its *All Ireland Standards for Community Work* (2016) statement, the AIEB indicates that this value is concerned with the promotion of a just and sustainable society, thereby stopping short of a specific emphasis on environmental justice. To date, the International Association for Community Development (IACD) has not placed urgent emphasis on environment or sustainability in its own standards statement or set of values. Instead, it refers more generally to "rights, solidarity, democracy, equality, environmental and social justice" (2021, p. 2) and states that "practitioners need an understanding of political, social and ecological sciences to give them wider insights into the interconnected realities of people's lives, of the social, political, cultural, economic and environmental contexts within which people live" (ibid). Similarly, the National Occupational Standards for Youth Work (England), which are also used as occupational standards for youth work on the island of Ireland, do not make reference to the environment or climate change and instead focus on human relationships, social systems approaches and young people's rights (National Youth Agency, 2023). These statements tend to suggest that youth and community work thinking and practice have yet to significantly evolve beyond Anthropocentrism, despite the activities of many community workers, activists, and researchers.

Conclusion

This chapter reflects on the development and co-delivery of two new modules on the MSW and BSocSc (Youth and Community Work) qualifying programs in University College Cork. The aim of the modules is to introduce green and ecosocial thinking to students on social professional qualifying programs in the university and to encourage students to think in resistant ways about climate change and how to combat it in their thinking and practice. One of the main challenges in introducing the modules is the continued influence of anthropocentric thinking and practice in the social professions, both in Ireland and more broadly. It is difficult to introduce new and diverse ways of thinking to qualifying programs which are "human centred and defined by the existing service structure, national legislation and accustomed ways of working" (Ranta-Tyrkko & Narhi, 2021, p. 616). Overall, while there is a dearth of environmental awareness and ecosocial practices in social work and community work settings, we recognize that this may not be due to the unwillingness of practitioners to consider the relevance of the environment in professional practice. It may be due to a lack of opportunity for practitioners to discuss ideas for ecosocial practice in supportive and encouraging work environments. While the rhetoric of sustainability and climate change strategies does now feature in the strategic planning of large employing organizations in Ireland such as the Health Service Executive, how these strategies will be operationalized in practice needs urgent attention and support if they are to be effectively and meaningfully implemented by the social professions on the ground.

Writing this chapter has given the authors an opportunity to consider the opportunities and challenges of teaching and learning ecosocial work in an Irish university. What has emerged from these considerations is a realization that for our teaching to be effective, it must have an interdisciplinary focus that connects students from allied social professions to each other in the first instance, to the wider academy and to like-minded colleagues in professional practice. We argue that it is only by building such connections that practitioners' capacity to be "climate activists" (HSE, 2023, p. 36) can be realized. We agree with Thich Nhat Hanh that it is only through a collective awakening that draws on the shared knowledge, skills, and commitment of every section of our professions that we can build our capacity to grapple against the worst impacts of the climate emergency for the planet and its people.

References

Angus, I. (2018). The Earth System has passed six of nine planetary boundaries. *Resilience*.

Arevalo, L. (2017). Agitating through an alternate economy: Social work, sustainability, and the circular economy. In M. Rinkel & M. Powers (Eds.), *Social work promoting community and environmental sustainability: A workbook for social work practitioners and educators* (Vol. 1). International Federation of Social Work (IFSW). http://ifsw.org/product/books/social-work-promoting-community-and-environmental-sustainability-free-pdf/

Attlee, C. (1920). *The social worker.* London: George Bell and Sons (reproduced by Forgotten Books). https://archive.org/details/socialworker00attliala

Besthorn, F. H., & McMillen, D. (2002). The oppression of women and nature: Eco-feminism as a framework for an expanded ecological social work. *Families in Society, 83*(3), 221–232.

Boetto, H. (2017). A transformative eco-social model: Challenging modernist assumptions in social work. *The British Journal of Social Work, 47*(1), 48–67.

Boetto, H., Närhi, K., & Bowles, B. (2022). Creating 'Communities of Practice' to enhance ecosocial work: A comparison between Finland and Australia. *The British Journal of Social Work, 52*(8), 4815–4835.

Bretzlaff-Holstein, C. (2018). The case for humane education in social work education. *Social Work Education, 37*(7), 924–935.

Bronfenbrenner, U. (1979). *The ecology of human development: Experiments by nature and design*. Harvard University Press.

Climate Change Advisory Council. (2021). *Annual review 2021*. Climate Change Advisory Council.

Climate Change Advisory Council. (2022). Annual Review 2022. Dublin: Climate Change Advisory Council.

Community Work Ireland. (2016). *All Ireland Standards for community work*. Community Work Ireland.

Couzens Hoy, D. (2005). *Critical Resistance: From Poststructuralism to Post-Critique. Boston: MIT.*

Crowley, P., & Hughes, A. (2021). *The impact of COVID-19 pandemic and the societal restrictions on the health and wellbeing of the population, and on the health service capacity and delivery: A plan for healthcare and population health recovery. Version 2*. Dublin: National QI Team, Health Service Executive.

De Bruin, K., & Yakut, A. (2022). *The Global Emissions Impact of Irish Consumption*, ESRI Working Paper 740. Economic and Social Research Institute. https://www.esri.ie/publications/the-global-emissions-impact-of-irish-consumption

Dominelli, L. (2011). Climate change: Social workers' roles and contributions to policy debates and interventions. *International Journal of Social Welfare, 20*(4), 430–438.

Dominelli, L. (2012). *Green social work: From environmental crises to environmental justice*. Polity Press.

Doyle, G. (2022). In the garden: Capacities that contribute to community groups establishing community gardens. *International Journal of Urban Sustainable Development, 14*(1), 15–32.

Drolet, J., Wu, H., Taylor, M., & Dennehy, A. (2015). Social work and sustainable social development: Teaching and learning strategies for 'Green Social Work' curriculum. *Social Work Education, 34*(5), 528–543.

Environmental Protection Agency. (2022). *Water quality monitoring report on nitrogen and phosphorus concentrations in Irish waters*. EPA.

Forde, C., Lynch, D., & Lathouras, A. (2021). Introduction - community development in social work education: Themes for a changing world. *Community Development Journal, 56*(4), 561–565. https://doi.org/10.1093/cdj/bsab027

Freire, P. (2014). *Pedagogy of the oppressed*. Bloomsbury.

Gardner, H. (2006). *Multiple intelligences: New horizons*. Basic Books.

Gorman, J. (2016). Stories of resistance and resilience: Developing a community work approach to climate change and climate justice. *Journal of Radical Community Work, 2*(1) http://www.rcwjournal.org/ojs/index.php/radcw/issue/view/2/showToc

Hanh, T. N. (2006). *Call for a collective awakening: Speech to UNESCO*. https://plumvillage.org/about/thich-nhat-hanh/letters/call-for-a-collective-awakening-speech-to-unesco

Harley, A., & Scandrett, E. (2019). *Environmental justice, popular struggle and community development*. Policy Press.

Health Service Executive. (2023). *HSE climate action strategy 2023–2050*. HSE.

Howe, D. (2009). *A brief introduction to social work theory*. Palgrave Macmillan.

Intergovernmental Panel on Climate Change. (2022). *Sixth assessment report climate change: Impacts, adaption and vulnerability* https://www.ipcc.ch/report/ar6/wg2/about/factsheets

International Association for Community Development. (2021). *Towards shared international standards for community development practice*. IACD.

International Association for Social Work. (2020). *Code of Ethics* https://www.iasw.ie/publications-for-social-workers

International Federation of Social Workers. (2024). Global Definition of Social Work. https://www.ifsw.org/what.is.social.work/global.definition.of.social.work/

Irish Association of Social Work. (2022). *Social workers: Stronger together for social change. Strategic plan 2022–2024*. IASW.

Jones, P. (2010). Responding to the ecological crisis: Transformative pathways for social work education. *Journal of Social Work Education, 46*(1), 67–84.

Jones, K. R., Merritt, L. N., Brown, A. N., Davidson, S., Nulliner, D. L., Smart, J., Walden, L., & Wingez-Yanez, N. (2017). Case studies in sustainable social work: MSW students explore principles in practice. *Journal of Sustainability Education, 12(1), 2151–7452.*

Klemmer, C. L., & McNamara, K. A. (2020). Deep ecology and ecofeminism: Social work to address global environmental crisis. *Affilia, 35*(4), 503–515. https://doi.org/10.1177/0886109919894650

Kornhaber, M., Ferrios, E., & Veenema, S. (2004). *Multiple intelligences: Best ideas from research and practice*. Pearson.

Lavalette, M. (2011). *Radical social work today: Social work at the crossroads*. Polity Press.

McDade, S. A. (1995). Case Study Pedagogy to Advamce Critical Thinking. *Teaching of Pyschology, 22(1), 9–10.*

McKinnon, J., & Bay, U. (2013). Social work enabling sustainable ecological living. *Australian Social Work, 66*(2), 153–155.

Mehrotra, G. R., Hudson, K. D., & Self, J. M. (2019). A critical examination of key assumptions underlying diversity and social justice courses in social work. *Journal of Progressive Human Services, 30*(2), 127–147.

Melekis, K., & Woodhouse, V. (2015). Transforming social work curricula: Institutional supports for promoting sustainability. *Social Work Education, 34*(5), 573–585.

Met Eireann. (2018). *Ophelia Report.* https://www.met.ie/cms/assets/uploads/2018/10/OpheliaReport.pdf

Met Eireann. (2020). *The Status of Ireland's Climate 2020.* https://www.met.ie/cms/assets/uploads/2021/08/The-Status-of-Irelands-Climate-2020.pdf

Met Eireann. (2024). *Human-caused climate change brings increased storm rain fall.* https://www.met.ie/human-caused-climate-change-brings-increased-storm-rainfall

Mullaly, B. (2007). *The new structural social work* (3rd ed.). Oxford University Press.

National Youth Agency. (2023). *Youth work in England: Policy, practice and the National Occupational Standards*. NYA.

Noble, C. (2016). Green social work – The next frontier for action. *Social Alternatives, 35*(4), 14–19.

Nolan, C., & Gorman, J. (2018). Just transition champions: Reflections on climate justice-focused continuous practice development. *Practice Insights, 20*, 17–18.

Pascoe, S., Waterhouse-Bradley, B., & McGinn, T. (2023). Social workers' experiences of bureaucracy: A systematic synthesis of qualitative studies. *British Journal of Social Work, 53*, 513–533. https://doi.org/10.1093/bjsw/bcac106

Pawar, M. (2019). Social work and social policy practice: Imperatives for political engagement. *The International Journal of Community and Social Development, 1*(1), 15–27. https://doi.org/10.1177/2516602619833219

Payne, M. (2005). *The origins of social work: Continuity and change*. Palgrave Macmillan.

Perkins, D. (1998). What is understanding? In M. S. Wiske (Ed.), *Teaching for understanding: Linking research with practice*. Jossey-Bass Publishers.

Philip, D., & Reisch, M. (2015). Rethinking social work's interpretation of 'Environmental Justice': From local to global. *Social Work Education, 34*(5), 471–483.

Rambaree, K., Powers, M. C. F., & Smith, R. J. (2019). Ecosocial work and social change in community practice. *Journal of Community Practice, 27*(3–4), 205–212.

Ramsay, S., & Boddy, J. (2017). Environmental social work: A conceptual analysis. *British Journal of Social Work, 47*, 68–86.

Ranta-Tyrkkö, S., & Närhi, K. (2021). Striving to strengthen the ecosocial framework in social work in Finland. *Community Development Journal, 56*(4), 608–625. https://doi.org/10.1093/cdj/bsab030

Reisch, M. (2013). Social Work and the Neoliberal Challenge: The US Response to Increasing Global Inequality. *Social Work: The International Journal, 21, 287–298.*

Shulman, L. S. (2007). *The many faces of engagement. Transcript of Keynote Address.* Presented at 13th annual conference of the coalition of urban and metropolitan universitieS, 21 October 2007, Baltimore, Maryland, USA, pp. 11–20.

Teixeira, S., & Krings, A. (2015). Sustainable social work: An environmental justice framework for social work education. *Social Work Education, 34*(5), 513–527.

Toreti, A., Bavera, D., Acosta Navarro, J., Arias Muñoz, C., Avanzi, F., Barbosa, P., de Jager, A., Di Ciollo, C., Ferraris, L., Fioravanti, G., Gabellani, S., Grimaldi, S., Hrast Essenfelder, A., Isabellon, M., Jonas, T., Maetens, W., Magni, D., Masante, D., Mazzeschi, M., McCormick, N., Rossi, L., & Salamon, P. (2023). *JRC technical report: Drought in Europe GDO report.* Brussels: European Commission.

Towey, N. (2018, Oct 14). Storm Ophelia caused almost €70m worth of damage, *Irish Times.* Storm Ophelia caused almost €70m worth of damage – The Irish Times.

Tsui, M., & Chueng, F. C. H. (2004). Gone with the wind: The impacts of managerialism on human services. *British Journal of Social Work, 34*, 437–442.

Ungar, M. (2002). A deeper, more social ecological social work practice. *Social Service Review, 76*, 480–497.

University College Cork. (2023a). *UCC Strategic Plan 2023–2028*. Cork: University College Cork. Securing Our Future - UCC Strategic Plan 2023–2028 by discoverUCC - Issuu.

University College Cork. (2023b). *UCC sustainability and climate action plan 2023–2028*. Cork: University College Cork.

Wehbi, S., & Straka, S. (2011). Revaluing student knowledge through reflective practice on involvement in social justice efforts. *Social Work Education, 30*(1), 45–54.

Weinberg, M., & Banks, S. (2019). Practising ethically in unethical times: Everyday resistance in social work. *Ethics and Social Welfare, 13*(4), 361–376. https://doi.org/10.1080/17496535.2019.1597141

Wiske, M. S. (Ed.). (1998). *Teaching for understanding: Linking research with practice*. Jossey Bass.

Zapf, M. K. (2009). *Social work and the environment: Understanding people and place*. Canadian Scholars' Press.

Chapter 6: Linking the Human Rights and Ecosocial Paradigms in Social Work Education: An Example from Practice

Ingo Stamm and Lisa Dörfler

Introduction

The implementation of ecosocial work is still an open task for many social work study programs. One concrete challenge is to combine ecosocial work with already established paradigms or approaches[1] in social work, such as human rights-based social work. Like ecosocial work, the human rights approach in social work incorporates different perspectives and is linked to a variety of fields of social work education, for example, social work ethics and international social work. Human rights are commonly seen as one of the guiding principles of social work; together with social justice, they form the ethical basis of the profession and inform social work practice. Human rights underlie a theory or concept of social work that guides social work action and has specific action methods (Prasad, 2023). There is a broad consensus that human rights are highly relevant for social work as a discipline and profession and therefore for social work education (Androff, 2016; Staub-Bernasconi, 2016, 2018). The overarching question asked by this chapter is how to combine human rights-based social work with ecosocial work in educational contexts. This question has rarely been discussed in the literature on social work education.

[1] We treat the terms "approach" and "paradigm" as synonymous in this chapter. Both terms can be understood to refer to premises, concepts, or theories that have a foundation in shared knowledge.

I. Stamm (✉)
University of Jyvaskyla, Jyväskylä, Finland
e-mail: stamm.ingo@fh-muenster.de

L. Dörfler
Paritätischer Gesamtverband, Berlin, Germany
e-mail: lisa.doerfler@paritaet.org

C. Forde et al. (eds.), *Teaching and Learning in Ecosocial Work*,
https://doi.org/10.1007/978-3-031-58708-5_9

To find answers to the question, we introduce an example from practice that centers on both the human rights and ecosocial work approaches and their intersections. The example we have chosen is a pilot seminar we conducted in spring 2022 at the Alice Salomon University of Applied Sciences Berlin (ASH), Germany, in the context of the MA study program "Social Work as a Human Rights Profession." The focus of the seminar was on the role of human rights in the sustainability discourse and the meaning of the natural environment within the human rights debate. First, we will explain the theoretical basis of the pilot seminar. This descriptive section is intended to provide a foundation that will enable a better understanding of our subsequent discussion of the seminar. It does not claim to delineate all the aspects of either paradigm, and it cannot show the potential conflicts between the two. Rather, it focuses on their common ground and intersections. Next, we will describe the teaching concept and rationale. Third, we will critically evaluate the seminar and reflect on its structure and outcomes, including the seminar's implications for social work practice, teaching, and learning. Our conclusions have to be seen as tentative, since our experience to date of combining the human rights and ecosocial paradigms in social work education is mostly limited to this single instance.

Theoretical Background

In the current *Global Standards for Social Work Education and Training* (International Association of Schools of Social Work [IASSW] and International Federation of Social Workers [IFSW], 2020), human rights are directly linked to environmental issues. The document states that students need to be able to apply human rights principles in order to understand how social issues affect different forms of justice, including environmental justice (p. 19). This is one indication of the link between human rights and the ecosocial paradigm. However, the *Global Standards* only provide a broad framework. They do not further explain how these links can be implemented in social work education. Before we go on to describe the practical teaching example, we will introduce the human rights and ecosocial paradigms and show what theoretical links can be made (see also Stamm, 2023).

Human Rights-Based Social Work

The seminar on which this article is based was built on both the human rights paradigm and the ecosocial paradigm of social work. In the German-speaking countries, the human rights paradigm of social work was established theoretically and conceptually by Silvia Staub-Bernasconi (1995, 2019), one of the pioneering social work scholars on human rights. Staub-Bernasconi was also the driving force in the foundation of the MA program "Social Work as a Human Rights Profession" (see section "Theoretical Background"). Drawing on human rights, she developed the concept of a triple mandate for social work. According to this concept, there is not only the double mandate for social work—that is, the classic state mandate based on legislation and the constitution, plus the client's mandate—but also a third mandate that comes from the profession itself. This third mandate comprises two dimensions: the science-based nature of professional practice and the ethics-based nature of the profession's national and international codes of practice (Staub-Bernasconi, 2016, 2018).

There is a current debate in social work about the importance of the physical environment for the realization and protection of human rights, as well as about how environmental crises—first and foremost the climate crisis—threaten the human rights of vulnerable people (Dörfler & Stamm, 2022; Tester, 2013). Simultaneously, the human rights discourse shows that environment-related issues can already be found in not only collective human rights but also liberal defensive rights; economic, social, and cultural rights; procedural rights; and minority rights, as well as at the level of regional human rights documents (Lohmann, 2012; Woods, 2016). For example, the right to life, enshrined in the United Nations (UN) Covenant on Civil and Political Rights, and the rights to food, health, and adequate housing, incorporated into the UN Covenant on Economic, Social, and Cultural Rights (see Wronka, 2012), presuppose the necessity of an intact environment. A breakthrough regarding the link between human rights and environmental protection occurred in 2021: resolution 48/13 of the UN Human Rights Council recognized for the first time the official human right to a clean, healthy, and sustainable environment for all people. A year later, this new human right was confirmed by a resolution of the UN General Assembly (UN, 2021, 2022). Both resolutions emphasize that a healthy environment is crucial to the realization of all other human rights. This means that the concept of a third mandate (i.e., from the profession itself)—which according to Staub-Bernasconi relies heavily on human rights—also implies an ethical obligation to consider the health of the physical environment. Human rights depend on a healthy environment and vice versa, and therefore both play an important role for social workers.

The human rights paradigm not only provides an ethical basis for the profession but also has practical implications. What the human rights orientation means at different levels of social work practice is pointed out by Prasad (2018, 2023). In her view, human rights can serve as a frame of reference for social workers. For example, if mandates or tasks from state actors are seen to disrespect or even violate human rights, then social workers can refer to human rights and the third mandate when they either refuse to fulfil or deliberately act against such tasks in order to support the people with whom they work. Relevant examples would include refugees' right to freedom of movement within a nation-state or their right to privacy in collective accommodation. Furthermore, human rights and other UN documents can be used to define core values of social work, such as participation and human dignity (see Androff, 2016). Likewise, they can be used as an analytical tool in relation to the living conditions of both vulnerable groups and individual social work clients, for example, the right to adequate housing for homeless people or asylum seekers.

Furthermore, using certain instruments of the human rights system can be seen as part of social work's methods at the structural level. Human rights treaties ratified by states are a legal tool to assert rights. Some treaties also include individual complaints procedures: for example, the UN Covenant on Economic, Social, and Cultural Rights can be used by social workers. When the national legal process is concluded, citizens, whose governments have ratified the Optional Protocol to the Covenant, can present their complaints to the UN treaty bodies. Social workers and NGOs can support individuals during this process, aiming to obtain a ruling that leads to structural improvements. Additionally, participation in reporting procedures and shadow or alternative reports, which are a method for nongovernmental organizations (NGOs) to provide UN monitoring bodies with additional information to supplement governmental treaty reports, is part of the toolbox of human rights-based social work. Such alternative reports are able to point out problems with governments' implementation of or noncompliance with human rights, and social workers can use them to support their concerns and demands (Prasad, 2018, pp. 37–44). There are also several examples showing how to extend human rights-related methods in social work to include elements of ecosocial work. For example, the German civil society organizations that are part of a network called the National Coalition, which includes large social welfare agencies as well as smaller NGOs, have coined the term "ecological children's rights." They argue that some of the most fundamental children's rights—for example, the right to life, enshrined in Article 6 of the UN Convention on the Rights of the Child—can only be fully realized if governments take the climate crisis more seriously than hitherto (National Coalition, 2019; Schramkowski & Stamm, 2023). However, these ideas have not yet been used in social work practice, and to our knowledge they are rarely discussed in social work education.

Ecosocial Paradigm

The second theoretical basis of our seminar was the ecosocial paradigm of social work. Key elements of this paradigm involve including the natural environment as an additional focus of social work, understanding people as part of the natural environment, putting social and ecological problems and goals together and recognizing the responsibilities of social work, looking critically at prevailing modes of production and consumption, forming new transdisciplinary and international alliances, and critically examining the role of social work in the ecological crisis (see chapter "Introduction"). The global definition of social work and social work ethics also ascribes ecosocial responsibility to social work. For example, the IFSW and IASSW's (2014) global definition states that "social work engages people and structures to address life challenges and enhance well-being" (n.p.). Since the climate crisis will be a life challenge for many social work service users across the globe and will have a negative impact on people's well-being, social workers will have to consider climate and environmental aspects in their work. The commentary notes for the global definition elaborate further:

> Social work's legitimacy and mandate lie in its intervention at the points where people interact with their environment. The environment includes the various social systems that people are embedded in and the natural, geographic environment, which has a profound influence on the lives of people. (IFSW & IASSW, 2014, n.p.)

This is a demand for social work practice to expand its narrow understanding of the environment as merely social and includes the physical environment as well. In reference to necessary social changes, the commentary further explains that "social change initiatives recognize the place of human agency in advancing human rights and economic, environmental, and social justice" (n.p.). In addition to advancing human rights, social work should also work toward environmental justice—to which we can add climate justice. Environmental justice is a broad concept that demands that all people, without discrimination, should be free from any environmental harm and have access to the environment's positive effects (Erickson, 2018). Climate justice is a narrower concept that refers to climate change impacts, climate protection, and climate mitigation programs. It seeks to shed light on the social and cultural effects of the climate crisis, and it is used to discuss the relationships between nation-states or groups within a society as well as intergenerational justice (Liedholz, 2022). At the level of international social work organizations, there are several other documents that delineate the connection between social and natural environmental factors and the relevance of social work. These include *Globalization and the Environment* (IFSW, 2014), the *Global Social Work Statement of Ethical Principles* (IFSW, 2018) and the *Global Agenda of Social Work and Social Development* (2010–2020, now in its second decade; IFSW et al., 2012). All these documents share an emphasis on the significance of ecosocial work—or at least certain elements of it, such as sustainability or environmental justice—and the importance of human rights for social work. However, further ideas about how to link these approaches cannot be found.

Summary

The human rights approach and the ecosocial approach share much common ground. Human rights and a healthy environment are mutually dependent, and the ecosocial paradigm can be linked to human rights-based social work at different levels, for example, with regard to social work ethics and social work methods. Numerous documents issued by the IFSW and IASSW already emphasize the equal importance of human rights and elements of the ecosocial paradigm. Moreover, as the current *Global Standards for Social Work Education and Training* (IASSW & IFSW, 2020) stipulate, human rights and principles such as environmental justice need to be brought together in social work education. However, if we wish to understand how all this can be implemented in social work education, more practical experience and reflection is necessary. The pilot seminar we describe in the next section should therefore be seen as part of future knowledge regarding the link between ecosocial work and human rights-based social work in social work teaching.

Teaching Concept and Rationale

In Germany, the discussion of ecosocial work and the importance of the climate crisis for social workers has recently accelerated (Dörfler & Stamm, 2022; Liedholz, 2020; Pfaff et al., 2022; Stamm, 2021). However, as in many other countries, connecting human rights and ecosocial approaches is a completely new field in German social work education. The connection has been highlighted in niche areas, such as in courses on international social work, youth work, and social work ethics (Verch, 2020). The example from practice we introduce in this section was one of the first teaching units to explicitly center the link between human rights and the environment. It was offered as an optional, 1-day seminar on the ASH MA study program "Social Work as a Human Rights Profession." ASH is one of Germany's largest universities of applied sciences for social work, health, and education studies. The MA program is a nonconsecutive, extra-occupational and cooperative program offered by different schools of social work in Berlin (ASH, Evangelische Hochschule Berlin, and Katholische Hochschule für Sozialwesen Berlin) and the German Institute for Human Rights. Its main objective is to train professionals to incorporate and implement human rights issues in teaching, in domestic and/or international practical projects, and within their own organizations (see MRMA, 2023). The additional 1-day seminar was announced to the current student cohort as well as alumni of the program across German-speaking countries. Altogether 18 participants signed up for the seminar, all of them current or former social work students. Since it is an extra-occupational study program, most of the current students worked in parallel as practitioners. Further information on the students' backgrounds was not assessed. The language of the pilot seminar was German. It was titled "Social Work as a Human Rights Profession and Sustainability." The main questions that structured the seminar were as follows: what role do human rights play in the principle of sustainability? What is the environmental dimension of human rights within the UN and at the supranational level? Lastly, what can a human rights-based social work profession learn from ecosocial approaches and vice versa? Since the discourse on the climate and environmental crises is not yet part of the curriculum but is already having a negative impact on the realization of human rights, the seminar was offered in order to start a debate in the context of social work as a human rights profession. The learning goals remained at the levels of understanding and applying, according to Bloom's taxonomy of educational objectives (Anderson & Krathwohl, 2001). Due to the brevity of the seminar, the learning goals were as follows: students would have a basic knowledge of the history of the ecosocial paradigm; they would understand the environmental dimension of human rights; they would be able to tentatively apply elements of the ecosocial paradigm in social work practice; and they would be capable of making conceptual links between human rights-based social work and ecosocial work during such application.

The 1-day seminar took place online (via Zoom) to allow people from all over Germany, Austria, and Switzerland to participate and to reduce mobility during the COVID-19 pandemic. To avoid any privacy issues, it was not recorded. It consisted of three main parts. The first presentation gave an overview of human rights in relation to the environment, the Sustainable Development Goals, and recent developments and discussions about the human right to a clean, healthy, and sustainable environment. In addition, the importance of combining human rights and an ecosocial approach was discussed with participants. Making service users aware of their fundamental human rights can also have an empowering effect (see Prasad, 2018). Including environmental topics within this approach helps social workers to focus on the natural living conditions of the people and groups with whom they work and the social and health impacts of those conditions. It also legitimizes social workers' participation in environmental and climate discourses. In this context, the discussion also integrated further examples of social work ethics drawn from international social work organizations (the IFSW and IASSW).

The second part introduced the ecosocial paradigm of social work and current concepts, practice research, and trends. It discussed whether and how the ecosocial and human rights paradigms might complement each other. The afternoon session focused on social work practice. After a brief overview of recent practice research in Finland and Germany, the session elaborated on possibilities for action at different levels of social work intervention. In the subsequent group work, the participants were able to discuss concrete ideas about how to combine human rights and environmental concerns in practical work. Regarding ecosocial work practice, we focused on different levels of intervention. Boetto et al. (2020) differentiate between the personal, individual, group, community, and structural levels. While the personal level deals with our own perceptions of nature and lifestyles, the individual level focuses on one-to-one work with individual clients. Based on the idea of a holistic approach, it also includes nature-based interventions and outdoor activities, for example. The group level of ecosocial work focuses on both client group work and social work organizations and institutions (Boetto et al., 2020). Ecosocial work at the community level aims to strengthen local networks and the participation of residents and organizations in decision-making processes. To reduce the exploitation of nature and people, social work should also expand its interventions at the structural level. Working toward social, political, and economic change can have far-reaching effects, minimizing the negative impact of the climate crisis on marginalized people and enhancing social and environmental justice (p. 60). In a group discussion in the afternoon during the second part of the seminar, students suggested numerous ideas stemming from their own experiences of practice, personal engagements with environmental issues, and theoretical knowledge. These included gardening projects, community work, environmental justice, and arts-based projects. Further, the participants discussed how social workers could link their work with lobbying for environmental concerns and influencing policymakers. These ideas and interventions addressed all of the levels described by Boetto et al. (2020).

Due to time limitations, we were unable to conduct a more systematic categorization or in-depth discussion of the emergence and feasibility of these ideas.

Finally, we discussed the question of how to take action within the university to highlight the responsibility of places of higher education in terms of the climate crisis. Students came up with the idea of organizing days of action, for example, regarding the conditions under which clothing is produced. Giving a talk or planning an exhibition about the social and ecological consequences of fast fashion might raise awareness of this issue at the university. It might also make the link between human rights-based social work and ecosocial work more tangible for university students and staff. To raise this to the level of action, it might be combined with a clothing swap party or an upcycling workshop for old clothes. These ideas are already being implemented in some German universities, but they have not been systematically assessed. Further linking these suggestions to social work practice would mean adhering to ecological and social standards when buying textiles for work purposes, and it would raise the awareness of service users and colleagues about the topic.

Evaluation and Reflection

Our evaluation of the seminar included not only our own critical reflections but also the views of participants as co-informants, based on a questionnaire distributed to all participants to evaluate their experiences. The overall assessment of the seminar among participants who filled out the questionnaire afterward was very good. However, less than half of participants filled out the questionnaire, which limits the informative value of the feedback. Many participants stated that ecosocial work was a completely new approach to them. Only one had extensive prior knowledge and therefore took a more active part in the discussion. Responses to the question of how relevant the seminar was for students' current questions and concerns varied. Half of participants said it was "very relevant," the other half only "relevant." The overview of concepts such as green social work and ecosocial work, and of recent research from a variety of countries across the globe, was found to be very helpful. At the end, participants also mentioned a wish for further seminars and formats to discuss and exchange ideas on the topic of ecosocial work. Some comments in the feedback also suggested that the link between human rights and the environment might be incorporated into future study projects or dissertations on the MA program.

Overall, the participants' feedback indicated that the learning goals had at least partially been reached. However, the first point of critical reflection is to acknowledge the limitations of a 1-day seminar. The learning goals were quite broad, and a lot of prior knowledge is needed if one is to make connections between and integrate two essential paradigms in social work. The presentations in the first part of the seminar could therefore have been reduced to focus only on certain elements or streams of the ecosocial paradigm, for example, only dealing with environmental justice. For the group discussion in the afternoon too, a narrower focus might have been helpful. However, in sum the seminar was successful, even though the limitations were evident.

For organizational reasons, the seminar has not yet been repeated. However, the medium-term goal is to integrate the link between human rights and the environment as a mandatory element of the MA study program. In the future, it might even be worthwhile adding ecosocial work to the program's introductory seminars as a crosscutting topic. For example, a stronger embeddedness in modules on international law and the historical development of human rights and their environmental dimension would be a fruitful addition. Further, modules on social work ethics, poverty and children's rights could include ecosocial aspects, to name just a few possibilities.

Conclusion

The main question of this chapter was how to combine human rights-based social work with ecosocial work in educational contexts. Using an example from practice, we wanted to demonstrate that the two paradigms can be linked in social work education. Our seminar successfully highlighted the interrelatedness of human rights, the natural environment, and sustainability. Human rights cannot be fulfilled without a healthy environment, and a healthy environment cannot be achieved without the fulfilment of basic human rights and underlying principles such as nondiscrimination and participation. Our learning goals were reached, and during group discussions, the students made their own links based on their personal experience. However, our example has also revealed some of the challenges.

Both approaches include a number of different perspectives, and without prior knowledge, it can be hard to grasp the commonalities or to draw conclusions for social work practice. The ecosocial paradigm in particular is new to many social work students, and therefore starting with certain elements of the paradigm—for example, environmental justice and its link to human rights—might be a better choice for a 1-day seminar. Group discussions regarding the practical implementation of the two paradigms could also be narrowed down, for example, to one specific human right and its environmental aspects. Another idea would be to start with concrete cases and reflect on how they include both a human rights and an ecosocial work dimension. In general, our experience makes clear that the ecosocial paradigm (or elements of it) needs to become much more integrated into social work education. This might be done in specialized modules or courses such as the seminar we conducted or as a new crosscutting topic that can be integrated into various fields of social work education. Either way, the ecosocial work paradigm will reshape the education of social workers in the future.

References

Anderson, L. W., & Krathwohl, D. R. (Eds.). (2001). *A taxonomy for learning, teaching, and assessing: A revision of Bloom's taxonomy of educational objectives*. Longman.

Androff, D. K. (2016). *Practicing rights: Human rights-based approaches to social work practice*. Routledge.

Boetto, H., Bowles, W., Närhi, K., & Powers, M. (2020). Raising awareness of transformative eco-social work: Participatory action research with Australian practitioners. *International Journal of Social Welfare, 29*(4), 300–309. https://doi.org/10.1111/ijsw.12445

Dörfler, L., & Stamm, I. (2022). Nachhaltigkeit und Soziale Arbeit als Menschenrechtsprofession. In Y. Liedholz & J. Verch (Eds.), *Nachhaltigkeit und Soziale Arbeit: Grundlagen, Bildungsverständnisse, Praxisfelder* (pp. 81–92). Budrich.

Erickson, C. L. (2018). *Environmental justice as social work practice*. Oxford University Press.

International Association of Schools of Social Work & International Federation of Social Workers. (2020). *Global standards for social work education and training*. Retrieved 5 June 2023 from https://www.iassw-aiets.org/wp-content/uploads/2020/11/IASSW-Global_Standards_Final.pdf

International Association of Schools of Social Work, International Federation of Social Workers, & International Council on Social Welfare. (2012). *The global agenda for social work and social development: Commitment to action*. Retrieved 5 June 2023 from https://www.ifsw.org/wp-content/uploads/ifsw-cdn/assets/globalagenda2012.pdf

International Federation of Social Workers. (2014). *Globalisation and the environment*. Retrieved 5 June 2023 from https://www.ifsw.org/globalisation-and-the-environment/

International Federation of Social Workers. (2018). *Global social work statement of ethical principles*. Retrieved 5 June 2023 from https://www.ifsw.org/global-social-work-statement-of-ethical-principles/

International Federation of Social Workers & International Association of Schools of Social Work (Eds.). (2014). *Global definition of social work: Commentary notes for the global definition of social work*. Retrieved 5 June 2023 from https://www.ifsw.org/what-is-social-work/global-definition-of-social-work/

Liedholz, Y. (2020). *Berührungspunkte von Sozialer Arbeit und Klimawandel: Perspektiven und Handlungsspielräume*. Budrich.

Liedholz, Y. (2022). Konzepte von Klimagerechtigkeit im Kontext Sozialer Arbeit. In T. Pfaff, B. Schramkowski, & R. Lutz (Eds.), *Klimakrise, Sozialökologischer Kollaps und Klimagerechtigkeit: Spannungsfelder für Soziale Arbeit* (pp. 69–80). Beltz Juventa.

Lohmann, G. (2012). Umweltzerstörung. In A. Pollmann & G. Lohmann (Eds.), *Menschenrechte: Ein interdisziplinäres Handbuch* (pp. 438–443). J. B. Metzler.

MRMA. (2023). *Master of Social Work – Soziale Arbeit als Menschenrechtsprofession. Studienumfang*. Retrieved 5 June 2023 from https://www.mrma-berlin.de/Studium/Studienprofil/

National Coalition Germany. (2019). *Implementation of the UN convention on the rights of the child in Germany, 5th/6th supplementary report to the United Nations*. Network for the Implementation of the UN Convention of the Rights of the Child. Retrieved 5 June 2023 from https://umsetzung-der-kinderrechtskonvention.de/wp-content/uploads/2019/11/NC_ErgaenzenderBericht_ENG_Web.pdf

Pfaff, T., Schramkowski, B., & Lutz, R. (Eds.). (2022). *Klimakrise, Sozialökologischer Kollaps und Klimagerechtigkeit: Spannungsfelder für Soziale Arbeit*. Beltz Juventa.

Prasad, N. (2018). Statt einer Einführung: Menschenrechtsbasierte, professionelle und rassismuskritische Soziale Arbeit mit Geflüchteten. In N. Prasad (Ed.), *Soziale Arbeit mit Geflüchteten: Rassismuskritisch, professionell, menschenrechtsorientiert* (pp. 9–29). Budrich.

Prasad, N. (Ed.). (2023). *Methoden struktureller Veränderung in der Sozialen Arbeit*. UTB Verlag Budrich.

Schramkowski, B., & Stamm, I. (2023). Die Klimakrise als universelle Kinderrechtsgefährdung: Mögliche Implikationen für die Kinder- und Jugendhilfe. *Sozialmagazin, 7-8*, 25–30.
Stamm, I. (2021). *Ökologisch-kritische Soziale Arbeit: Geschichte, aktuelle Positionen und Handlungsfelder*. Budrich.
Stamm, I. (2023). Human rights-based social work and the natural environment: Time for new perspectives. *Journal of Human Rights and Social Work, 8*, 42–50. https://doi.org/10.1007/s41134-022-00236-x
Staub-Bernasconi, S. (1995). Das fachliche Selbstverständnis Sozialer Arbeit – Wege aus der Bescheidenheit: Soziale Arbeit als 'Human Rights Profession'. In W. R. Wendt (Ed.), *Soziale Arbeit im Wandel ihres Selbstverständnisses: Beruf und Identität* (pp. 57–104). Lambertus.
Staub-Bernasconi, S. (2016). Social work and human rights: Linking two traditions of human rights in social work. *Journal of Human Rights and Social Work, 1*(1), 40–49. https://doi.org/10.1007/s41134-016-0005-0
Staub-Bernasconi, S. (2018). *Soziale Arbeit als Handlungswissenschaft: Soziale Arbeit auf dem Weg zu kritischer Professionalität*. Budrich.
Staub-Bernasconi, S. (2019). *Menschenwürde, Menschenrechte und Soziale Arbeit: Vom beruflichen Doppelmandat zum professionellen Tripelmandat*. Budrich.
Tester, F. (2013). Climate change as a human rights issue. In M. Gray, J. Coates, & T. Hetherington (Eds.), *Environmental social work* (pp. 102–118). Routledge.
United Nations. (2021). *Resolution adopted by the Human Rights Council on 8 October 2021, 48/13: The human right to a clean, healthy and sustainable environment*. Retrieved 5 June 2023 from https://undocs.org/A/HRC/RES/48/13
United Nations. (2022). *Resolution adopted by the General Assembly on 26 July 2022, A/76/L.75. The human right to a clean, healthy and sustainable environment*. Retrieved 5 June 2023 from https://digitallibrary.un.org/record/3982508?ln=en
Verch, J. (2020). Perspektiven zu einem nachhaltigkeitsbildenden Professionsverständnis und einer nachhaltigkeitskulturellen Ethik in der Sozialen Arbeit: Einleitende Überlegungen zu einer sich entwickelnden Beziehung von Nachhaltiger Entwicklung. In B. Völter, H. Cornel, S. B. Gahleitner, & S. Voß (Eds.), *Professionsverständnisse in der Sozialen Arbeit* (pp. 170–183). Beltz & Gelberg.
Woods, K. (2016). Environmental human rights. In T. Gabrielson, J. M. Meyer, & D. Schlosberg (Eds.), *The Oxford handbook of environmental political theory* (pp. 333–345). Oxford University Press.
Wronka, J. (2012). Overview of human rights: The UN conventions and machinery. In L. M. Healy & R. J. Link (Eds.), *Handbook of international social work: Human rights, development, and the global profession* (pp. 439–446). Oxford Scholarship Online. https://doi.org/10.1093/acprof:oso/9780195333619.001.0001

Chapter 7: Eco-Diversity at the Margins: Permaculture as an Ecosocial Pedagogy for Social Work

Alex Skinner and Deborah Lynch

Introduction

Globally, ecosocial work exists as an emerging field of practice in response to a broad suite of rapidly evolving, overlapping, and interrelated crises, which threaten both "human and non-human life" (Panagiotaros et al., 2022, p.1). There is a rapidly growing body of work calling for social work to engage with ecosocial transition (Boetto, 2017, 2019; Boetto et al., 2020; Matthies & Närhi, 2016; Panagiotaros et al., 2022). The challenge for social workers and educators is no longer articulating the *why* we need to shift in this direction but to demonstrate the *how*. Social work education must engage with the lofty challenge of (re-) imagining and (re-) building an ecosocial world while also equipping students with the practical tools, knowledge, and guidance to implement ecosocial practice from the ground up. As social work educators, one of whom is also a permaculturist, we posit that permaculture provides an existing theoretical framework that can be critically and reflexively adapted to this challenge.

To this end, the chapter will focus on engagement with permaculture not only as a method of practice but as a critical pedagogy for ecosocial work education. To further this, as social work educators, we must develop a body of practical solutions, practice models, case studies, and examples of this work. Ramsay and Boddy (2017) have argued for a "partnership approach between social work and permaculture" (p. 206). This chapter aims to progress and further that notion, toward the embedding of permaculture in social work education as a transformative pedagogy. We will explore the ethics and some key principles of permaculture to illustrate how this aligns with social work's role in an ecosocial transition.

A. Skinner (✉) · D. Lynch
School of Nursing, Midwifery and Social Work, University of Queensland, Brisbane, QLD, Australia
e-mail: alex.skinner@uq.edu.au

C. Forde et al. (eds.), *Teaching and Learning in Ecosocial Work*,
https://doi.org/10.1007/978-3-031-58708-5_10

Permaculture is a way of holistic systems thinking that foregrounds observation of natural processes from an eco-centric perspective. Social workers have adopted "person-in-environment" theories for many years. These theories have generally focused on the social environment, but there is a contemporary shift toward re-centering practice with an eco-centric rather than anthropocentric perspective. See, for example, "posthumanism" (Braidotti, 2013; Ife, 2018;) "ecofeminism" (Mies & Shiva, 2014; Bell et al., 2019), and "deep ecology" (Besthorn, 2012; Naess, 1995). These worldviews have long existed in countercultural spaces and are certainly not new. However, what is new is their rapid entry into mainstream discourse as a response to the existential threat of anthropogenic catastrophic climate change and environmental collapse. As each unfolding crisis highlights our connection and interdependence with the natural world, these ideas are foregrounded, not just in social work but in many industries and institutions globally through the United Nations (2015) Sustainable Development Goals (SDGs). There is an impetus for social work to adapt more rapidly to this changing practice context.

The crises we face are rapid, multifaceted, and interrelated, and so are the responses. There are many differing approaches to social work's response and considerable effort applied to formulate a cohesive definition of environmentally focused practice in the Anthropocene. This has led to the development of overlapping and contested conceptualizations, including "ecosocial work," "green social work," and "environmental social work." There are competing demands for social work to adopt a unified definition. Here we draw upon our first permaculture principle which is *Use and Value Diversity* (Holmgren, 2002) to argue that multiple conceptualizations, multiple practices, and multiple examples are useful. In a permaculture garden, as in a natural ecosystem, resilience is created through diversity. In monocultural industrial agriculture, a single crop is grown, with limited genetic diversity, in order to create efficiency. However, this monocultural approach is also vulnerable to changes in the environment. A change in rainfall or an outbreak of a particular pest can destroy entire crops. In contrast, nurturing a diversity of crops, insects, animals, and microclimates means that while particular elements of the system may fail, there is more likelihood that others will thrive, and the system as a whole is thus much more resilient in the face of adversity. As social workers facing such complex and fast-moving challenges, we should tend toward a diversity of practices to provide more overall viability and stability of the movement toward ecosocial transition.

In light of this principle, we do not offer permaculture as *the* direction for social work practice but humbly present permaculture as one branch of practice that offers both conceptual and practical utility. We begin by discussing "ecosocial transition" in the context of social work education before exploring permaculture and ecosocial work in relation to our own teaching and learning practices. We introduce the principles of permaculture and extend and deepen our analysis and pedagogical approach through an ecosocial case study and other project examples.

Social Work Education in Ecosocial Transition

Social work in ecosocial transition is an emerging area of both undergraduate and postgraduate curriculum development in social work programs internationally as the profession strives to articulate and embed the practice dimensions of the ecosocial approach (Beltrán et al., 2016; Boddy et al., 2018; Drolet et al., 2015; Papadimitriou, 2022; Ranta-Tyrkkö & Närhi 2021; Reu & Jarldorn, 2022; Sanhueza et al., 2022). Ongoing questions arise for social work educators about the kinds of thinking and skills that social work students need for deeper transformational practices in the ecosocial transition. Our students are encouraged by Bell's (2022) assertion that "all constructive efforts towards transformative change count, whether little or large" (p. 22) as she argues that this will sustain effort and the momentum for change. Nevertheless, grounding ecosocial practice concepts within and across social work curricula remains an ongoing challenge for educators (Papadimitriou, 2022; Reu & Jarldorn, 2022).

In our experience as practitioners and educators, students are often highly cognizant of the ecological challenges of our times and willing and able to adapt their practice. Students often express an interest in ecosocial work and related concepts but find it difficult to explain or understand in practice. Our challenge as educators is to bring these concepts to life and forge links between conceptual and day-to-day practices.

Approaches to environmental social work education vary and according to Jones (2013 cited in Boetto and Bell (2015)) can be categorized in three different ways: *Bolt-On* refers to adding content relating to ecological sustainability to the existing curriculum; *Embedded* refers to permeating ecological sustainability throughout the existing curriculum; and *Transformative* means changing the fundamental orientation of social work education to reflect a holistic understanding of the place of humans in the natural world.

Initially, when faced with the challenges of incorporating this type of knowledge into the social work curriculum, we sought to "bolt-on" or at most "embed" this knowledge into the existing curriculum, rather than seek a transformative approach. Reshaping social work curricula can be a slow and protracted process due to institutional and professional governance structures and processes. This slow change process is at odds with the need for engaging proactively with a rapidly unfolding ecological crisis. Other factors that stand in the way of a transformative approach are competing demands among educators. For example, the plan to "green" the curriculum could be perceived to be in competition with the other important tasks of "queering" and "Indigenising" or "decolonizing" the curriculum, as well as incorporating more critical perspectives around disability, mental health, domestic and family violence, and other issues of importance to social workers.

However, to adopt a holistic view, one can see the threads of connection between all of these issues and recognize that an ecosocial perspective is not in competition with these other practice perspectives but rather is intertwined with all of these. From a critical perspective, our analysis would show that the notion that these issues are competing for scarce resources is more symptomatic of how we as academics have accepted the compartmentalization and specialization of knowledge and the neoliberal model of competition for survival. This is the antithesis of what an ecological perspective provides, and thus we offer this perspective not in competition with other diverse perspectives but as part of a rich ecosystem of thought. The further we engage with ecosocial concepts, the clearer it is to us that social work education does in fact require transformative change, and this change is already underway. In the current global social and environmental context, we would argue that it is increasingly difficult to maintain the artificial separation of the social from the environmental that arose as a part of the project of modernity. In order to reconnect and reorient our practice toward ecocentrism, we draw upon permaculture thinking.

In our context at an Australian university, the most logical place to begin to embed permaculture and ecosocial work is through the community development curriculum. We have done this in four broad ways. Firstly, we have reengaged with place. At our university, there is a community garden run on permaculture principles. We have moved our final tutorials from the classroom to take place in this garden as we understand the importance of engagement physically as well as intellectually with these concepts. Secondly, we have included some broader lecture content around the current ecological crises and how these link to social issues, as well as how these links are made in the profession of social work. Thirdly, we have embedded group assessment where students engage with international ecological issues. This includes students choosing a topic from the International Federation of Social Workers' (IFSW) workbook on international social work practices that advance community and environmental sustainability (Rinkel & Powers, 2019) to present to the class, as well as facilitating group activities. Finally, we have created structured tutorial activities around understanding links between environmental and social issues, and we facilitate discussion around this.

Permaculture: An Overview

Permaculture is a term coined by Australians Bill Mollison and David Holmgren in the 1970s. Permaculture, meaning permanent agriculture, or permanent culture, began its life as an integrated system of design which is based on observing natural processes and working alongside nature to create abundant sustainable systems. It has since grown from a fringe collection of gardeners into a global sustainability movement (Spangler et al., 2021).

At its heart, permaculture is a design system for creating sustainable environments "which function to benefit life in all its forms" (Mollison, 1988, p. xi). On the small scale, this can mean learning from natural processes to create sustainable gardens, for example, cleverly designing gardens using plants for multiple uses in such a way that minimizes water usage. On the large scale, permaculture is a way of seeing and being in the world that can lead to whole systems change.

Permaculture is of interest to social workers and community development workers as they are concerned with social justice and holistic well-being, which includes people's physical and emotional well-being, as well as community well-being and resilience. Gardening, and more particularly community gardening, has a beneficial effect on the psychosocial well-being of individuals, which includes reducing stress and increasing social connection (Theodorou et al., 2021; Spano et al., 2020). Getting your hands dirty is not just emotionally satisfying, but there is also a growing body of scientific evidence around the connection between soil microbes, human gut health, and mental health (Dawud et al., 2022). Local food production can also increase community food security, independence, and resilience. In addition to this, there are a myriad of less quantifiable but important cultural and spiritual benefits to permaculture practice, such as a connection to place, connection to natural cycles and seasonal rhythms, and connection to the processes of life and death and growth and decay. Beyond this, permaculture should be of interest to radical and critical ecosocial workers as:

> Permaculture movements, because of their combination of local, situated design practices and underlying social and political philosophies, provide an interesting case of imaginaries that make it possible to reimagine the relations between humans, non-human species and the natural environment. (Roux-Rosier et al., 2018, p.1)

Permaculture's three ethics *Earth Care, People Care, Fair Share* (Holmgren, 2002, 2018) are strongly aligned with social work's values and ethics, and these do not necessarily offer much that is novel to social work's teaching and learning. These ethics are embedded in the Global Social Work Statement of Ethical Principles (IFSW, 2018) as "promoting social justice" and "working towards access and the equitable distribution of resources and wealth" (p. 4–5). Permaculture's 12 principles also align with these ethics and provide a robust conceptual framework which is of utility in ecosocial work. Table 1 presents Holmgren's design principles (2018) which we will apply to social work practice later in the chapter.

Permaculture has a reputation as a grassroots, fringe movement. There is no central accrediting body although there has been much debate in permaculture circles about the need for this. There are accredited permaculture courses available in Australia through our vocational *Technical and Further Education* (TAFE) training system. Generally, permaculture training takes the form of a Permaculture Design

Table 1 Holmgren's permaculture design principles (2018)

1. Observe and interact	7. Design from patterns to details
2. Catch and store energy	8. Integrate rather than segregate
3. Obtain a yield	9. Use small and slow solutions
4. Apply self-regulation and accept feedback	10. Use and value diversity
5. Use and value renewable resources and services	11. Use edges and value the marginal
6. Produce no waste	12. Creatively use and respond to change

Source: authors. (Adapted from Holmgren 2018)

Certificate (PDC), which is often delivered as a 2-week intensive course. The term permaculture has been used broadly by practitioners with varying degrees of expertise. There are differing perspectives on the practicalities of permaculture. Permaculture has a somewhat countercultural history and can be perceived as a "hippy" movement. Permaculture writing is abundant, although this writing exists mostly outside formal academia (Krebs & Bach, 2018; Morel et al., 2019; Spangler et al., 2021). Permaculture therefore does not possess a strong body of rigorous academic literature. Much of the dialog, discussion, and critique of permaculture happen in less formal settings such as in person, websites, gray literature, forums, and blog posts.

More problematic from a social work perspective are some of the critiques of permaculture originating from Indigenous communities and the Global South. While these critiques are voiced mostly in informal dialogs, they are worth exploring here. A key criticism is that many of the concepts that permaculture is built on are in fact traditional ideas that have been present in many cultures for thousands of years (Morel et al., 2019; Prein, 2002). This critique sees permaculture as another way that Indigenous knowledge has been colonized and commodified. To their most strident critics, the founders of permaculture have done nothing other than categorize and rebrand traditional knowledge, repackaging it for easy consumption by a predominantly white audience. A further critique is that permaculture is an expensive lifestyle movement popular in wealthy countries and is often inaccessible for poorer farmers. While similar principles have guided historical preindustrial agricultural practices in the Global North, developments such as the use of powerful machinery and synthetic pesticides and fertilizers as well as low-cost refrigeration and transportation networks have led to farmers growing large quantities of single crops. Permaculture and other organic or "natural" farming methods have been critiqued as less efficient and therefore seen as "a hobby" for wealthier people who are not reliant on maximizing food production for their livelihoods.

These are serious concerns from the perspective of ecosocial work, but they are not insurmountable. These critiques contain within them some of the strengths of permaculture. Firstly, permaculture as a movement that is relatively decentralized, unregulated, and informal has significant benefits. As discussed earlier, when facing a multitude of challenges, a diversity of responses is needed. Permaculture in its decentralized form offers flexibility and responsiveness which is often less possible within large formal professions such as social work which have become more conformist and risk averse. Secondly, permaculture's "borrowing" of concepts and techniques from other cultures has increased their visibility and, while obviously problematic, has helped to foreground these concepts in affluent communities of the Global North. When approached with cultural humility and a critical perspective, these concepts can be seen as a dialogical bridge between cultures. This leads us to explore the permaculture concept of *observant humility*.

Observant Humility

One permaculture principle is *Observe and Interact* (Holmgren, 2002, 2018). This principle highlights the importance of careful observation before interacting. It also includes a commitment to making the "smallest intervention necessary" (Holmgren, 2002). Upon settling on a block of land, an astute permaculturist will observe the land for a lengthy period of time, before digging the first hole or planting the first tree. This observation period could be for an entire year, to see how the land changes throughout the seasons. Through careful observation and respect for "what is," a permaculturist will aim to understand the processes that are already occurring, and if and when practitioners do intervene, it will be with a holistic view of potential ramifications through the system. Permaculture can trace some of its philosophical origins to Masanobu Fukuoka's book *The One-Straw Revolution* published in 1978. He was a Japanese Zen Buddhist farmer, who advocated "do-nothing farming." Fukuoka's radical approach was to do as little as possible while carefully observing his farm. This careful, holistic approach of observation with minimal intervention resulted in a type of farming with very few external inputs, very little turning of the soil, and high yields.

This kind of observant humility is very relevant in ecosocial work and community development and is in direct contrast to the high speed, results oriented, neoliberal context in which students, practitioners, and educators often find themselves. Social work educators must therefore provide tools for observation, reflection, and analysis and carve out spaces for thoughtful, reverent observation. With this kind of a lens, a social worker first tries to appreciate *what is* without jumping to *what should be*. This requires a conscious effort of observation and engagement. Educators can guide students toward a thoughtful analysis of what is occurring in a family or in a community or with an individual. One way to do this is through the use of case studies and examples that can challenge student's perceptions of issues. Students in our course are invited to observe and connect with place through physical means such as hosting tutorials outdoors in our community garden or in a circle outside. We build upon existing work that is happening in the ecosocial space. One way that we have done this is through students choosing a topic from the IFSW workbook (Rinkel & Powers, 2019) and presenting this topic to other students. Another way that we involve students in humble observation is through engagement in class activities and case studies such as the example below.

Deepening the Analysis: An Ecosocial Case Study

Azin is a single parent with two children, aged 4 and 6. Azin moved to Brisbane 10 years ago as a refugee from Afghanistan and resettled in Australia after several years in a refugee camp in Pakistan. She is a keen gardener and used to enjoy creating art in textiles and pottery, though she has not done much of this in recent years.

She used to live in the inner suburbs, but due to increasing rent, Azin has recently moved further out to an outer Brisbane suburb with limited public transport. She does not know many people in her local area. She worked casually in a hospitality job at a restaurant, but this work dried up during the COVID-19 pandemic and subsequent lockdowns. She is now on a parenting payment, and a requirement of this payment is that she must attend fortnightly meetings in person with her job network provider, which is located two suburbs away. She is a keen cook but often buys fast food to feed herself and the children as there is a fast food outlet nearby which has cheap options. Her local shop is expensive and has experienced some shortages due to the pandemic as well as recent flooding. She sometimes travels to a supermarket further away for shopping. Azin is reliant on her car to get to her jobseeker appointments and often to get the children to school as well. This is becoming increasingly difficult for her to manage with rising petrol prices, as well as putting aside money for car registration payments and repairs. She also has difficulties affording her electricity bill. While Azin is very good at budgeting, she is surviving on little and sometimes has to make difficult choices. Her rental house is located adjacent to an industrial area which is heavily concreted. Her house is not well insulated and gets extremely hot in the summer. She can't afford air-conditioning. Azin says that she has been feeling down lately and sometimes finds life overwhelming. She has spoken with her general practitioner about this and has been diagnosed with anxiety and depression and prescribed an antidepressant medication, which she takes daily.

In using this case study, we foster discussion around global level forces in the everyday context. We take a critical questioning approach which includes a learning activity in groups which is then brought back to the class for discussion. Here are some examples of questions that form the discussion and analysis.

Examples of permaculture questions

- *Principle 1: Observe and interact.* How may we "observe and interact" with Azin? Can we identify what is going on and find the smallest changes that we could make for the most benefit?

Principle 2: Catch and store energy. Where does Azin's energy go such as time, money? Is this caught and stored within the local community, or does this go elsewhere? Are there ways that we can keep more of this energy circulating locally?

Principle 3: Obtain a yield. What kind of a yield can Azin obtain that may support her? What are some potential yields for Azin? For example, how can we support Azin to have an early "win" by selling some pottery, thus obtaining a "yield" from her labor?

Principle 7: Design from patterns to details. Taking a step back, are there any patterns that we can identify here? What are the individual issues for Azin and what are the larger forces impacting on Azin's situation? Are there patterns that occur across space (e.g., other people in similar situations in the geographical area)? Are there patterns that occur across time? (e.g., times of abundance and scarcity with income and bills)? Do these patterns point to any kinds of solutions?

Principle 11: Use edges and value the marginal. Does Azin have knowledge, skills, and perspectives that come from being an "outsider" or being on the "edges"? Can these be used productively? How does Azin perceive her situation, and what can we learn from her?

Through a permaculture lens, students are able to identify structural issues, as well as potential interventions or solutions. They are also able to better grasp the interconnectedness of environmental issues and social issues. From this holistic and structural analysis flowed more holistic and structural responses.

It is our intention as educators to foster observant humility among social work students. We believe that these types of processes offer a widening and a deepening of analysis. This humble observation is the first step toward the ethic of "first, do no harm." We aim to engage with these concepts, not just in an intellectual way but in an emotional way and in an embodied way. It is our hope that a deep reverence and humility toward natural processes fosters a deep reverence and humility toward social processes as well. We hope that this lens offers a space, a pause for reflection, which may assist social workers to engage more thoughtfully, before leaping to action. To sum it up, the task of an ecosocial worker is "don't just do something, sit there."

Permaculture thinking includes trying to see the relationships between elements and noticing the energy flows within a system. One permaculture principle of particular utility is *Catch and Store Energy* (Holmgren, 2018). A functioning permaculture system aims toward being a closed loop system where there are limited external inputs and needs are met from within the system. In a permaculture garden, excess plant material that is not consumed is recycled back into the system as mulch or compost. Water that has been used once already in the kitchen may be diverted to ponds with reeds for filtration and then flow on to other crops. Permaculturists often have a keen eye for these energy flows, and this lens can be useful in social work and community development, for example, in assisting practitioners to codesign sustainable responses to social issues in an environment of constrained resources.

Through a permaculture lens, the simultaneous import and export of the same item is a wasteful venture. Permaculturists are often seeking to redefine "waste" by looking at multiple uses for the excess outputs of our wasteful systems and finding ways that these can be put to use. When social work students start to think like permaculturists, they start to notice these outputs and think creatively about how they may be used. This is a lens that can be practiced and honed. In the previous case study example of Azin, examining energy flows allows students to unearth structural injustices. For example, if Azin drives to another suburb to purchase groceries from a large supermarket, then her limited "energy" leaves her local area, either as emissions or as profits for the supermarket which benefits nonlocal shareholders. As there are only limited "inputs" into Azin's system in the form of parenting payments, it is very difficult for Azin to improve her situation. This kind of energy analysis clearly points toward finding ways to challenge these systemic injustices both in terms of assessing the "inputs" and also keeping a close eye on the "outputs."

Extending the Analysis: Some Practice Examples

One useful way to assist students to understand these concepts is to provide real world practical examples of this permaculture systems thinking. A permaculturist will seek to have multiple uses for each element in a design. An early permaculture design principle from permaculture's cofounder Bill Mollison is "each element performs multiple functions" (Mollison & Slay, 1994). For example, in a permaculture garden, lemongrass may be grown for use as a tea but also to stabilize a hillside and aid in water retention as well as be chopped for use as mulch for the vegetable garden and also as a physical barrier to stop wallabies[1] from grazing on other plants. Sometimes, when applied to social services, this thinking of multiple uses can yield surprising and beneficial outcomes. Here is an example from our own practice:

In an urban neighborhood center, we discovered some older community members experiencing social isolation. We also found some community members who needed practice in learning English. By linking these people together for conversations, we are using a resource that the older people have in abundance, which is time, to benefit those who need the support. In return, through this conversation and mutual exchange, the people learning English are providing purpose and connection for the older community members.

Another example from a neighborhood center is the development of a social enterprise café, in which each element serves multiple uses. In this small café in an inner suburb of Brisbane, Australia, coffee was provided on a "pay as you feel" basis. This created a space where all people were welcome, whether they could afford to pay or not. The café employed people who would otherwise find it difficult to find employment. The coffee grounds and food scraps from the café were composted on site, and the compost was used to grow food for community meals. In turn, gardening groups using this compost shared gardening and cooking knowledge from a range of cultures. Excess produce was cooked on site and then frozen into free meals that people experiencing hardship could take with them or heat up and eat on site. The barista benefited from employment and training, and the community members benefited by having a space where they could be, without any pressure to pay money or "consume." From a permaculture viewpoint, there were many "outputs" from this system, which required very little in terms of external "inputs" such as money and staffing. The above examples demonstrate a strong synergy between permaculture systems thinking and community development methodologies, including Asset Based Community Development (ABCD), an approach characterized by the identification and strengthening of community assets (Asset-Based Community Development Institute, 2017).

Another example of the intersection of ecosocial work and permaculture is a gardening and cooking project at an urban permaculture farm in Brisbane, Australia, which hosts organic markets, educational workshops, and events. The project involved a partnership with a local hospital. Residents of the adult inpatient mental

[1] Small Australian marsupials or mammals with pouches

health service from the hospital were invited once a week to the farm to grow and harvest vegetables. Participants then used these vegetables to make homemade pizzas in an earth oven on the farm and shared these with each other and the broader farm community. This small project allowed residents some respite from their sterile surroundings such as fluorescent lighting and hospital food. Participants had access to a natural environment, connection with one another, and the development of practical skills. The project also included the more intangible benefits of fostering a deeper connection to the land and developing a more holistic relationship with food. In alignment with community development principles, the project offered a non-stigmatizing environment for hospital residents to engage with other community members. Consistent with critical, strengths-based, anti-oppressive social work practice, hospital patients were invited to the space as active participants rather than passive welfare recipients and were able to step outside the role of a patient and to share and learn in equal relationship with other volunteers and community members. These practice examples illustrate the multiple benefits of permaculture from a social work perspective.

Not Just About Gardens: "Radical" Reimagining

> Human-induced climate change, including more frequent and intense extreme events, has caused widespread adverse impacts and related losses and damages to nature and people, beyond natural climate variability. Some development and adaptation efforts have reduced vulnerability. Across sectors and regions the most vulnerable people and systems are observed to be disproportionately affected. The rise in weather and climate extremes has led to some irreversible impacts as natural and human systems are pushed beyond their ability to adapt (Intergovernmental Panel on Climate Change [IPCC], 2022, p.1).

We are currently in the midst of the sixth great extinction event with global average temperatures rising, an increase in adverse weather events, and a rapid loss of biodiversity threatening the very systems that support life on our planet. We are, at this point in human history, experiencing a great shift from a carbon economy to a post-carbon economy. Business as usual is no longer an option, as continuing on our current trajectory will render our planet vastly less habitable (IPCC, 2022). Radical change is therefore inevitable, whether we are involved in it or not. This current predicament entails a radical reimagining of human society, which has been conceptualized as an "ecosocial transition" (Matthies, 2016). This ecosocial transition can be described as an *economic* shift from a capitalist to a postcapitalist world. It can be described as an *energy* shift, from fossil fuel dependence to renewable energy dependence. It can be described as a *relational* shift from an extractive relationship with the natural world to a regenerative relationship with the natural world. It can also be described as a *philosophical* shift from anthropocentrism to ecocentrism. Ecosocial transition can be described as a shift of scale from *global* to *local* and, finally, as a *distributive* shift from inequality to equity. These shifts are necessary, albeit highly challenging, and success is certainly not guaranteed. We believe, however, that our conceptualization presents aspirational signposts for practice.

In 2022, the theme of World Social Work Day and the People's Summit was "Co-building a New Eco-Social World: Leaving No One Behind" (IFSW, 2022). This theme acknowledges the urgency of ecosocial transition and embeds the values of participation "co-building" and social justice "leaving no-one behind." All types of human organizational structures are now engaged in the rapid shifts necessary, particularly around decarbonization. This includes nation states, multinational corporations, and nongovernment organizations as well as global bodies such as the United Nations and the World Bank. Social work has a vital role to play in this shift, in ensuring that these changes take place in a way that is participatory and just. In this rapidly changing environment, there are significant risks to the most vulnerable populations, and social work and community work must be a strong voice to ensure that this transition does not further exacerbate inequality.

One of the key strengths of postmodern social work is its critical perspective on power (Healy, 2022) which connects with collectivist approaches and community development ideas (Forde & Lynch, 2015). When combined with permaculture thinking, we have a potent force for change. The permaculturalist Geoff Lawnton who founded the Permaculture Research Institute (2023) is well known for expressing "All the world's problems can be solved in a garden." Indeed, perhaps many of the world's problems could be solved in a garden, but there are structural barriers to people having access to a garden. In our neoliberal economic paradigm, urban and suburban land is valued so highly, and food production is valued so slightly that urban and suburban agriculture is not, generally, economically viable. Most of our food production takes place a great distance away from where it is consumed. We are now at a critical juncture in terms of climate change and energy descent where there is a public mandate and growing political will to reassess and reorganize this system. Social workers who have the tools and expertise to advocate for these kinds of changes are an invaluable part of this shift. A transition is happening one way or another, and ecosocial workers have a responsibility to be a strong voice for a "just transition."

Two permaculture principles with particular relevance to addressing these significant challenges are *Use and Value Diversity* and *Use Edges and Value the Marginal*. Here we look at the complex interplay between neighboring ecosystems. These areas of convergence often produce a large amount of diversity and productivity. This is known to permaculturists as the *edge effect*. As educators, we are interested in the "edge" between permaculture and ecosocial work, both with compelling ecosystems of thought. It is the interplay between these that helps us to create a diversity of responses. In addition to this, interdisciplinary approaches can be useful to create novel responses to "wicked problems." As an example, a research partnership between social work and agriculture can produce results that would not be possible through the lens of each discipline alone. This interdisciplinary practice can be useful for ideas from the margins to gain more mainstream credibility.

In parallel to this *edge effect*, permaculturists, like social workers, also value the marginal. Holmgren (2002) argues that those on the margins may be able to produce radical solutions as they are less embedded or invested in the systems that maintain the status quo. As social workers, we are often at "the margins" or are working with those who are. This principle reminds us to value and uplift the perspectives of those people when working toward systemic change. A diversity of perspectives from the edges creates a more diverse, resilient, and robust society. As educators, our task is to assist students to put aside their assumed professional expertise and to open themselves to other perspectives.

Ecosystems Thinking as Critical-Consciousness Raising

It is clear that the situation is challenging. It is clear that the effects of climate change are occurring now and are not reversible. There is, however, a vast spectrum of thought around how far along we are and therefore what action we should take. At one end of the spectrum, we have the notion that we can solve most of our issues if we "tinker around the edges" and continue, much as we have been, in our present way of life, aided by some advances in technology. At the other end, we have the growing "deep adaptation" movement, which argues that environmental and societal collapse is now inevitable, and our key task is to collapse well (Bendell & Read, 2021). Many social workers and social work scholars and educators are positioned somewhere between these extremes. There is support among social work scholars for the notion that our current neoliberal, colonial, and extractivist paradigm must be critiqued and dismantled. In order to progress toward a new ecosocial future, a radical reimagining must occur. Indeed, this radical reimagining is occurring already in many different forms from community led "rewilding" projects, to alternative economies, to reclamation of the commons in the form of guerrilla gardening and tactical urbanism, to citizen-led participatory democratic processes. When viewed together, these projects, processes, and movements add up to what Joanna Macy (2009) calls the "Great Turning" (which we now interpret as ecosocial transition). She describes this as "… a name for the essential adventure of our time: the shift from the Industrial Growth Society to a life-sustaining civilization" (para.1). While the challenges are indeed great, the authors of this chapter align themselves toward radical hope. This hope is reflected in the words of author Arundhati Roy at the World Social Forum held in Porte Alegre (Brazil) in 2003: "Another world is not only possible, she is on her way. On a quiet day, I can hear her breathing." As the above examples indicate, different solutions do exist currently, but the challenge is shifting our attention and strengthening these practices. As a good gardener cares for soil so that plants may thrive, our task as educators is to nurture the conditions for these practices to multiply and flourish.

One of the key challenges of our time is shifting from globalized economies to a reinvigoration of the local. There are many organizations and collectives working toward this aim. In Australia, one such group is the New Economy Network Australia with a focus on a "well-being economy," a concept linked to Kate Raworth's (2017) "Doughnut Economics" which is currently informing city level policy in Amsterdam, Copenhagen, Portland (Oregon, USA), and Birmingham (UK) (Wahlund & Hansen, 2022). Doughnut Economics is an economics that is focused on keeping growth within sustainable outer limits while raising the standard of living for those at the lower ends of the socioeconomic spectrum, conceptualized as the inside hole of the doughnut. This offers a visualization to engage social work students in deeper conversation about well-being as opposed to economic growth. This economic shift requires a dramatic reorganization of global trade (Norberg-Hodge, 2019). In our current economic system, it has been viable, for example, for Australia to export roughly the same amount of bottled water to the UK that we import or, in one of the author's experience, for a macadamia farmer in Australia to ship macadamia nuts to China to be cracked and then shipped back to Australia to be sold as local produce.

As we know, these kinds of wasteful energy flows have been made possible by cheap energy and global economic systems based solely on profit. The COVID-19 pandemic, global political instability, and rising prices of oil have highlighted the fragility of these systems and the urgent impetus for change. When students begin to view the world like a permaculturist does, as a living system, and start to glimpse the connections among ecosystems, they become more aware of the injustice of these systems.

Concluding Thoughts

We understand and acknowledge the rich and deep philosophies and bodies of knowledge from Indigenous communities around the world, on being, doing, and knowing in relationship with the natural world. This is a much larger conversation than we could do justice to within the scope of this chapter. As non-Indigenous academics, we would not profess to teach these traditions but instead offer to enter into dialog with those knowledge holders to cocreate a "third space" of understanding. In our Australian context, we would encourage students to engage with authors and thinkers such as *Bruce Pascoe*,[2] *Tyson Yunkaporta*,[3] and *Miriam-Rose Ungunmerr-Baumann*.[4] There is much here to inspire and shape future thinking.

In terms of current and future directions for social work practice, we believe that there is no longer a need to spend energy on arguing that the climate is changing,

[2] Bruce Pascoe is an Aboriginal Australian and Enterprise Professor of Indigenous agriculture at the University of Melbourne in Australia. His book *Dark Emu: Black Seeds: Agriculture Or Accident?* published in 2014 explores approaches to sustainability by Aboriginal communities through land cultivation and management practices in precolonial Aboriginal society. This presents an alternative perspective to the view of Australian Aboriginal peoples as hunter-gatherers.

[3] Tyson Yunkaporta is a Senior Lecturer in Indigenous knowledges at Deakin University in Melbourne, Australia. His work grounds Indigenous knowledges in ecological sustainability and custodianship of the land.

[4] Miriam-Rose Ungunmerr-Baumann is an Aboriginal artist and educator from the Daly River region in the Northern Territory in Australia. Her peoples' spiritual connection and respect for nature are expressed through the practice of "deep listening," known as dadirri (Miriam Rose Foundation, 2023).

when there is evidence that this is so. Similarly, there is no longer a need to advocate for social work to adapt to a changing environment or to argue that we must incorporate the physical, living world in our work. Given the rapid changes that are facing the world, we simply do not have the time to wait. The argument has been made, and now it is time for the doing. However, a key challenge of adopting such a large-scale, holistic approach is how we can possibly comprehend and analyze social and environmental systems of such complexity? And how can we effectively teach this knowledge to students? We need to continue to adapt and develop a broad array of theories, methods, techniques, and skills in order to meet the significant challenges of our time. What we are hearing from students is "yes I am onboard, now what is my role in this, and how do I do it?". To this end, we believe in a multifaceted, multi-scalar approach to ecosocial work pedagogy, working within and around our institutions, in the ways that we can make that impact. The gap our students have identified is the link between the abstract concepts and concrete, practical examples of how to do this work.

Permaculture systems thinking offers one practical framework that can be taught to social work students in a way that enriches their practice and enhances rather than competes with other practice theories and methods. We believe that students need to see, feel, and do this work in order to have the confidence to put these concepts into practice. To this end, we support a proliferation of real life case studies, tutorial activities, and embodied learning including taking our classrooms into the garden. We hope that the case studies and examples here provide some value to this process.

References

Asset-Based Community Development Institute. (2017). *Asset-based community development: The essentials*. Asset-Based Community Development Institute.

Bell, K. (2022). Post-anthropocentric social work: Transformative philosophy for contemporary theory and practice. [Advance online publication]. In D. Hölscher, R. Hugman, & D. McAuliffe (Eds.), *Social work theory and ethics: Ideas in practice* (pp. 1–36). Springer.

Bell, K., Kime, K., & Boetto, H. (2019). Gender, environmental degradation and ecofeminism. In M. Rinkel & M. Powers (Eds.), *Social work promoting community and environmental sustainability: A workbook for social work practitioner and educators* (Vol. 3, pp. 117–137). International Federation of Social Workers.

Beltrán, R., Hacker, A., & Begun, S. (2016). Environmental justice is a social justice issue: Incorporating environmental justice into social work practice curricula. *Journal of Social Work Education, 52*(4), 493–502. https://doi.org/10.1080/10437797.2016.1215277

Bendell, J., & Read, R. (Eds.). (2021). *Deep adaptation: Navigating the realities of climate chaos*. Polity Press.

Besthorn, F. H. (2012). Deep Ecology's contributions to social work: A ten-year retrospective. *International Journal of Social Welfare, 21*(3), 248–259. https://doi.org/10.1111/j.1468-2397.2011.00850.x

Boddy, J., Macfarlane, S., & Greenslade, L. (2018). Social work and the natural environment: Embedding content across curricula. *Australian Social Work, 71*(3), 367–375. https://doi.org/10.1080/0312407X.2018.1447588

Boetto, H. (2017). A transformative ecosocial model: Challenging modernist assumptions in social work. *The British Journal of Social Work, 47*(1), 48–67.

Boetto, H. (2019). Advancing transformative ecosocial change: Shifting from modernist to holistic foundations. *Australian Social Work, 72*(2), 139–151.

Boetto, H., & Bell, K. (2015). Environmental sustainability in social work education: An online initiative to encourage global citizenship. *International Social Work, 58*(3), 488–462. https://doi.org/10.1177/0020872815570073

Boetto, H., Bowles, W., Närhi, K., & Powers, M. (2020). Raising awareness of transformative ecosocial work: Participatory action research with Australian practitioners. *International Journal of Social Welfare, 29*(4), 300–309.

Braidotti, R. (2013). *The posthuman*. Polity Press.

Dawud, L. M., Holbrook, E. M., & Lowry, C. A. (2022). Evolutionary aspects of diverse microbial exposures and mental health: Focus on 'old friends' and stress resilience. In J. Savitz & R. H. Yolken (Eds.), *Microorganisms and mental health. Current topics in behavioral neurosciences* (Vol. 61, pp. 93–118). Springer. https://doi-org.ezproxy.library.uq.edu.au/10.1007/7854_2022_385

Drolet, J., Wu, H., Taylor, M., & Dennehy, A. (2015). Social work and sustainable social development: Teaching and learning strategies for 'Green Social Work' curriculum. *Social Work Education, 34*(5), 528–543. https://doi.org/10.1080/02615479.2015.1065808

Forde, C., & Lynch, D. (2015). *Social work and community development. A critical practice perspective*. Palgrave.

Fukuoka, M. (1978). *The One-Straw Revolution: An introduction to natural farming*. New York Review Books Classic.

Healy, K. (2022). *Social work theories in context: Creating frameworks for practice*. Bloomsbury Academic.

Holmgren, D. (2002). *Permaculture: Principles and pathways beyond sustainability*. Holmgren Design Services.

Holmgren, D. (2018). *Retro Suburbia: The downshifter's guide to a resilient future*. Melliodora Publishing.

Ife, J. (2018). *Social work and the Post-Human* [Paper Presentation]. 2018 ANZSWWER Symposium, Adelaide, South Australia.

Intergovernmental Panel on Climate Change [IPCC]. (2022). *Climate change 2022: Impacts, adaptation and vulnerability – summary for policymakers. IPPC Working Group 2 contribution to the sixth assessment report of the IPCC.* World Meteorological Organization (WMO), United Nations Environment Programme (UNEP).

International Federation of Social Workers (IFSW). (2018). *Global social work statement of ethical principles.* Retrieved 4 September 2022 from https://www.ifsw.org/global-social-work-statement-of-ethical-principles/

International Federation of Social Workers (IFSW). (2022). *World Social Work Day 2022. Co-building a new ecosocial world: Leaving no-one behind.* Retrieved 4 September 2022 from https://www.ifsw.org/social-work-action/world-social-work-day/world-social-work-day-2022/

Krebs, J., & Bach, S. (2018). Permaculture – Scientific evidence of principles for the agroecological design of farming systems. *Sustainability, 10*(9), 3218. https://doi.org/10.3390/su10093218

Macy, J. (2009). *The great turning. Center for Ecoliteracy.* Retrieved 4 September 2022, from, https://www.ecoliteracy.org/article/great-turning

Matthies, A. (2016). The conceptualisation of ecosocial transition. In A. Matthies & K. Närhi (Eds.), *The ecosocial transition of societies: The contribution of social work and social policy* (pp. 17–35). Routledge. https://doi.org/10.4324/9781315615912

Matthies, A., & Närhi, K. (Eds.). (2016). *The ecosocial transition of societies: The contribution of social work and social policy*. Routledge. https://doi.org/10.4324/9781315615912

Mies, M., & Shiva, V. (2014). *Ecofeminism Critique. Influence. Change*. Bloomsbury Publishing.

Miriam-Rose Foundation. (2023, Feb 25). *Empowering Indigenous futures through art, education and opportunity.* Miriam-Rose Foundation. https://www.miriamrosefoundation.org.au/

Mollison, B. (1988). *Permaculture – A designer's manual*. Tagari Publications.

Mollison, B., & Slay, R. (1994). *Introduction to permaculture*. Tagari Publications.

Morel, K., Leger, F., & Ferguson, R. (2019). Permaculture. In B. D. Fath (Ed.), *Encyclopedia of ecology, 4* (pp. 559–567). Elsevier.

Naess, A. (1995). Deepness of questions and the Deep Ecology movement. In G. Sessions (Ed.), *Deep Ecology for the 21st century: Readings on the philosophy and practice of the new environmentalism* (pp. 204–212). Shambhala.

Norberg-Hodge, H. (2019). *Local is our future: Steps to an economics of happiness*. Local Futures.

Panagiotaros, C. V., Boddy, J., Gray, T., & Ife, J. (2022). (Re-) imagining social work in the Anthropocene [Advance online publication]. *The British Journal of Social Work*, 1–17. https://doi.org/10.1093/bjsw/bcac075

Papadimitriou, E. (2022). Coverage of environmental issues in undergraduate curricula in social work in four European countries: The UK, Switzerland, Germany and Greece [Advance online publication]. *Social Work Education*, 1–18. https://doi.org/10.1080/02615479.2022.2028763

Pascoe, B. (2014). *Dark Emu: Black seeds: Agriculture or accident?* Magabala Books.

Permaculture Research Institute. (2023, Feb 25). Permaculture Research Institute. https://www.permaculturenews.org/

Prein, M. (2002). Integration of aquaculture into crop-animal systems in Asia. *Agricultural Systems, 71*, 27–146.

Ramsay, S., & Boddy, J. (2017). Environmental social work: A concept analysis. *The British Journal of Social Work, 47*(1), 68–86. https://doi.org/10.1093/bjsw/bcw078

Ranta-Tyrkkö, S., & Närhi, K. (2021). Striving to strengthen the ecosocial framework in social work in Finland. *Community Development Journal, 56*(4), 608–625. https://doi.org/10.1093/cdj/bsab030

Raworth, K. (2017). *Donut economics: Seven ways to think like a 21st century economist*. Random House.

Reu, P., & Jarldorn, M. (2022). Social work students' perceptions of eco-social work in the curriculum [Advance online publication]. *Australian Social Work*. https://doi.org/10.1080/0312407X.2022.2102431

Rinkel, M., & Powers, M. (Eds.). (2019). *Social work promoting community and environmental sustainability: A workbook for social work practitioner and educators* (Vol. 3). International Federation of Social Workers.

Roux-Rosier, A., Azambuja, R., & Islam, G. (2018). Alternative visions: Permaculture as imaginaries of the Anthropocene. *Organization, 25*(4), 550–572. https://doi.org/10.1177/1350508418778647

Roy, A. (2003, Jan 23–28). [Conference Presentation]. *2003 World Social Forum*, Porte Alegre, Brazil.

Sanhueza, G., Matus Sepúlveda, T., & Moreno, J. (2022). Innovating social work education in Chile: addressing the UN sustainable development goals, enhancing our profession. [Advance online publication]. *Social Work Education*. https://doi.org/10.1080/02615479.2022.2115028

Spangler, K., McCann, R. B., & Ferguson, R. S. (2021). (Re-)Defining permaculture: Perspectives of permaculture teachers and practitioners across the United States. *Sustainability, 13*(10), 5413. https://doi.org/10.3390/su13105413

Spano, G., D'Este, M., Giannico, V., Carrus, G., Elia, M., Lafortezza, R., Panno, A., & Sanesi, G. (2020). Are community gardening and horticultural interventions beneficial for psychosocial well-being? A meta-analysis. *International Journal of Environmental Research and Public Health, 17*(10), 3584. https://doi.org/10.3390/ijerph17103584

Theodorou, A., Panno, A., Carrus, G., Carbone, G. A., Massullo, C., & Imperatori, C. (2021). Stay home, stay safe, stay green: The role of gardening activities on mental health during the Covid-19 home confinement. *Urban Forestry & Urban Greening, 61*, 127091. https://doi.org/10.1016/j.ufug.2021.127091

United Nations. (2015). *UN Sustainable Development Goals (SDGs)*. Retrieved 25 February 2023 from https://sdgs.un.org/goals

Wahlund, M., & Hansen, T. (2022). Exploring alternative economic pathways: A comparison of foundational economy and Doughnut economics. *Sustainability: Science, Practice and Policy, 18*(1), 171–186. https://doi.org/10.1080/15487733.2022.2030280

Chapter 8: Social Work and Environmental Sustainability Toolkit: Building a Practical Guide to Apply Sustainable Practices in Social Work

Rita Luís, Joana Gomes, and Afonso Borga

Introduction

Our planet faces numerous challenges regarding (un)sustainability. Fighting against the worsening of increasingly adverse climate conditions requires the involvement of all stakeholders in society, mobilizing groups and communities towards a common goal: sustainable development. Therefore, it is necessary to design and implement sustainable responses that do not put at risk the Earth's natural limits and social work increasingly needs to have the capacity to develop a proactive social innovation agenda that allows social work professionals to incorporate environmental concerns as part of their endeavor. With this foundation, three aspiring social workers, in collaboration with both social work and nonsocial work experts, have developed a digital toolkit encompassing sustainable practices in the field of social work.

The *Social Work and Environmental Sustainability Toolkit: A practical guide* was published in 2022 in E-book format in Portuguese and can be consulted online for free on the website of the Portuguese Association of Social Workers (https://www.apss.site/publicacoes). The toolkit is a document designed to serve as a supportive educational resource that promotes collaboration, critical thinking, and reflexivity for social workers, because all social workers can be ecosocial workers. It provides a theoretical and practical framework, promotes interdisciplinarity, and encourages the sharing of information, new approaches, and projects on the six topics covered.

This chapter introduces the objectives and the premise that led to the construction of the toolkit. The six topics covered in the toolkit are then listed. In these six topics, we present a full translation of the original document, contextualizing the themes in line with an eco-social approach in social work. In each topic, we delve

R. Luís (✉) · J. Gomes · A. Borga
ISCTE University Institute of Lisbon, Lisbon, Portugal

C. Forde et al. (eds.), *Teaching and Learning in Ecosocial Work*,
https://doi.org/10.1007/978-3-031-58708-5_11

into a theoretical framework that examines the connection between social work and a specific topic. Within each of these topics, we present a practical case that illustrates the implementation of ecosocial practices in social intervention. Additionally, we highlight the social worker's role in these practical cases. The only change from the original version is that just one case study is presented per topic in this chapter (in the Portuguese version, we present two to three case studies).

The six topics addressed are "Environmental Social Work in the Community", "Social Work and the Circular Economy," "Food Security, Sustainability, and Social Work", "Social Work and Environmental Education", "The Human-Animal Relationship as an Element in Social Work", and "Social Work in Environmental Disaster". Each topic is not intended to deepen the theme addressed but to frame it in an ecosocial perspective that allows the understanding of the relationship between social work and environmental sustainability. The six topics are only examples. There are many other fields where social work can promote environmental sustainability that remain to be explored in future research.

This chapter, along with the original Toolkit, is intended to inspire and motivate social workers to get involved in ecosocial practices and contribute to a more sustainable world. It is important that all social workers become aware of their role in building a more sustainable world, in line with the principles and values of the profession, helping to fulfil the principle "Leave no one behind" (United Nations, 2015). Therefore, it is necessary that sustainability issues are addressed in the initial training of students. The inclusion of the six topics mentioned above in the social work curriculum is fundamental, not only from a theoretical perspective that deepens the relationship between social work and environmental sustainability/sustainable development but also from a practical perspective that shows students and professionals that it is possible to reconcile social issues with environmental concerns.

How Was the Toolkit Developed?

The gathering of information on different projects, research, and interventions on ecosocial practices has been operationalized with the organization of two cycles of webinars on the general theme: Social Work and Environmental Sustainability. The organization was handled by the three students who developed the toolkit. They took charge of the entire process, including research, guest invitations, and webinar moderation. The Portuguese Association of Social Workers and the association Novas Raizes offered logistical support to facilitate the execution of this initiative.

The first cycle of webinars was realized between April and May 2021, and covered three topics: "Environmental Social Work in the Community", "Social Work and the Circular Economy," and "Social Work and Environmental Education". The second cycle of webinars was realized between September and October 2021 and also covered three topics: "Food Security, Sustainability, and Social Work", "The Human-Animal Relationship as an Element in Social Work", and "Social Work in Environmental Disaster". In total, there were 6 webinars with 6 different sub-themes, in which 20 experts participated. After that, the information collected from the webinars and literature review was compiled into a document, which became the *Social Work and Environmental Sustainability Toolkit: A practical guide.*

The construction of a toolkit of sustainable practices for social work and social intervention is the result of a cooperative process that was made possible through interdisciplinary work. Professionals, specialists, and researchers came together to promote and raise awareness among social work professionals and students about environmental sustainability issues and ecosocial practices.

Goals of the Toolkit

- To deepen the knowledge about the relationship between social work and environmental sustainability
- To raise awareness in the social sector about the importance of promoting environmental sustainability connected with social intervention
- To create a tool that enhances the positioning of social work in the field of environmental sustainability

Six Topics

The structure of the chapter is segmented into six distinct topics introduced above. The first topic: "Environmental Social Work in the Community" addresses the importance of using an ecosocial approach and model in community intervention, associating environmental sustainability with social work practice in a multidisciplinary and multilevel framework. In the second topic: "Social Work and the Circular Economy," we address the importance of transitioning to a circular model of society and present some examples of how this transition can be appropriated by social work in social intervention logics and projects. The next topic: "Food Security, Sustainability, and Social Work" links a traditional field of social work—food support—to the need to adopt logics that promote not only environmental sustainability, but also the empowerment of people receiving food support and put into practice the optimization of food resources to promote a better nutritional balance and less food waste. The fourth topic: "Social Work and Environmental Education" addresses the importance of education in promoting environmental sustainability through programs that promote awareness of the subject and critical thinking and through approaches such as ecopedagogy to guide an ecosocial approach in social work. The fifth topic: "The Human-Animal Relationship as an Element in Social Work" highlights the significance of the human-animal relationship and its implications for animal social work. The sixth and final topic: "Social Work in Environmental Disaster," conceptualizes what a disaster is and underscores the significance of social workers' involvement across different phases and levels of disaster intervention.

Environmental Social Work in the Community

The recognition of the importance of promoting environmental sustainability in social work is essential as it is an ethical duty for social workers to promote well-being, social justice, and dignity not only to present but also future generations. The community level in social work has the potential to raise awareness about environmental responsibility in community members, and developing advocacy skills and strategies is essential to work towards a healthier environment and ensure that environmental issues are approached regularly (Rinkel & Powers, 2019). Thus, social workers should use an ecosocial perspective to enhance social change through empowerment and to promote the participation and resilience of people and communities in the process (Peeters, 2012; Luís, 2020).

Pinto (2014) identifies three levels for practice concerning the application of the principles of sustainable development in social work: the individual level, the meso level, and the macro level. At the meso level, we can find the intervention in organizations and community work, highlighting the intervention of the social worker as a professional who acts in proximity with individuals and groups in communities.

Communities in vulnerable situations will be the ones that will suffer the greatest impact in terms of the consequences resulting from climate change, since they are also the ones with the fewest resources to deal with these consequences (Dominelli, 2018). Thus, social workers and community workers can be fundamental interlocutors and facilitators in community participation processes, adopting a multilevel approach that connects the micro, meso, and macro levels (Pinto, 2014) to facilitate a clearer identification of the wishes and concerns of the community and the development of sustainability programs.

Environmental Social Work in the Community must be oriented by an ecosocial approach (Matthies et al., 2001) in which social workers should guide their practice by an ecosocial model (Rocha, 2015) that suggests a long-term vision considering the balance between the social, economic, and environmental dimensions, alongside the response to people's immediate needs. Through an ecosocial model for social work, we highlight the following aspects of social work in a community context:

- Fostering a holistic view that conjugates the human dimension, the community and local development dimension, and, finally, the sociopolitical dimension. (Rocha, 2015).
- Relating local problems faced by communities to global causes such as climate change, and using this insight for advocacy.
- Inserting environmental concerns into community-level social diagnostics and intervention plans.
- Promote consultation sessions with the community and create opportunities and spaces for dialogue with communities (Rocha, 2015).
- Promote multidisciplinary work and networking with other actors present in the territory, bringing the connection between environmental sustainability and community work to the agenda (Rocha, 2015).

- Promote projects that respond to the needs arising from the community's environmental sustainability such as community gardens, environmental education workshops, tree planting, and reforestation.

In summary, social work should focus on working within communities in co-designing and co-building social and environmental protection, leading to capacity-building, and policy development to combat complex and intersecting crises, committing to building a new ecosocial world in which all rights are ensured and no one is left behind (IFSW, 2022).

Practical Case: Bela Flor Agroforestry

An agroforestry is a production system that imitates what nature normally does, with the soil always covered by vegetation, many types of plants together, helping each other, without problems with "pests" or "diseases," dispensing with the use of poisons.

Bela Flor Agroforestry is located in the center of Lisbon and unites residents of the neighborhood and volunteers of all ages around an initiative that supplies the community with horticultural products. From a plot of land that had nothing but rubbish, the agroforestry has become a space for the community to socialize, and this space is maintained by the people of the community and the parish council/local government.

An agroforestry in the city allows for many ecological and social functions to be performed in a limited space with the potential to bring greater resilience to the population, in some cases representing an important contribution to daily nourishment. Mukhlis et al. (2022) affirm that an agroforestry can have benefits such as improving carbon sequestration, promoting biodiversity through the incorporation of different species of plants/crops which may provide homes for various wildlife, providing alternative incomes for the community, promoting economic and social resilience through the implementation of a diverse agroecosystem including trees (timbers, fruits) and livestock and improving household food security through diversified food sources.

The Role of the Social Worker: Bela Flor Agroforestry

Social workers engaged in community work with the population of the area surrounding Bela Flor are instrumental in the process of planning, building, and maintaining the Agroforestry. Due to the newness and lack of experience with the concept, local populations may be hesitant to understand and actively participate in the implementation and maintenance of an agroforestry and a social worker can mobilize a participatory and communal process, communicating with the target audience, involving the community in the planting and management of the agroforestry, listening to their feedback, and building the project accordingly. In this sense the professional has an important role as a mediator with the population: as a point of reference for any situation or as a promoter and facilitator of the population's interaction with this space since he/she has empirical knowledge about their motivations and needs. Additionally, the social worker will play an essential role in establishing partnerships and communicating with organizations. At Bela Flor Agroforestry, social workers facilitate all of these processes and play a fundamental part in maintaining the community enrolled with the project, utilizing an ecosocial approach and combining environmental and social concerns.

Social Work and the Circular Economy

The circular economy is based on the premise that the planet's resources are limited and that consumption patterns should be adapted to this reality, aiming to represent an alternative solution to promote sustainable development. It is based on three principles: eliminate waste and pollution, circulate products and materials, and regenerate nature. By applying circular economy models, we can drive systemic change that creates long-term resilience, generates commercial and economic opportunities, and delivers environmental and social benefits (Ellen MacArthur Foundation, 2017).

Promoting the circular economy in social work can be a process of empowerment: working with individuals and communities to create long-term sustainable solutions (economically, socially, and environmentally) can restore, renew, or refurbish products, creating local circular economy-based businesses and generating employability opportunities for people (Arevalo, 2017).

Gomes (2020) identifies the circular economy as a strategy that can be appropriated by social work for social intervention. She takes the textile industry as an example of the potential that a circular logic can have by bringing employment, giving another purpose to discarded textile materials, and creating an environmental and economic return. She draws examples from the community to show the positive impact that can be generated on the social and environmental levels.

Initiatives that promote the use of social/local currencies—currencies used as money created especially to promote social or environmental projects, mainly at the local level (Elena, 2022)—are also identified as good tools to promote circular economy and consequent local/territorial development (Marques, 2021a). Another way to contribute to the circular economy can be by raising awareness of issues such as responsible consumption and production (Matenga et al., 2019) and advocacy for the right to choose to consume sustainable products.

Therefore, social workers should incorporate and accompany the circular economy, namely, through:

- Promoting a paradigm shift at the economic level, driven by principles of social justice and environmental values
- Raising awareness about the importance of the transition to a more sustainable economic model
- Exploring the community dimension, creating networks and partnerships to develop circular economy initiatives and projects
- Facilitating the involvement of vulnerable populations in the creation and/or implementation of circular economy responses
- Enabling training and education on skills to be developed for work in circular economy structures (e.g., household appliance repair, sewing, etc.)

Practical Case: Vintage for a Cause

Vintage for a Cause is a social business and a circular economy brand that reuses textile waste, mostly in the pre-consumption phase. It is based on collaborative design, challenging different designers to create limited editions/collections signed by them. These pieces are built on the prism of functionality durability and the least possible waste of resources.

This social enterprise, founded in Porto in 2012 develops sewing, stitching, and tailoring workshops open to women over 50 years old, where they have the opportunity to develop and learn new skills. This can also apply to the elderly female population (over 65) as an active aging project. For those in active age, the project can create employability opportunities, as it only hires women over 50. This commitment stems from the fact that this target audience is especially vulnerable after their active role in society given that unemployed women over 50 remain one of the groups most vulnerable to poverty in Portugal (EAPN Portugal, 2023).

The Role of the Social Worker: Vintage for a Cause

Social workers can use an ecosocial approach in analyzing and evaluating specific situations and refer unemployed women over the age of 50 to access training or even employment in Vintage for a Cause. In a community and social innovation dimension, the example set by Vintage for a Cause can inspire the creation of more similar projects, and the social worker can have a role in enhancing and facilitating these processes. To support social workers gaining skills to facilitate these processes, an investment can be made in the training and education of new and experienced social workers in the area of social entrepreneurship using circular logics that promote environmental, social, and economic sustainability.

Food Security, Sustainability, and Social Work

Access to food is defined as a human right by the Universal Declaration of Human Rights (1948), which asserts that no one should be hungry and all people should have access to adequate nutrition (United Nations, 1996). To fulfill this human right, it is necessary that conditions of availability, stability, accessibility, sustainability, and adequacy of food are ensured (Vizinho et al., 2021). However, nowadays, we are being confronted with several challenges related to food consumption and production, linked with issues like climate disruption increasingly threatening food security around the world and agrofuels that increase food prices when replacing fossil fuels.

According to the Food and Agriculture Organization of the United Nations (2023), food insecurity means not having regular access to sufficiently safe and nutritious food for normal growth and development and an active and healthy life. This can be due to unavailability of food and/or lack of resources to obtain food. Food insecurity, hunger, and malnutrition are directly linked to issues of poverty and social inequalities (Aguiar & Padrão, 2022). These global issues affect different regions of the globe differently, considering different geographical, economic, and social characteristics. In Portugal, according to Rocha (2021a), moments of socioeconomic crisis seem to be linked to a higher prevalence of food insecurity, as food is the most flexible expense and easy to cut down on in moments of monetary scarcity. In these moments, there is also a greater public investment in emergency food responses, often managed by nonprofit organizations. However, in Portugal there are few integrated food aid responses designed/funded for a medium-/long-term intervention, outside the emergency context.

Promoting sustainability in relation to food insecurity requires social work professionals to adopt a critical perspective in their intervention, understanding the underlying structural issues and seeking to rethink, in an empowerment logic, the food assistance models (Schwemmlein, 2017). Thinking about the concept of food security focusing on the promotion of active and healthy lifestyles is also a challenge for social work. Poverty can be one of the main obstacles to achieving healthy and sustainable food, as economic and literacy issues can impact on the choice of lower-quality groceries (Vilar-Compte et al., 2021). Sufficient food is not enough; it must be adequate, with encouragement for the consumption of healthy food produced locally, with the aim of preventing diseases related to food and nutrition in all age groups (FAO & IFAD, 2019). Social workers can facilitate information and nutritional education programs for health and food security for the people, groups, and/or communities with whom they work.

From the rural development point of view, we also face several challenges. Environmental sustainability of food systems should start at the production stage; however, the model of society that we live in gives primacy to imports and large-scale productions that cause social, environmental, and economic consequences. Here it is important for social workers to be aware of the global causes of local problems, since currently a population that suffers food insecurity is workers in the agriculture industry and small-scale farmers subject to the consequences of seasonality and climate change (Schwemmlein, 2017). The defense of sustainable small-scale agriculture is one of the keys to a more sustainable world (FAO & IFAD, 2019).

Social workers can also act in the binomial between food scarcity and food waste by raising awareness on minimizing predictable food waste and promoting the valorization of non-predictable food waste, enabling the transition from food waste to consumable goods for people experiencing food insecurity (Morone et al., 2019). An example of action in the binomial between food scarcity and food waste is the Refood project, which channels food waste from restaurants, bakeries, and commercial supermarkets to people in need through a network of volunteers.

Practical Case: ACTUAR—Association for Cooperation and Development

ACTUAR is an international nongovernmental organization (NGO) committed to building and developing sustainable food systems, based on an ecological ethic of protecting biodiversity and the environment and respecting human rights.

In Portugal, ACTUAR is focused on the design of policies, strategies, and programs for the promotion of sustainable food systems, through the implementation of projects in collaboration and partnership with national and local public bodies, farmers' organizations, consumers, and social, environmental, and local development organizations. ACTUAR also works in the CPLP (Community of Portuguese Language Countries), with the following associates: São Tomé and Príncipe, Cape Verde, Angola, Mozambique, Guinea-Bissau, Brazil, and Portugal.

The Role of the Social Worker: ACTUAR—Association for Cooperation and Development

The ACTUAR team is multidisciplinary and includes social workers who perform duties such as advocacy for the promotion of sustainable food systems at various territorial levels and based on networking and active partnerships with government entities, universities, and other relevant actors. Therefore, social workers in many fields of practice can play an important role in promoting food security in line with environmental sustainability principles in various ways such as:

- Advocating and promoting sustainable small-scale agriculture in both urban and rural settings
- Developing activities and projects around the promotion of the consumption of healthy and sustainable food
- Creating partnerships with food retailers to reduce food waste and channel it to food insecure families
- Acting at a political level, seeking to rethink the models of food assistance

Social Work and Environmental Education

According to Ogando (2022) environmental education is not only important for providing information about the environment but also for increasing awareness, critical thinking, problem-solving, and decision-making skills, empowering children to be more prepared to face the effects of climate change and to contribute to policymaking, mitigation, and adaptation. Davis (1998) states that children, teachers, and communities can work through environmental education so that collectively and democratically they can move towards the resolution of environmental questions, issues, and problems.

Social work can deepen its relationship with the development of activities with schools and communities with the perspective of environmental education as political education of intervention and participation, directed towards the construction of a fairer and more sustainable society. From this point of view, it is important to promote an education that enables respect for the environment, in contrast to a model that is devastating the relationships of human beings among themselves and between them and the environment (Freire, 2001). According to Câmara et al. (2018), in a framework for schools, Environmental Education for Sustainability should prepare young people to exercise a conscious, dynamic, and informed citizenship towards current environmental problems. It is important to mention that although Câmara et al. (2018) focus on the education of children and young people, environmental education can be applied to other target audiences with whom social workers work, such as communities, the older population, or people with disabilities.

To establish a coherent approach to environmental education, social workers can engage with ecopedagogy. This is a social and political movement that emerges from various segments of society, including educators, ecologists, workers, and entrepreneurs, all sharing a deep concern for the environment. What sets ecopedagogy apart from traditional education is its holistic perspective. Instead of just focusing on nature conservation or the consequences of human activities on the environment, ecopedagogy delves deeper. It revolves around the idea of a new model of ecologically sustainable civilization that implies a change in economic, social, and cultural structures (Gadotti, 2001).

In this context, environmental education can be an important means of promoting social change towards a more sustainable society, and social workers can play an important role:

- In the promotion of environmental education activities in community and school contexts, enabling students to acquire and understand relevant information regarding environmental sustainability.
- In the training of strategic publics in environmental education, who work with or are in situations of greater risk of vulnerability. This can include a broad spectrum, professionals, policymakers, community leaders, and populations most susceptible to the effects of climate change, such as those residing on islands or in developing countries.

Despite being an enabler of social change towards a more environmentally sustainable society, environmental education and its connection to social work needs further exploration both at the level of research and inclusion in social work education curricula.

Practical Case: "School Gardens—Cascais Ambiente"

The School Gardens is a project based in the municipality of Cascais (Lisbon district) and aimed at schools that wish to create a vegetable garden or already have one and need support or technical assistance. To reinforce the garden space and classroom material, the "Garden at School" program launched the "Biological Garden in Schools Notebook." This notebook comprises three volumes. The first volume presents the objectives and elucidates the functioning of school gardens. The second volume provides activity sheets designed for students, while the third volume offers activity sheets for professors. With this new material, students and educators can, with total autonomy, explore new knowledge through practical activities in the garden. For example, in the subject of Environmental Studies, students learn the life cycle of plants through observation from seed to fruit. If they divide the amount of compost by the area of the garden, they are learning mathematics.

The Role of Social Worker: "School Gardens—Cascais Ambiente"

This project is in line with the food justice movement as it seeks to ensure nutritious food for all and to eliminate the lack of access to healthy food in communities. The social worker has the role of promoting projects such as school gardens due to their importance in providing more social opportunities and positive emotions (Pollin & Retzlaff-Fürst, 2021). These types of projects are relevant because the gardens can act as safe spaces where social workers can simultaneously educate for environmental practices and establish trusting relationships with children, young people, and even families or communities, providing a place to be, learn, and share (Bailey et al., 2018).

The social worker can also be involved in the development of materials that enhance the use of these spaces. The three notebooks created are designed to raise awareness about the importance of environmentally sustainable practices and food production.

The Human-Animal Relationship as an Element in Social Work

According to the American Veterinary Medical Association, the "human-animal relationship" is mutually beneficial and dynamic, with behaviors essential to the health and well-being of both. There is ample evidence that owners tend to view their pets as members of the family (Cohen, 2002). Due to this degree of importance given to pets, social workers should recognize the interconnectedness of all life in our ecosystem, seeking to address the complex relationships between humans, animals, and the environment (Boetto et al., 2017; Coates, 2003; Matthies et al., 2001 cited by Rambaree et al., 2019). Despite this, Gomes et al. (2021) recognize that the human-animal relationship and its intersection with social work is little explored in Portugal, either in the training offer or in the production of scientific knowledge.

Pinto (2021) states that a precise clarification of the terminology used in relation to the consideration of the animal on social work does not exist. It is a very polysemic field, where there are various understandings of what the terms indicate and a proliferation of proposals to define what intervention with nonhuman animals in social work is. Several nomenclatures have emerged, among them the concept of animal social work (Pinto, 2014; Arkow, 2020).

Animal social work consists of providing social support to individuals experiencing poverty or economic hardship, using a holistic approach that considers the well-being of their pets as well. In such situations, social workers should initially gather information about all the pets and humans in a family, communicate interest and concern for the whole family, and demonstrate an integrated approach to care that can help in planning appropriate interventions and preventive care (Arkow, 2020). Specifically, professionals should analyze with the user what the benefits and responsibilities of owning a pet are and support the most vulnerable users with cheaper veterinary care and animal food banks (Pinto, 2014). Likewise, Labrecque and Walsh (2011) state that it is important that pets are kept with their owners, as taking them away can have negative consequences for the person. The author gives the example of an experience with homeless women who endured loss due to the inability to maintain their companion animals while staying in homeless shelters.

Practical Case: Animalife

Animalife, a national nonprofit animal awareness and social support association, was founded in October 2011 and is based in three Portuguese cities. It is a nonprofit organization that operates on a voluntary model and has the objective to reduce the serious problem of animal abandonment, acting upstream on one of the main causes of abandonment in Portugal: the economic and social vulnerability of families. Thus support is provided, whenever possible, for feeding, vaccination, and deworming, and sterilization support is provided whenever possible.

The support provided by Animalife is organized into three programs, one of which is the Support Program for Homeless People (PSSA) that was created in 2016. This program aims to support people in their process of reintegration into society by ensuring that they can do so with their pets; offer a better quality of life for these animals and their owners, halting the rise in the number of abandoned pets; and carry out electronic identification of animals using microchipping to check whether the person already had the animal with them when they moved to the street or whether it was adopted by the person already on the street.

The Role of the Social Worker: Animalife

Social workers' main interventions include providing veterinary care and food assistance to low-income pet owners, rescuing and rehabilitating abandoned and mistreated animals, and promoting responsible pet ownership through education and awareness-raising campaigns. However, social workers not only identify the needs relating to the pet but also identify social issues associated with the pet owner. By using their professional skills and benefiting from the strong link between humans and animals, social workers can establish a better relationship with the client, making it easier to help them or refer them to other support structures.

Social workers working at Animalife reinforce ecosocial practices, as this work recognizes not only the human-animal relationship as beneficial but also moves them away from an anthropocentric approach (Boctto et al., 2017). The goal of the professionals who work at Animalife is not only to protect disadvantaged people but also to identify animals as beings with value and rights.

Social Work in Environmental Disaster

A disaster is an event that causes great destruction and loss to people and can result from natural events, human-induced or human intentional actions. The phenomena that can cause a disaster are diverse and are divided into four types, (1) natural disasters, (2) environmental disasters resulting from indirect human intervention, (3) environmental disasters resulting from direct human intervention, and (4) disasters directly related to human action: terrorist acts and military conflicts (Alston et al., 2019).

According to the code of ethics for social workers in Portugal (APSS, 2018), social workers have an ethical duty to intervene in disaster situations. In such situations, social workers have the opportunity to put into practice holistic and ecological approaches that should be guided by professional knowledge, ethics, and values, with a commitment to engage and support diverse communities (IFSW, 2022).

One model that can be useful to social workers in practice is the Community-Based Risk and Disaster Management Model (Marques, 2021a, b), which is structured in three phases:

1. Pre-disaster: During this phase, the key elements are prevention, mitigation, and preparedness. To strengthen disaster response and prevention efforts, social workers can integrate risk assessment into social diagnoses particularly during home visits and community work (Pascual & Fernández, 2017; Marques, 2021b).
2. Disaster: In this phase, social workers get into action to try to contain the situation and identify the most urgent needs. There must be multidisciplinary work between first responders, volunteers, and professionals involved. One of the actions social workers can take is locating and reuniting families, with a special focus on the most vulnerable individuals, such as children, the elderly, and animals (Pascual & Fernández, 2017; Marques, 2021b).
3. Post-disaster: During this phase, social workers should focus on community rehabilitation and repairing the consequences of the disaster. To facilitate recovery and preparedness for future disasters, social workers can evaluate the response provided during the disaster, aiming to improve coordination for subsequent responses (Pascual & Fernández, 2017).

Beyond this model, it is crucial for social workers to possess the necessary knowledge and skills to intervene effectively at various levels. This understanding plays a pivotal role in fostering collaborative partnerships that drive positive change and facilitate opportunities for sustainable development and ecosocial practice. By committing to the construction of a new eco-social world, social workers actively contribute to building a more inclusive and environmentally conscious society (IFSW, 2022). This has to happens in a multilevel process:

Macro Level In way of intervening on a macro level, the NGO statement on the war in Iraq to the UN Commission on Human Rights (2003) was developed. This document was a statement by various nongovernmental organizations (including the International Federation of Social Workers) condemning and calling for the immediate cessation of the illegal invasion of Iraq by the United States and the United Kingdom in 2003 (cited in IFSW, 2005). Unfortunately, the war only ended in 2011, having created extensive damage to the biophysical environment, resulting in the loss of date palms, the drying of the marshlands, and worsening of soil and water salinity. These ecological consequences had profound impacts including triggering the displacement of populations and creating a scarcity of essential resources such clean water and food (ICRC, 2021).

Meso Level An example of this was the Photovoice project that took place in the municipality of Mafra in Portugal. Photovoice is a research methodology which uses images and words to give individuals a voice and bring about change. Social workers used this methodology to make changes, such as revising the Forest Fire Management Plan and including a specific budget for community programs to adapt and recover from forest fires.

Micro Level An illustrative example is the intervention adopted by a social worker in the face of the earthquake that devastated the Bam city in Iran on 26 December 2003. The professionals directed their efforts towards a survivor of the earthquake, guiding the individual's attention towards a variety of rational solutions and presenting a range of alternative strategies for the client to consider. For instance, one approach involved helping the client understand the reality by identifying various gestures of support and affection from family and friends to help him cope with his loss (Javadian, 2007).

Practical Case: "Are You Ready" Emergency Kit—Prevention Campaign

The "Are You Ready" campaign was held at the EBI Arrifes school in São Miguel Island, Azores, with the aim of teaching Azorean children and young people some simple behaviors to prepare for possible disasters. Specifically, they were taught which different types of disaster can happen and what people can do in each situation, how to organize a kit/backpack for disaster situations, and how to build up an emergency plan with their family.

The Role of Social Workers: "Are You Ready" Emergency Kit—Prevention Campaign

Social workers educate and reach out to communities about disaster preparedness and response. They can develop and deliver educational materials and provide training to build knowledge and skills. The goal of the social workers in this type of intervention is to empower people to take an active role in building resilience and preparedness in their communities (Boetto et al., 2017).

With this project, the social workers chose schools as a means of education for preparing for and responding to potential disasters. Schools were chosen not only because they are trusted educational institutions that serve as a meeting point for sharing information and raising awareness but also because they are important communication networks, since through children and young people, this information can reach parents and family members more quickly. This is a very good example of how eco-social work can cross over into practice because it provides guidance and education on practices for sustainable living (Boetto et al., 2017).

Practicality of the Toolkit

The toolkit combines a practical approach together with a theoretical contextualization of the topics, and has since its publication garnered considerable interest among social workers, social work students, and social work educators across Portugal and Europe, resulting in invitations to participate in lectures and conferences aimed at discussing the contents of the E-book. This chapter encompasses a significant portion of the original E-book, now translated into English. This translation enhances its potential to connect with a wider audience of social workers, thereby igniting a drive for environmentally sustainable practices within the social sector.

For educators and students, the toolkit serves as a valuable resource to incorporate into their process of teaching/learning. It provides a simple and practical framework for introducing the intersection of social work and environmental sustainability. The inclusion of real-world case studies allows teachers to illustrate concrete examples while explaining the topic in the classroom and providing students with greater contact with reality and understanding of how environmental social work occurs in practice. Its user-friendly format and comprehensible language make it an accessible and informative resource for case studies, research, and professional development.

For professionals in the field of social work, the toolkit serves as an accessible resource for ongoing learning and development. It offers practical tools and insights that can help professionals to adapt to new social and environmental challenges. This guide serves as a source of inspiration, encouraging professionals to explore innovative solutions and make a positive impact on the well-being of individuals and communities across the world.

Reflections and Conclusion

This chapter provides a practical perspective that can encourage the appropriation of an ecosocial approach and enhance ecosocial work (Rambaree et al., 2019; Närhi & Matthies, 2016). This is the first of many steps that can continue to affirm an ecosocial work view and reflect on a holistic and multifocal social work that considers action at the micro, meso, and macro levels to promote a more socially and environmentally sustainable society.

The six subjects covered merely represent a selection of the multitude of areas where social workers can incorporate eco-social principles. Our anticipation is that social workers who engage with this chapter can gain insights for instigating transformations within their communities, institutional frameworks, and policy landscapes towards a more environmentally sustainable response to social issues. It is important to acknowledge that there remains a substantial unexplored terrain concerning the interplay between social work and the pursuit of environmental sustainability and further research is needed.

Specifically, the practical cases presented here create some reflection and discussion about how social workers can in their practice connect social work and environmental and sustainability issues along with the challenges associated with this. These practical cases offer valuable insights: The "Bela Flor Agrofrorestry" initiates a compelling conversation about difficulties involved in engaging the community in such a project and underscores how social workers can facilitate their involvement in the agroforestry initiative. "Vintage for a Cause" illustrates the potential for generating social impact within a project centered on environmental concerns. It highlights how social workers can identify and refer individuals from vulnerable target audiences. In the practical case of "ACTUAR," discussion centers on the importance of social interventions that promote sustainable food systems. It emphasizes how social workers can contribute as valuable assets to multidisciplinary teams. The "School Gardens—Cascais Ambiente" discusses how a project can serve a dual purpose for social workers. Firstly, they educate on environmental practices, and, secondly, they build trusting relationships with service users. The "Animalife" underscores the importance of including animals in social interventions. In this context social workers extend their support not only to the pets but also to the humans involved. In the final practical case, "'Are You Ready' Emergency Kit—Prevention Campaign", the discussion centers on the importance of education in disaster prevention and preparedness. In this scenario, social workers play a vital role in empowering individuals and fostering community resilience with a strong emphasis on readiness for potential disasters.

In conclusion, this chapter serves as a call to action—a call for social work students, professionals, and educators to adopt an ecosocial approach in their reality. To achieve this, utilizing the material presented in this chapter and the practical guidance in the *Social Work and Environmental Sustainability Toolkit* can highlight the potential of this approach in promoting a fairer, more conscious, and more just world. It is crucial to recognize the interconnection between social and environmental issues, and, by adopting an ecosocial approach, we can work towards a more sustainable future for all.

Bibliography

Aguiar, O. B., & Padrão, S. M. (2022). Direito humano à alimentação adequada: fome, desigualdade e pobreza como obstáculos para garantir direitos sociais. *Serviço Social & Sociedade, 143*, 121–139.

Alston, M., Hazeleger, T., & Hargreaves, D. (2019). *Social work and disasters: A handbook for practice* (1st ed.). Routledge.

American Veterinary Medical Association. (n.d.). *Human-animal interaction and the human-animal bond.* AVMA. https://www.avma.org/resources-tools/avma-policies/human-animal-interaction-and-human-animal-bond. Accessed 20 August 2023.

Animalife. (n.d.). *Animalife.* https://www.animalife.pt/pt/home. Accessed 31 August 2023.

Arevalo, L. (2017). Agitating through an alternate economy: Social work, sustainability, and the circular economy. In M. Rinkel & M. Powers (Eds.), *Social work promoting community and environmental sustainability: A workbook for global social workers and educators* (Vol. 1, pp. 35–45). International Federation of Social Work (IFSW).

Arkow, P. (2020). Human–animal relationships and social work: Opportunities beyond the veterinary environment. *Child and Adolescent Social Work Journal, 37*(6), 573–588. https://doi.org/10.1007/s10560-020-00697-x
Associação de Profissionais de Serviço Social (APSS). (2018). *Código deontológico dos Assistentes Socais em Portugal*. APSS.
Bailey, S., Hendrick, A., & Palmer, M. (2018). Eco-social work in action: A place for community gardens. *Australian Social Work, 71*(1), 98–110. https://doi.org/10.1080/0312407X.2017.1384032
Boetto, H., Bell, K., & Kime, K. (2017). Holistic ecosocial work: A model for transformative change through being, knowing and doing. In M. Rinkel & M. Powers (Eds.), *Social work promoting community and environmental sustainability: A workbook for global social workers and educators* (Vol. 2, pp. 46–58). International Federation of Social Work (IFSW).
Borga, A., Gomes, J., & Luís, R. (2022). *Toolkit serviço social & sustentabilidade ambiental* [E-book].
Brazão, M. (2021, April 22). *Circular economy Portugal* [Webinar]. https://www.facebook.com/APSS.SERVICOSOCIAL/videos/300605164805523?locale=pt_PT. Accessed 4 August 2023.
Brito, M. (2021, October 21). *Cascais Ambiente* [Webinar].
Calado, R. (2021, May 6). *Repais Café* [Webinar]. https://www.facebook.com/APSS.SERVICOSOCIAL/videos/587838785943330/
Câmara, A. C., Proença, A., Teixeira, F., Freitas, H., Gil, H., Vieira, I., Pinto, J., Soares, S., Gomes, M., Amaral, M., & Castro, S. (2018). *Referencial de Educação Ambiental para a Sustentabilidade para a Educação Pré-Escolar, o Ensino Básico e o Ensino Secundário*. Ministério da Educação, Direção Geral de Educação.
Carvalho, M. I. (2021, October 14). *Apresentação sobre o conceito Serviço Social em Catástrofes* [Webinar].
Cohen, S. P. (2002). Can pets function as family members? *Western Journal of Nursing Research, 24*(6), 621–638. https://doi.org/10.1177/019394502320555386
Davis, J. (1998). Young children, environmental education and the future. In N. Graves (Ed.), *Education and the environment* (pp. 141–154). World Education Fellowship.
Dominelli, L. (2018). Green social work in theory and practice: A new paradigm for the profession. In L. Dominelli (Ed.), *The Routledge handbook of green social work* (pp. 8–10). Routledge.
EAPN Portugal. (2023). *Poverty Watch Portugal 2023*. https://www.eapn.pt/wp-content/uploads/2023/10/Poverty-Watch-Report-Portugal-2023.pdf. Accessed 21 August 2023.
Elena. (2022, September 26). *Social currencies that promote local economies*. Comunitaria. https://comunitaria.com/en/social-currencies-that-promote-local-economies/
Ellen MacArthur Foundation. (2017). *Let's build a circular economy*. https://ellenmacarthurfoundation.org/. Accessed 21 August 2023.
FAO & IFAD. (2019, May 29). *United Nations decade of family farming 2019–2028 – Global action plan*. Retrieved July 30, 2023, from https://reliefweb.int/report/world/united-nations-decade-family-farming-2019-2028-global-action-plan?gclid=Cj0KCQjwmvSoBhDOARIsAK6aV7i6iJwsC4W9WlfAU47TkPlhvqz5DStAedNmhVNN5YK6z81DCvtCKz0aAh8JEALw_wcB
Folgado, G. (2021, April 22). *Associação locals* [Webinar]. https://encurtador.com.br/dinEZ. Accessed 4 August 2023.
Food and Agriculture Organization of the United Nations. (2023). [Website]. https://www.fao.org/hunger/en/
Franco, R. (2021, April 22). *Ecobairro da boavista* [Webinar]. https://encurtador.com.br/dinEZ. Accessed 4 August 2023.
Freire, P. (2001). *Pedagogia da Autonomia: Saberes necessários à prática educativa* (25th ed.). Paz e Terra.
Gadotti, M. (2001). *Pedagogia da Terra*. Peirópolis.
Gomes, J. (2020). Iniciativas de economia circular têxtil: Um quadro de aplicação para o Serviço Social ambiental. *Espacios Transnacionales: Revista Latinoamericana-Europea de Pensamiento y Acción Social, 8*(15), 40–51. https://dialnet.unirioja.es/servlet/articulo?codigo=7747655
Gomes, J. (2021, October 21). *Kit Emergência* [Webinar].

Gomes, J. F. P., Marques, E. J. S. T., & Schmitt, A. R. V. (2021). Os Desafios Contemporâneos do Serviço Social Ambiental: Contributos Para Uma Reflexão Epistemológica Sobre As Potencialidades Das Intervenções Assistidas Com Animais. In E. J. S. T. Marques & A. R. V. Schmitt (Eds.), *Serviço Social: Aplicação da Ciência e os seus antagonismos* (1st ed., pp. 1–20). Atena Editora.

Gonçalo, R., & Silva, M. (2021, May 20). *Animalife* [Webinar]. https://encurtador.com.br/rsFI7. Accessed 8 August 2023.

Hunter, H. (2021, September 30). *Re-food* [Webinar].

International Committee of the Red Cross (ICRC). (2021, July, 21). *Iraq's perfect storm—A climate and environmental crisis amid the scars of war*. Retrieved July 30, 2023, from https://www.icrc.org/en/document/iraqs-perfect-storm-climate-and-environmental-crisis-amid-scars-war

International Federation of Social Workers (IFSW). (2005, October 13). *NGO joint statement on the war in Iraq to the UN Commission on Human Rights*. Retrieved July 18, 2023, from https://www.ifsw.org/ngo-joint-statement-on-the-war-in-iraq-to-the-un-commission-on-human-rights/

International Federation of Social Workers (IFSW). (2022, May 24). *The role of social workers in advancing a new eco-social world*. Retrieved July 27, 2023, from https://www.ifsw.org/the-role-of-social-workers-in-advancing-a-new-eco-social-world/

Javadian, R. (2007). Social work responses to earthquake disasters: A social work intervention in Bam, Iran. *International Social Work, 50*(3), 334–346. https://doi.org/10.1177/0020872807076047

Labrecque, J., & Walsh, C. A. (2011). Homeless women's voices on incorporating companion animals into shelter services. *Anthrozoös, 24*(1), 79–95. https://doi.org/10.2752/175303711X12923300467447

Luís, R. (2020, December 22). *Serviço Social e Desenvolvimento Sustentável* [Master's thesis]. Retrieved July 5, 2023, from https://repositorio.iscte-iul.pt/handle/10071/21328

Marques, E. (2021a, May 6). *Serviço Social e Economia Circular* [Webinar]. https://www.facebook.com/APSS.SERVICOSOCIAL/videos/587838785943330/. Accessed 8 July 2023.

Marques, E. (2021b, October 14). *Apresentação sobre o conceito Serviço Social em Catástrofes* [Webinar].

Matenga, L., Mathende, T. L., & Nhapi, T. G. (2019). Constraints and opportunities in the mitigation of adverse environmental impacts in mineral resources harnessing—Some Marange diamond perspectives. In S. Mugova & P. Sachs (Eds.), *Opportunities and pitfalls of corporate social responsibility* (CSR, Sustainability, Ethics & Governance) (pp. 215–240). Springer.

Matthies, A. L., Närhi, K., & Ward, D. (Eds.). (2001). *The eco-social approach in social work*. Sophi.

Morone, P., Koutinas, A., Gathergood, N., Arshadi, M., & Matharu, A. (2019). Food waste: Challenges and opportunities for enhancing the emerging bio-economy. *Journal of Cleaner Production, 221*, 10–16. https://doi.org/10.1016/j.jclepro.2019.02.258

Mukhlis, I., Rizaludin, M. S., & Hidayah, I. (2022). Understanding socio-economic and environmental impacts of agroforestry on rural communities. *Forests, 13*(4), 1–13. https://doi.org/10.3390/f13040556

Munera, L. (2021, October 21). *Fundación Vida Sostenible* [Webinar].

Närhi, K., & Matthies, A. L. (2016). The ecosocial approach in social work as a framework for structural social work. *International Social Work, 61*(4), 490–502.

Ogando, P. (2022, July 18). *A pending task: 5 reasons why we need environmental education*. Retrieved September 1, 2023, from https://www.unicef.org/lac/en/stories/a-pending-task-5-reasons-why-we-need-environmental-education

Pascual, M. G., & Fernández, A. L. (2017). Building sustainable communities through social work in emergency/disaster situations: Learning from the Association of Social Work Professionals of Spain. In M. Rinkel & M. Powers (Eds.), *Social work promoting community and environmental sustainability: A workbook for global social workers and educators* (Vol. 1, pp. 223–238). International Federation of Social Work (IFSW).

Patrão, A. (2021, October 14). *Apresentação do projeto de Phtotovoice no concelho de Mafra* [Webinar].

Peeters, J. (2012). Social work and sustainable development: Towards a social–ecological practice model. *Journal of Social Intervention: Theory and Practice, 21*(3), 5–26. https://doi.org/10.18352/jsi.316

Pinto, C. (2014). Serviço Social e Desenvolvimento Sustentável: Missões Entrecruzadas. In M. I. Carvalho (Ed.), *Serviço Social, Teorias e Práticas* (pp. 143–162). Pactor.

Pinto, C. (2021, May 20). *Consideração dos animais como um elemento do Serviço Social* [Webinar]. https://www.facebook.com/APSS.SERVICOSOCIAL/videos/192684666040864/. Accessed 18 July 2023.

Pollin, S., & Retzlaff-Fürst, C. (2021). The school garden: A social and emotional place. *Frontiers in Psychology, 12*, 1–11. https://doi.org/10.3389/fpsyg.2021.567720

Rambaree, K., Powers, M. C. F., & Smith, R. J. (2019). Ecosocial work and social change in community practice. *Journal of Community Practice, 31*(2), 205–212. https://doi.org/10.1080/10705422.2019.1660516

Raposo, S. (2021, May 6). *CRAES* [Webinar]. https://encurtador.com.br/hSZ26. Accessed 15 July 2023.

Rede DLBC Lisboa. (2021, September 30). *Food wave* [Webinar].

Rinkel, M., & Powers, M. (2019). *Social work promoting community and environmental sustainability: A workbook for global social workers and educators*. International Federation of Social Workers.

Rocha, H. B. (2015). *Serviço social e ambiente: A sustentabilidade ecológica das comunidades socialmente vulneráveis* [Doctoral thesis]. Retrieved July 5, 2023, from http://hdl.handle.net/10071/11964

Rocha, H. B. (2021a, April 22). *Serviço Social e ambiente na comunidade* [Webinar]. https://www.facebook.com/APSS.SERVICOSOCIAL/videos/300605164805523/. Accessed 8 July 2023.

Rocha, S. (2021b, September 30). *Actuar—Associação para a cooperação e Desenvolvimento* [Webinar].

Schwemmlein, K. (2017, November 22). *Serviço Social, desenvolvimento sustentável e soberania alimentar* [Master thesis]. Retrieved July 5, 2023, from https://repositorio.iscte-iul.pt/handle/10071/15386

Schwemmlein, K. (2021, September 30). *Serviço Social, desenvolvimento sustentável e soberania alimentar* [Webinar].

Silva, H. (2021, May 6). *Vintage for a cause* [Webinar]. https://www.facebook.com/APSS.SERVICOSOCIAL/videos/587838785943330/. Accessed 8 July 2023.

Strand, B. E. (2021, May 20). *Veterinary social work* [Webinar]. https://encurtador.com.br/qLN69. Accessed 8 July 2023.

United Nations. (1948). *Universal Declaration of Human Rights*.

United Nations. (1996, December 16). *International covenant on economic, social and cultural rights*. Retrieved July 30, 2023, from https://treaties.un.org/doc/treaties/1976/01/19760103%2009-57%20pm/ch_iv_03.pdf

United Nations. (2015). *Transforming our world: The 2030 Agenda for Sustainable Development*.

Vilar-Compte, M., Burrola-Méndez, S., Lozano-Marrufo, A., Ferré-Eguiluz, I., Flores, D., Gaitán-Rossi, P., Teruel, G., & Pérez-Escamilla, R. (2021). Urban poverty and nutrition challenges associated with accessibility to a healthy diet: A global systematic literature review. *International Journal for Equity in Health, 20*(1), 1–19. https://doi.org/10.1186/s12939-020-01330-0

Vizinho, A., Mamede, C., Santos, E., & Cruz, P. (2021). *O APOIO ALIMENTAR EM PORTUGAL: Uma análise das necessidades alimentares e dos recursos mobilizados para a sua satisfação na perspetiva do cumprimento do Direito Humano à Alimentação e Nutrição Adequada*. EAPN Portugal.

Case Study 3: Stormkop: Nurturing Minds, Inspiring Action for a Sustainable Future

Luna Haine

In the vibrant city of Antwerp (Belgium), the NGO *Stormkop* is located in a vast barn next to an ancient dry dock at the Scheldt River. The organisation has created an environment where children and adults are encouraged to touch, see, hear, and feel. The dynamic philosophy and approach of this NGO is constantly evolving, tailored to the needs of the diverse groups they serve. Today, Stormkop NGO is an organisation where young people and children can participate in planned camps during school holidays and on class days organised by them, where they spend a day in the magical world of Stormkop. The literal translation of 'Stormkop' is 'Storm head.' This name was chosen because in this organisation, they aim to create a storm in the minds of their target audience. They want to fill their heads with thoughts, interests, questions, and more and then work with these to stimulate reflection.

Stormkop offers an inspiring story on how to engage children on the issues of climate change and pollution. They use art, science, and philosophy to create awareness around pollution and plastic waste. In this organisation, I undertook the internship in my second year of the professional bachelor of social work. In this way, I witnessed how Stormkop not only cleans up but inspires, raises awareness, and spurs action, making a better future for our planet possible.

Stormkop was born out of a passion for art education with a focus on 'process over product'. The organisation began in 2012 as an educational initiative called 'De Steen der wijzen' or '*The Philosopher's Stone*' which focused on art, philosophy, and science. With the support of a dedicated team and the city of Antwerp, Stormkop eventually found shelter in a former dock workers' workshop in 2018. There they developed a unique approach in an even more unique place. Stormkop receives financial support from the city of Antwerp as well as an income from activities such as camps, a restaurant, a bar, class days, and rental of spaces. Together with some project funding, this supports Stormkop's continued existence.

L. Haine (✉)
Antwerp, Belgium

C. Forde et al. (eds.), *Teaching and Learning in Ecosocial Work*,
https://doi.org/10.1007/978-3-031-58708-5_12

Stormkop's philosophy revolves around the keywords 'Touch, See, Hear and Feel'. These principles are materialised through the countless possibilities of their building, a gigantic ancient storehouse. The organisation works from the heart and adapts their approach based on the changing groups of children they serve. Their themes include art, philosophy, science, sustainability, climate, mobility, and adventure.

Nowadays, Stormkop's main focus is on climate issues that children encounter in their daily lives, with a focus on recycling beachcombed plastic. In recent times, it's often forgotten that the youth are the future. It's crucial to engage in conversations with them, not just about them. They should have a more prominent voice in discussions related to topics like climate, mobility, mental health, and more. Authorities and governments must not overlook this fact. Change needs to stem from the children through active collaboration. They are closer to the problems and will be the ones dealing with the consequences in the future. Stormkop's goal is to encourage a long-term vision and ensure that the youth are actively involved in shaping the future.

They have introduced four new pillars: vulnerability (fragile like children), bravery (have courage, hold on), assembly (connection), and uprising (rebellion, rioting, artistic citizenship), which further shape their vision and are applied in activities for various audiences (Koos Hogeweg, Lot Saldien, Personal communication, 2021). These pillars are similar to those we learned on the social work degree itself: proximity, advocacy (and political action), process work, generalist work, and connecting work. While attending college, courage, connection, and rebellion were also exemplified. These values and attitudes closely align with the pillars of Stormkop, and I found them particularly relevant during my participation in the school strike for climate that took place in Belgium from February to May 2019.

During my internship (within the social work degree), I participated in raising awareness and action around pollution and plastic waste in the Scheldt (the river: De Schelde) estuary. My role consisted of supervising class days and camps for children, focusing on the theme of pollution. Stormkop drew inspiration from the courageous 'Schelde jutters'. They clean up the areas where the river, The Scheldt, meets the beach and nature; this is called 'jutten'. This led to organising day trips in which school children first discussed trash and plastic in our daily lives, its origins, and its devastating effects. Then the children set out in groups armed with buckets, boots, and gloves on a 'Schelde jutting' trip. Here, over many years, all kinds of trash pile up, from little green toy men to Lego bricks and balloons. But what stands out the most are the tiny plastic granules scattered everywhere, the infamous nurdles. These pea-sized granules are the building blocks for almost all plastic products and derivatives. After the clean-up walk, the children are given creative assignments, depending on the class day. They create works of art from the collected plastic – from still lives to musical instruments and even robots.

My choice of an internship at Stormkop was based on stories about the 'organization where creativity can flourish,' a remote place where people come together to create, learn, and share. The challenge for me was the lack of a familiar and bureaucratic structure; they are more flexible and creative. After getting used to this working style, I discovered that this was precisely where the strength of the organisation lies. It was a challenge for me to go with the flow of this unique organisation. Although I initially struggled with the unstructured nature of the organisation, I learned to be more flexible in my approach. Stormkop taught me to deal with change and unpredictability, which is a valuable skill.

The most striking aspect of Stormkop is how they approach children, interns, employers, and employees as equal participants. Hence the organisation has a (complete) horizontal hierarchy. This way of working has stuck with me because it proved to be very powerful in engaging others in social and environmental issues. The children who leave Stormkop are more aware of 'global warming' than most people on the planet. They understand that the tourist splendor of Antwerp hides a dark side that is often ignored. These children now know that pollution is unacceptable. Moreover, they learn that it is not right to point the finger at individual behaviour, as in not picking up litter (as most children don't litter themselves). Stormkop confronts them with bigger questions: 'What happens to the trash a boat loses?' 'How are companies that process these nurdles controlled?', 'What do you think about the plans for a new production facility in the harbor that produces the main component for plastics?' Stormkop prompts children to think about broader responsibilities in the fight against pollution and provides them with the knowledge and inspiration to make a difference.

My personal commitment to the issues of global warming and pollution did not originate at Stormkop, but during my internship, it grew and became more fundamental. From the front row, I experienced how plastic pollution was viewed from the perspective of art, science, and philosophy. I saw how paintings and sculptures were made from trash that children's hands had picked up. I heard about recycling, microplastics, and the impact of our lifestyle on nature and the planet. For example, imagine the fact that we eat the aggregate of a bank card in plastic every year! This made me feel very close to myself in this organisation. Six-year-old Luna, who once tinkered with trash on the playground, was now at Stormkop amongst mountains of trash to be cleaned up, triaged, and worked on afterwards.

My internship definitely accomplished a bigger awareness of my own consumption behaviour. Somehow, I started educating others on the impact of plastic pollution. This experience also made me eager to integrate environmental issues into my current and future job(s) as a social worker. During the classes in the university after my internship, I remember feeling a need to work more around climate and environmental issues. I felt that ecological issues are not a far-away issue for social workers. In no time our profession would be dealing with victims of ever closer natural disasters, climate migration, famine, and poverty or the lack of education which in turn could be a consequence of climate issues. The combination of an internship at Stormkop and a course on global challenges in the third year of social work in which we discussed these issues made me feel better prepared and well informed as a brand new social worker.

I personally think that the link between social work and the future of our planet should be emphasised even more in both education and practice. Through organisations like Stormkop, there is a place for environmental education and climate action. This is what our society and the social work profession need. The prospect of enrolling in a graduate programme focused on ecosocial work strikes me as a fantastic opportunity and an essential milestone. If I were to contemplate resuming my studies with the option to specialise in ecosocial work, my motivation would be strong, and I wouldn't hesitate for long when making that choice. When teaching and learning ecosocial work, an important focus should be to 'also' offer opportunities, possibilities, and positive projects or hopeful events. Ecology is a dark topic, and it is vital to maintain a spot of light in this darkness. It is hope that fuels student engagement. Our planet is changing. The social and ecological issues we face in our profession are changing. The study of social work, in turn, should change too. Preparing for ecosocial challenges means looking these squarely in the eye and beginning to critically re-examine the content of courses and graduate programmes.

In conclusion, the story of Stormkop offers an inspiring insight into the power of arts education and action in fighting environmental issues, notably plastic pollution. This organisation, which once began as an educational initiative with a passion for art, philosophy, and science, has evolved into a dynamic place where children and adults are encouraged to see and understand the world around them in new ways. Stormkop not only cleans up but inspires, opens eyes to the shocking reality of pollution, and encourages awareness and recycling. Social work should learn something from Stromkop's 'burning fire' that makes them take action. They turn their frustrations and optimism for the future into deeds, action, and riot.

During my classes, I was often told how important it is to get out on the streets, to make your voice heard, but also to set a good example. This is exactly what Stormkop does with Scheldt jutting. Putting words into action was what I missed during my social work studies. We could have united and worked with students and teachers equally to find topics that were raised during the lectures and make a difference in practice as well. In a million ways, this can be done through a (river) clean-up campaign, a goods raiser action, a rally or demonstration, a page or canvas highlighting socioecological issues on social media, or collaboration around art and creativity to increase awareness and activate citizens committed to a sustainable future, just like Stormkop does. Social work programmes can call for reconsideration of consumption patterns and behaviours and strive for a more conscious and sustainable society. Social work programmes can offer inspiration and hope in the struggle to protect and cherish our planet for future generations, just like Stormkop does.

Chapter 9: Teaching for Ecosocial Work: What Can Ecosocial Policy Offer?

Joe Whelan

Introduction

Environmental and ecological justice have generally not been a central concern for social work at a structural level. The reasons for this are complex and devolve in part on the way in which social work is conceptualised and delivered. In this chapter, I want to suggest that a further reason for why social work has been slow to encompass climate work and climate justice devolves upon the proposition of 'not knowing how', that is, not knowing how to confront the overwhelming nature of a global and wholly existential phenomenon which takes the form of a 'hyperobject'. Social work is also a multidimensional activity meaning that a response to the climate crisis is required in multiple contexts including in social work practice and in social work scholarship and activism. A further important focal point is the role of social work education and the training of future practitioners. This chapter will focus on the latter and suggest that one method for confronting the climate crisis through social work education requires a turn to ecosocial policy. The method I employ for charting this path is to look toward what social policies and what policy programmes social work in the global north should seek to potentially teach about, champion and align with on the basis of what is environmentally sustainable and also with due regard to an already established social work identity, present in how the profession is defined and in a robust value base. In this respect and couched within the overarching framework of a degrowth aligned capability approach, universal basic services (UBS) and a participation income (PI) are explored. Ultimately, it is suggested that focusing on ecosocial policies in the context of social work education can give future practitioners something to aspire toward, to engage in activism for and to help shape practice.

J. Whelan (✉)
School of Social Work and Social Policy, Trinity College Dublin, Dublin, Ireland
e-mail: jwhelan9@tcd.ie

C. Forde et al. (eds.), *Teaching and Learning in Ecosocial Work*,
https://doi.org/10.1007/978-3-031-58708-5_13

Grasping a Hyperobject

In a report on forging a 'climate-resilient Europe' from the European Commission (2021) to the European parliament and others, the nature and effects of climate change are laid out in stark terms. In the first instance, the report acknowledges that climate change is happening now and is not something that is going to happen at some unspecified time in the future, the world having recently concluded the hottest decade on record. The report goes on to note that were all greenhouse gas emissions to stop overnight, the pervasive effects that have already begun to accumulate would not cease and would—and will—in fact, continue for decades. The report then acknowledges the increasing frequency of extreme weather events before going on to note what the enormous and devastating social and economic consequences of the current climate trajectory are likely to be. Taken as a whole and despite some assurances that Europe is 'rising to meet the climate challenge' (2021: 2), this makes for grim reading. Though grim, it is an also familiar picture that is replicated in various national and international reports in different ways. The world is not short on reports detailing the magnitude of the current crisis (see European Environmental Agency, 2021; United Nations, 2021; Intergovernmental Panel on Climate Change [IPCC], 2022 for just some examples). The frequency and abundance of such reporting conveys the sense of an awareness at high levels that the climate crisis happening around us right now, along with associated phenomena such as extreme weather events, glacial deterioration and a warming global climate, is well documented, widely discoursed and generally accepted. Yet, it can also be suggested that something about the climate crisis remains anomalous, intangible even, its immediacy difficult to grasp in the everyday. This is problematic of course because a sense of immediacy is generally required as a precursor to action in the face of crisis as recent experiences with the coronavirus pandemic aptly demonstrate (see Hick & Murphy, 2021 re emergency policy responses). In *Hyperobjects: Philosophy and Ecology after the End of the World*, Morton (2013) provides some useful conceptual tools that can be adapted to help us think through this anomaly, and I take up the concept of 'hyperobject' as a heuristic device here.

'Hyperobject' is undoubtedly an ambiguous term which is difficult to grasp. However, this is not surprising given that hyperobjects themselves are ambiguous and difficult to grasp. Hyperobjects include things like capitalism, the economy, society, the labour market, and the welfare state. We know about and somewhat implicitly understand what these things, these hyperobjects, are, yet they somehow remain intangible, unreachable, and untouchable. In describing how we interact with hyperobjects, in the first instance, Morton (2013: np) distinguishes between background and foreground objects, background objects being those safe 'neutral screen[s] that enables us to have a human drama in the foreground'. Morton (2013: np) uses the example of conversations about the weather—a staple social lubricant that almost all people have applied in both the most mundane and tense of social exchanges—to denote what is meant by backgrounds. In doing so, he demonstrates the intrusion of a 'hyperobject' into the realm of the everyday, the hyperobject in this case, being global warming:

> As you approach your car, a stranger calls out, 'Hey! Funny weather today!' With a due sense of caution—is she a global warming denier or not?—you reply yes. There is a slight hesitation. Is it because she is thinking of saying something about global warming? In any case, the hesitation induced you to think of it. Congratulations: you are living proof that you have entered the time of hyperobjects. Why? You can no longer have a routine conversation about the weather with a stranger. The presence of global warming looms into the conversation like a shadow, introducing strange gaps. Or global warming is spoken or—either way the reality is strange. A hyperobject has ruined the weather conversation…

For the purposes of this chapter, Morton's (2013) articulation of a hyperobject can be stretched and adapted a little and so used to frame both the questions and the thesis that will be explored. In this respect, taking up both Morton's (2013) articulation of a hyperobject and his example of same in the form of global warming and climate change, I want to suggest that a hyperobject, though it intrudes into and permeates our everyday, remains as yet intangible. This can mean that attempts to muster an environmentally conscious response are caught in a continuous struggle to attach themselves to a course or courses of action owing to the fact that in the context of a hyperobject like climate change 'Environmentalism seems to be talking about something that can't be seen or touched' (Morton, 2013: np). So then, the climate crisis as a hyperobject presents us with a peculiar 'double problem'. We both know and we don't know; it is all around us, and yet it is distant from us. We understand that there is a crisis; indeed, it pervades the everyday as demonstrated by Morton's (2013) example of the 'weather conversation'. We also understand that it must be addressed as demonstrated by numerous international reports, some of which are cited above. Yet, and despite empirical evidence which goes beyond words written in a report in the form of extreme weather events (Maglajlic & Ioakimidis, 2021), the crisis remains intangible, unknowable, at a distance, particularly, perhaps, for people not as yet directly impacted. This is the peculiarity of the climate crisis as a hyperobject which, on the one hand, envelopes us, surrounds us, and intrudes into the everyday while, on the other, remains distant, unknowable, and hard to grasp. Hyperobjects then are vast and need human apprehension.

Social Work and the Hyperobject of Climate: What's to Be Done?

If we were to use the framing of 'climate as hyperobject' for the purposes of a discussion about social work and about social work education in particular, before then asking how social work educators might go about teaching and preparing graduates to work in a world experiencing a severe climate crisis, we will arrive at a similar set of problems. Social work as a global entity (as a hyperobject in and of itself) and endeavour is aware of the hyperobject that is the climate crisis (see International Federation of Social Workers [IFSW], 2022b). This hyperobject also undoubtedly intrudes into social work classrooms and lecture halls as it should (Powers et al., 2019; Boetto et al., 2020; Lynch et al., 2021; Ranta-Tyrkkö & Närhi, 2021). Yet, and despite growing scholarship in this area, questions remain around how to confront climate change in the social work educative process in a way which accounts for the often-interpersonal nature of the work, while also encapsulating a political dimension. Focusing attention on the latter, questions such as 'what should we teach about in the context of the climate crisis?' and 'how should what we teach align with social work theories, values, methods and skills?' are pertinent. Moreover, and given the social justice mandate which underpins the profession, we might reasonably ask, in confronting the hyperobject of climate and climate crisis, 'how do we stay true to and even champion the emancipatory aspects of social work?' These are the questions at the heart of this chapter. Yet they are also big, indeed, existential questions with a range of potential answers. For this reason, in setting out one possible answer, I confine my analysis to the structural as it connects to the personal and specifically, therefore, to the areas of political economy and social policy. Geographically, I remain in the global north on the basis that this is where the most pervasive effects of the crisis are continually created—though they may be felt elsewhere—and also where the major changes in practice need to happen fastest. In doing so, I aim to explore the following key question to help answer those questions posed above: 'what sustainable social policies should social work educators embrace, advocate for and teach about?', the tentative answer to which is presented as a means of potentially progressing the broader task of ecosocial work and of ecosocial work education in the context of climate justice.

In attempting to begin to offer an answer and after offering a brief discussion on the nature of ecosocial work, I synthesise some select contributions from an emerging social policy literature which has begun to explore questions of environmental degradation and its potential repercussions, along with solutions to the former and the latter in the context of welfare states and policy responses. I base my suggestions for what social policies social work education should seek to incorporate into curricula on both how social work is globally defined and on what social policies overlap with and encapsulate the social work value base. In doing so, I explore the overarching potential of a postgrowth capability approach by drawing on the work of both Martha Nussbaum and Amartya Sen. From here, I explore an emerging literature in the context of social policy and political economy. With respect to offering concrete suggestions which go beyond the ungraspable nature of the hyperobject to include courses of action, I suggest a number of policy areas with which the value base underpinning social work is arguably naturally aligned and through which social work scholarship, but also practice and education, might chart a way forward in the context of setting a new green agenda.

Ecosocial Work and Justice

Climate change, environmental sustainability and ecosocial work have begun to capture the social work imagination across scholarship and practice and an array of contributions made in recent years demonstrates this (see Lucas-Darby, 2011; Philip & Reisch, 2015; Noble, 2016; Harris & Boddy, 2017; Holbrook et al., 2019; Powers et al., 2019; Boetto et al., 2020; Ranta-Tyrkkö & Närhi, 2021 for just some contributions over a number of years). The scope of this literature and, indeed, of the ideas that can be associated with ecosocial work is wide-ranging, and I draw here on a relatively recent special issue on ecosocial work, published in the *Journal of Community Practice*. Within the editorial for this special issue, Rambaree et al. (2019) grapple with what might be meant by the term ecosocial work and with what it can potentially encapsulate in a way that is instructive. In the first instance, they introduce the idea that ecosocial work is not and should not be viewed as a speciality within the profession but rather should be viewed as encapsulating the whole of the profession so that all social work is ecosocial work and thus should seek to acknowledge continuously the interconnectedness of all life. This is a useful idea that has been extended to social policy so that all social policy is seen as ecosocial policy (see Murphy, 2023), yet it remains an idea that is underdeveloped in the context of social work, at least with respect to the overarching issue of climate as noted by Maglajlic and Ioakimidis (2021: 2870) who, reacting in part to the lacklustre outcome of the COP26 summit in Glasgow, suggest that:

> The climate emergency seems to be mobilising the global social work community. Despite the welcome increase in climate change-related literature in social work journals, the fact of the matter is that, in much of the world, environmental justice had not been a central theme in social work practice and scholarship.

While it is true that striving for and toward environmental justice is just one facet of an overall ecological approach and a facet that Rambaree et al. (2019) would undoubtedly point to as being overly anthropocentric, it nevertheless remains a crucial component of an approach encapsulated within the more wide-ranging paradigm of 'ecological justice'. Ecological justice can be distinguished from environmental justice through a focus on interspecies justice, a position that Rambaree et al. (2019) favour. Environmental justice on the other hand is concerned with the unequally distributed consequences of environmental deterioration and warming climates. Yet one is intimately bound up with the other. This is because ecological justice is deeply affected by environmental degradation and warming climates which are now also inextricably linked with social and economic justice. Maglajlic and Ioakimidis (2021) for their part point to the already developing effects of the climate catastrophe from the creation of climate refugees due to wildfires in Greece to the effects of devastating flash-floods in the German regions of Rhineland Palatinate and North Rhine-Westphalia to demonstrate this. Through drawing attention to these examples, Maglajlic and Ioakimidis (2021) illustrate the very real practice of social workers acting and reacting in the face of conditions created by climate change: conditions that require a justice-orientated social work response. The usefulness of Rambaree et al.'s (2019) broader illustration of ecological justice for the purposes of this chapter comes via their acknowledgement that adequate responses require both interpersonal responses at the micro level and structural responses at the macro level, the latter of which they acknowledge in the following terms:

> …ecosocial work is inclusive of structural social work in addressing the social structures causing social problems. It, therefore, pays particular attention to the socio-economic and political structures of a society/community. (Rambaree et al.'s, 2019: 5)

For the purposes of this chapter then, a conceptualisation of ecosocial work which captures the importance of socioeconomic and political structures is essential. The ideas that are advanced in the paragraphs that follow reflect this and are concerned with movement at the level of policy and political economy. An analysis at the level of political economy is also relevant because where social work practices concerned with the natural environment have found a way into 'social-workesque' activities, early research has shown that these have tended to be individualising, (de)moralising and potentially stigmatising, effectively aiming to 'greenwash' the poor by shaming them into sustainable behaviour, thus eschewing the fundamentally structural nature of the climate challenge (Whelan, 2022a see also Malier, 2019 for example). The argument presented here is that the poorly formulated types of practice approaches described above happen in-part because they have no broad-based, humane policy platform on which to devolve and thus become individuating and obsessed with particular types of micro responses. These are practices which emphasise 'personal responsibility' as a matter of course while implicitly suggesting that poor people are responsible somehow for contributing to the climate crisis. Such an approach ignores the fact that an inability to act sustainably with respect to life choices may often have more to do with the socioeconomic reality that people face than with any inherent indifference to environmental issues (Malier, 2019). This logic cuts across social work education also where early research has shown that the inclusion of issues relating to the natural environment as part of the curriculum for social work programmes, while growing, needs further development (see Whelan, 2022a; Dominelli, 2014; Drolet et al., 2015; Teixeira & Krings, 2015; Harris & Boddy, 2017; Holbrook et al., 2019). This may mean that future practitioners receiving their education now are likely to incorporate climate issues into existing frameworks. In other words, though the environment—social and natural—may be rapidly changing, social work education is in danger of remaining stagnant in the face of such changes (Whelan, 2022a). An overarching policy framework, a set of ideas which in turn help us to grasp the hyperobject that is climate change, is necessary then, and it is here that I introduce the idea of a capability approach located within a postgrowth model of economy.

Postgrowth Social Policies: A Capability Approach

Developing ecosocial work knowledge and practices and situating these as a pillar of modern social work education require the social work community to acknowledge and confront the falling growth model of economy in the global north. Alongside being deeply unjust due to the fact that its effects are largely being felt in the global south (Gough, 2022), this model of economy is also unsustainable from the perspective of human and planetary survival (Whelan, 2022a). Rambaree, Powers and Smith (2019: 206) point to neoliberal capitalism which they tie to colonialism and acknowledge this reality in the following terms:

> Today, there is growing evidence that neoliberal capitalism, an outgrowth of colonialism, is among the root causes for climate change and other anthropogenic disturbances that are damaging ecosystem resources.

There is also a growing body of scholarship within social policy that chimes with this assertion in myriad ways (Walker, 2012; Fanning et al., 2020; Bailey, 2020; Koch, 2022; Gough, 2017, 2022; Dukelow, 2022). Fundamentally therefore, this set of global conditions either has to change or lead to further ecological destruction. The challenge of communicating this necessity in the context of a hyperobject is undoubtedly a steep one. Indeed, it is not hard to be minded of Jameson's now famous line, versions of which were given in both *The Seeds of Time* (1994) and in the essay *Future City* (2003) that at the current conjuncture '…it is easier to imagine the end of the world than the end of capitalism' (in Beaumont, 2014: 79). Moreover, in the remainder of this quote by Jameson (1994, 2003), he goes on to note that imagining capitalism has become synonymous with imagining the end of the world. Keeping in mind the idea of the hyperobject, it could be argued and, indeed Morton (2013) draws attention to this, that the capitalist mode of production is itself a hyperobject hence the difficulty that accrues to even imagining ways to confront it. Yet it is equally possible to be minded of Ursula K. Le Guin's (2014) speech at the National Book Awards in New York where she famously noted that:

> We live in capitalism. Its power seems inescapable. So did the divine right of kings. Any human power can be resisted and changed by human beings. (Le Guin, 2014: np)

In this respect and while the effects of global warming will continue to be experienced in the world even after our greatest efforts have been mustered, there is a literature which describes tangible ways of organising economies and societies that can, at the very least, help mitigate against the worst of these effects. As an example of this, the work of Hickel et al. (2021) is incredibly useful. In the first instance, Hickel et al. (2021) recognise that mainstream or 'green-growth' climate mitigation plans devolve on continued economic growth in all countries; a course of action which is then reconciled with desired outcomes based on speculative and as yet unrealised technological changes. They problematise this approach however by highlighting the science which suggests that such an approach will very likely not be adequate in the context of the existential crisis we face. Importantly, as opposed to just highlighting the challenges and thus investing the climate change hyperobject with even more intangibility, Hickel et al. (2021: np) describe a viable alternative, an alternative which undoubtedly aligns with the ethos of ecosocial work, in the form of a postgrowth model of economy which they illustrate in the following terms:

> When it comes to achieving strong social outcomes, what matters is not a continuous increase in commodity production, but access to livelihoods and provisioning. In high-income countries, delivering the latter does not require additional growth; rather, it requires a fairer distribution of income and wealth, and guaranteed access to universal public services. Post-growth scholarship demonstrates that by organizing the economy around principles of equity and sufficiency, societies can deliver high levels of human well-being with significantly less energy and resources than rich countries presently use.

While unpacking the finer details of Hickel et al.'s (2021) analysis is beyond the scope of this chapter, the quote given above is nevertheless offered as a starting point. In particular, the principles of fairer income distribution guaranteed access to universal services, and the prospect of organising economies and societies around equity and sufficiency is inherently connected to a project which seeks to foster a more sustainable, ecosocial welfare model. However, in taking up the work of Hickel et al. (2021), it is important to explore what a more sustainable, ecosocial welfare model—a model that can be efficacious in a post/degrowth world—could look like and consist of. In this respect, theorising a postgrowth capability approach which draws here on the work of Sen (1985, 1987, 1999, 2004, 2009), Drèze and Amartya Sen (2002) and Nussbaum (1997, 2007) arguably provides a potential policy anchor to the postgrowth economic arguments found in Hickel et al.'s (2021) work.

Taken together, the work of Nussbaum (1997, 2007) and Sen (1985, 1987, 1999, 2004, 2009) can, in the first instance, be seen as a project of moving on from an antiquated language of human rights which Nussbaum (1997: 275) notes is a language that is 'not especially informative, despite its uplifting character'. This suggests that a conceptual anchor which moves beyond ideas about rights, which tend to be inclusive/exclusive as opposed to integrated, is needed to fully capture the potentialities for human flourishing (Whelan and Flynn, 2024).

Nussbaum (2007) lists ten specific capabilities consisting of *Life; Bodily Health; Bodily Integrity; Senses, Imagination and Thought; Emotions; Practical Reason; Affiliation; Other species; Play; and Control over One's Environment* as being essential to human flourishing. Exploring the specifics of these ten capabilities goes beyond what is possible here though we may, for the purpose of this chapter, take them at face value. It is, however, interesting to note that Nussbaum (2007) does identify a relationship to other species and the natural environment as a key component of her capability framework. She elaborates on this by noting that part of a capability approach means:

> Being able to live with concern for and in relation to animals, plants, and the world of nature. (Nussbaum, 2007: 23)

In Nussbaum (1997, 2007) then, we find a valuing of human agency along with a role for the collective in ensuring that people can lead rich, meaningful lives and these principles are easily transposed to a postgrowth model of political economy. Sen, for his part, emphasises the 'person centeredness' of a capability approach, and in a publication with Drèze (Drèze & Amartya Sen, 2002: 6), this is described in the following terms:

> The approach … is essentially a 'people-centered' approach, which puts human agency (rather than organizations such as markets or governments) at the centre of the stage. The crucial role of social opportunities is to expand the realm of human agency and freedom, both as an end in itself and as a means of further expansion of freedom.

Sen in much of his work (1985, 1987, 1999, 2004, 2009), like Nussbaum (1997, 2007), describes a key role for human agency and for freedom. However, this freedom should not be mistaken for the type of freedom espoused by neoliberalism's canonical figures for whom the state is always only one overreach away from tyranny (see Hayek, 1960; Friedman, 1962). Rather this is a conceptualisation of freedom which resonates with Marx's (1867 [2013]) general assertion that the realm of freedom begins where and when necessity ends. The ideal of a capability approach couched in a postgrowth model of economy therefore is to remove the necessity for productivism through a reorientated social policy which reappraises and re-values how we work and use time while also interrogating what constitutes work and what activities are valuable and good for humans and for the natural environment (Dukelow, 2022). If we move next to locate the work of Nussbaum (1997, 2007) and Sen (1985, 1987, 1999, 2004, 2009) in an ecosocial policy approach, we find that this is an idea that has already been taken up (see Bonvin & Laruffa, 2018, 2022; Bonvin, 2012; Burchardt & Hick, 2018; Laruffa, 2020 for some examples). Following Bonvin and Laruffa (2022) who have made a useful recent contribution to this debate, the capability framework they propose suggests that we can move beyond the need only to ensure and support material well-being toward a focus on expanding the capabilities of persons in a way that values their freedom to live a valuable and fulfilling life (Whelan, 2022a). In doing so they suggest that:

> …the capability approach allows rejecting the identification of social progress with economic growth, opening instead a democratic debate on the concrete meaning of wellbeing and quality of life and on the best ways to promote them. (Bonvin & Laruffa, 2022: 10)

The crucial point here then is the suggestion that, as an overarching idea, taking a capability approach to the welfare of people and the natural environment aligns with the social work value base and with how social work is globally defined wherein it is suggested that 'social work engages people and structures to address life challenges and enhance wellbeing' (International Federation of Social Worker [IFSW], 2022a [2014]: np). If it is possible through a postgrowth capability approach to uphold this aspect of the global definition while also promoting environmental sustainability and a welfare ethic of care, then such an approach must be given due consideration within the social work community. However, the discussion must also be moved beyond the theoretical to capture discrete and feasible policy programmes so that the question set out in the introduction to this chapter: 'what sustainable social policies should social work educators embrace, advocate for and teach about?' may be given an answer. Therefore, in the final part of this chapter, I elucidate, briefly, the idea of universal basic services (UBS) followed by participation income (PI). Following this, I offer a short summary conclusion.

Universal Basic Services

If a capability approach located in a postgrowth model of economy and drawing inspiration from the work of Nussbaum (1997, 2007) and Sen (1985, 1987, 1999, 2004, 2009) is an overarching idea, it must, as noted, be achieved through discrete social policies. Keeping in focus the central mission of social work and of social work education, I want to suggest that, at its core, the idea of an environmentally sustainable capability approach means organising societies around the following key principles:

- Value people and their inherent humanity.
- Develop social policies which are aligned with an ethic of care and not with capacity for production.
- Promote an idea of well-being that is not reliant on production, consumption or the means to achieve these through the market.
- Encourage people to look after themselves, each other and the natural environment. (Adapted from Whelan (2022a))

If these are some of the ideals of a postgrowth capability approach, the next question becomes, how do we begin to go about realising these ideals through enacting congruent social policies? There are many potential responses here. I have chosen to focus on two: universal basic services (UBS) and participation income (PI) on the basis that both of these ideas, when communicated as potential policy programmes or agendas, naturally align with the mission of social work. UBS and PI can also be conceived of in a way that is environmentally sustainable, meaning that teaching about these approaches in the context of social work education is both congruent with the mission of the profession and with a turn toward ecosocial work as captured at the structural level. Moreover, because UBS and PI present tangible and 'doable' policy options, they are potentially efficacious in the context of confronting the hyperobject that is climate change. I move first to introduce the concept of UBS and I draw primarily on Coote (2022) and Coote and Percy (2020) who have usefully explored the concept of UBS with an eye to environmental sustainability.

Coote (2022) notes the collectivist ideal of pooling of resources at the heart of the ideas about the doing of welfare as part of the post-war settlement. Coote (2022) has also noted the steady decline of these ideas and the practices which underpin them. This is an ideological battleground which continues to characterise welfare debates and which will undoubtedly make Coote's (2022) implicit suggestion that we need, to some degree, to turn the clock back to reconsider the idea of a welfare commons consisting of things like health services, education, adequate housing provision, transport, childcare and income maintenance, a difficult message to successfully impart despite clear necessity. Nevertheless, in setting out a vision of UBS, Coote (2022: 2) suggests the following:

> The normative goal of UBS is to ensure that everyone has access to life's essentials—the things that every individual needs to participate in society and lead a life they value.

Eschewing a liberal and/or conservative view of freedom then, the central message contained in UBS is that we can flourish as a society—now and importantly, in future—if we act together and take collective responsibility to meet needs we all share, and this aptly captures the idea of redistributional justice inherent in a postgrowth model of economy. In practical terms, this will mean more and better collectively provided services which everyone has access to according to need and not the ability to pay, thus evoking the famous Marxian dictate of 'From each according to his ability, to each according to his needs' (Marx, 1875: (3) n.p.) and even evoking the Christian thought of Augustine of Hippo in *City of God* (426/2014) where he suggests that:

> …ought not all Christians, who make common property of their riches with a far nobler purpose, even that (according to what is written in the Acts of the Apostles) they may distribute to each one according to his need, and that no one may say that anything is his own, but that all things may be their common possession…

The fundamental ideas of UBS are clearly not new. However, and crucially, Coote (2022) also presents a version of UBS which is orientated to the future to take account of planetary boundaries so that resources are pooled collectively, yes, but also sustainably. The key, therefore, to UBS as one part of an environmentally sustainable social programme is that by moving away from productivism, they combine universal entitlement, something that is essential to a postgrowth model of economy, with sufficiency, something which is equally essential. Thus, UBS goes toward providing a secure social safety net within planetary boundaries (Coote, 2022). Using the examples of food, housing and childcare and drawing on earlier work, Coote (2022: 9) illustrates this in the following terms:

> Childcare, housing and food occupy different points on the collective-individual spectrum for securing life's essentials. Together, they illustrate the inescapable logic of exercising collective responsibility to meet human needs in ways that are universal and sufficient. Applying the UBS framework to provide in-kind benefits can bring substantial gains in terms of equity, efficiency, solidarity and sustainability.

In answering a question that is central to this chapter, 'what sustainable social policies should social work educators embrace, advocate for and teach about?', UBS as one facet of a set of social policies clearly aligns with social work and with the social work value base and involves the free distribution of social goods such childcare and healthcare. In terms of educating future practitioners, this suggests that teaching about UBS as part of a turn toward (or return to) the structural components of social work education has clear potential. However, it can be argued that UBS only represents one side of a coin, the other side necessarily being based on income. Next, therefore, I take up the idea of participation income (PI). In doing so, I draw attention to an article by Laruffa et al. (2022) and to work by Murphy (2023) who revises Atkinson's (1996) concept of PI with one eye toward the potential for environmental sustainability. I argue that, particularly when coupled with the idea of UBS, PI is an idea that is environmentally sustainable, that works in the context of social work, aligns with the social work value base and could therefore be included in a curriculum that teaches about and for ecosocial work.

Participation Income

In setting out his vision for a participation income (PI), Atkinson (1996: 68) notes the following:

> In my proposal, the basic income would be paid conditional on participation. I should stress at once that this is not limited to labour market participation…it would also include people engaging in approved forms of education or training, caring for young, elderly or disabled dependants or undertaking approved forms of voluntary work, etc.

Participation income then involves cash transfers to all persons on a universal basis in exchange for participation broadly defined. Setting their sights on a 'post-productivist' social landscape which is largely congruent with the postgrowth model of economy favoured here and wherein an ethic of care allows for the primacy of social reproduction over and above economic reproduction, Laruffa et al. (2022) take up Atkinson's (1996) blueprint of a participation income as given above and move toward the idea of a 'green conditional basic income'. Atkinson was clearly a pragmatist as demonstrated by his retention of conditionality. Yet he also deftly broadens what participation might mean. Atkinson (1996) clearly understood that the ideological legacy of Thatcher and Reagan has made all kinds of beneficent welfare proposals a 'hard sell'. Therefore, moving toward a green conditional basic income, as was the case when discussing the potential for the introduction of UBS, requires advocacy and action in order to foster a shift away from means-tested and sanction-based conceptualisations of income support (Whelan, 2021a, b, c). With an eye to environmental sustainability, this shift would need to be toward a policy agenda that could fundamentally alter and:

> …re-shape the focus of social policy on individuals' capability to 'take care of the world', thereby shifting the emphasis from production and employment to social reproduction and environmental reparation. (Laruffa et al., 2022: 2)

The old ideal of collective welfare in the form of a welfare commons must, as with UBS, resurface in the context of PI also. Laruffa et al. (2022) suggest that one outcome of moving toward a green conditional income could be the removal of the stigma of public burden in the context of human welfare, a burden which has garnered much scholarship in recent years in the context of austerity capitalism (see Whelan, 2021a, b, c, 2022b and a special themed section of *Social Policy and Society* on stigma, 2022, edited by Whelan and colleagues). Returning to the idea of a postgrowth capability approach, a shift in thinking about how and why income support as a form of welfare is administered is also necessary in the context of introducing a PI. In keeping with Atkinson's (1996) original blueprint, Laruffa et al. (2022) remain cognizant of the problems associated with universal basic income and so conditionality remains a feature of their conceptualisation of a green conditional basic income. However, such conditions align with an overarching postgrowth capability approach and are based on the concept of coproduction which involves an:

> …enabling approach that relies on appealing to claimants' intrinsic motivations through the quality of participation options available. (Laruffa et al., 2022)

The symmetry, therefore, between the green conditional basic income described above and the central mission of social work is arguably obvious. There is also clear alignment with a wide range of social work values from client self-determination to working in partnership and with a host of social work practice approaches, from strengths-based perspectives to empowerment approaches up to and including community development. Moreover, taken alongside the introduction of UBS and couched within a postgrowth capability approach, such social programmes offer a platform for social work as a whole to coalesce around and for social work education in particular to begin incorporating into curricula. This will take imagination on the part of social work educators but in real terms may mean something as simple as redefining core social policy-based concepts such as need, risk, health and wellness and the family to incorporate an emphasis on sustainable social practices and to critically interrogate how social work integrates sustainable approaches into these ideas in practice. Furthermore, because these are ideas which, despite being pitched at a high structural level, are reasonably concrete, they can empower social work at all levels to confront the hyperobject that is climate change in a tangible way through an advocacy and activism which describes the ecological and social justice the planet and its people need.

Conclusion

This chapter has suggested that climate change can be thought of as a hyperobject which, though it fosters awareness by default, is nevertheless difficult to grasp and therefore even more difficult to confront. Drawing on a conceptualisation of ecosocial work which acknowledges the necessity for engagement at the structural level, I have suggested that a turn to ecosocial policy potentially gives social workers something to grasp, advocate for and, crucially, for social work educators to teach about. The central thesis at the heart of this chapter therefore suggests that in order to permit a degree of focus, in order to place climate justice and the environment squarely on the social work agenda in a real, coherent and meaningful way across practice, education, scholarship and activism, aligning social work with sustainable social policies which naturally speak to the social work value base can provide a starting point. With respect to what these policies might look like, in the first instance I have suggested that an overarching framework characterised by a postgrowth capability approach aligns with the central mission of social work and with ecosocial work. Following on from this, I have suggested that universal basic services (UBS) and a pragmatic participation income (PI) have much to offer social work and, again, fit naturally with the broad mission of the profession while helping to progress a green agenda. Importantly, an endorsement of sustainable or ecologically sound social policies also provides a focal point for a type of social work education which interrogates the political and moral economy of the global north and offers alternatives. In this respect, advocating for and teaching about a post-growth capability approach which encompasses environmentally sustainable social programmes arguably address what has previously been a theoretical, conceptual and practical lacuna in the context of structural social work. In many ways and because the territory remains reasonably untrodden, this chapter represents a tentative offering designed to foster engagement and debate. While being far from the last word, it has hopefully offered some food for thought.

Bibliography

Atkinson, A. B. (1996). The case for a participation income. *The Political Quarterly, 67*(1), 67–70.

Augustine, A. (426/2014). *City of god*. Retrieved November 30th 2020, from https://www.gutenberg.org/files/45304/45304-h/45304-h.htm

Bailey, D. (2020). Re-thinking the fiscal and monetary political economy of the green state. *New Political Economy, 25*(1), 5–17.

Beaumont, M. (2014). Imagining the end times: Ideology, the contemporary disaster movie and contagion. In M. Flisfeder & L. P. Willis (Eds.), *Žižek and media studies* (1st ed., pp. 79–89). Palgrave Macmillan.

Boetto, H., Bowles, W., Närhi, K., & Powers, M. (2020). Raising awareness of transformative eco-social work: Participatory action research with Australian practitioners. *International Journal of Social Welfare, 29*, 300–309.

Bonvin, J.-M. (2012). Individual working lives and collective action. An introduction to capability for work and capability for voice. *Transfer: European Review of Labour and Research, 18*(1), 9–18.

Bonvin, J.-M., & Laruffa, F. (2018). Human beings as receivers, doers and judges. The anthropological foundations of sustainable public action in the capability approach. *Community, Work and Family, 21*(5), 502–518.

Bonvin, J.-M., & Laruffa, F. (2022). Towards a capability-oriented eco-social policy: Elements of a normative framework. *Social Policy and Society, 21*(3), 484–495. https://doi.org/10.1017/S1474746421000798

Burchardt, T., & Hick, R. (2018). Inequality, advantage and the capability approach. *Journal of Human Development and Capabilities, 19*(1), 38–52.

Coote, A. (2022). Towards a sustainable welfare state: The role of universal basic services. *Social Policy and Society, 21*(3), 473–483. https://doi.org/10.1017/S1474746421000385

Coote, A., & Percy, A. (2020). *The case for universal basic services*. Polity Press.

Dominelli, L. (2014). Promoting environmental justice through green social work practice: A key challenge for practitioners and educators. *International Social Work, 57*(4), 338–345.

Drèze, J., & Amartya Sen, A. (2002). *India: Development and participation*. Oxford University Press.

Drolet, J., Wu, H., Taylor, M., & Dennehy, A. (2015). Social work and sustainable social development: Teaching and learning strategies for 'green social work' curriculum. *Social Work Education, 34*(5), 528–543.

Dukelow, F. (2022). What role for activation in eco-social policy? *Social Policy and Society, 21*(3), 496–507. https://doi.org/10.1017/S1474746421000889

Dwyer, P. (2016). Citizenship, conduct and conditionality: Sanction and support in the 21st century UK welfare state. In M. Fenger, J. Hudson, & C. Needham (Eds.), *Social policy review* (Vol. 28, pp. 41–62). Policy Press.

Etzioni, A. (1997). *The new golden rule*. Profile Books.

European Commission. (2021). *Communication from the commission to the European parliament, the council, the European economic and social committee and the committee of the regions: Forging a climate-resilient Europe – The new EU Strategy on Adaptation to Climate Change*. Retrieved May 28th 2022, from https://eur-lex.europa.eu/legal-content/EN/TXT/?uri=COM%3A2021%3A82%3AFIN

European Environmental Agency. (2021). *Imagining sustainable futures for Europe*. Retrieved May 28th 2022, from https://www.eea.europa.eu/publications/scenarios-for-a-sustainable-europe-2050

Fanning, A. L., O'Neill, D. W., & Büchs, M. (2020). Provisioning systems for a good life within planetary boundaries. *Global Environmental Change, 64*, 102135. https://www.sciencedirect.com/science/article/abs/pii/S0959378020307184?via%3Dihub

Friedman, M. (1962). *Capitalism and freedom*. Chicago University Press.

Gough, I. (2017). *Heat, greed and human need: Climate change, capitalism and sustainable well-being*. Edward Elgar Publishing.

Gough, I. (2022). Two scenarios for sustainable welfare: A framework for an eco-social contract. *Social Policy and Society, 21*(3), 460–472. https://www.cambridge.org/core/journals/social-policy-and-society/article/two-scenarios-for-sustainable-welfare-a-framework-for-an-ecosocial-contract/BEB3596AE2F312375B564A8B9BFC0EEC

Harris, C., & Boddy, J. (2017). The natural environment in social work education: A content analysis of Australian social work courses. *Australian Social Work, 70*(3), 337–349.

Hayek, F. A. (1960). *The constitution of liberty*. Routledge.

Hick, R., & Murphy, M. P. (2021). Common shock, different paths? Comparing social policy responses to COVID-19 in the UK and Ireland. *Social Policy and Administration, 55*, 312–325.

Hickel, J., Brockway, P., Kallis, G., et al. (2021). Urgent need for post-growth climate mitigation scenarios. *Nature Energy, 6*, 766–768.

Holbrook, A. M., Akbar, G. L., & Eastwood, J. (2019). Meeting the challenge of human-induced climate change: Reshaping social work education. *Social Work Education, 38*, 955–967.

Intergovernmental Panel on Climate Change. (2022). *Climate change 2022: Mitigation of climate change*. Retrieved May 28th 2022, from https://www.ipcc.ch/report/sixth-assessment-report-working-group-3/

International Federation of Social Workers [IFSW]. (2022a [2014]). *Global definition of social work*. International Federation of Social Workers. Retrieved May 28th 2022, from https://www.ifsw.org/what-is-social-work/global-defini tion-of-social-work/

International Federation of Social Workers [IFSW]. (2022b). *Climate Justice programme*. Retrieved May 28th 2022, from https://www.ifsw.org/social-work-action/climate-justice-program/introduction/

Jameson, F. (1994). *The seeds of time*. Columbia University Press.

Jameson, F. (2003). Future city. *New Left Review*. Retrieved May 28th 2022, from https://newleftreview.org/issues/ii21/articles/fredric-jameson-future-city

Koch, M. (2022). Social policy without growth: Moving towards sustainable welfare states. *Social Policy and Society, 21*(3), 447–459. https://www.cambridge.org/core/journals/social-policy-and-society/article/social-policy-without-growth-moving-towards-sustainable-welfare-states/635A7EC691A0A16E40B888AFF39FB56B

Laruffa, F. (2020). What is a capability-enhancing social policy? Individual autonomy, democratic citizenship and the insufficiency of the employment-focused paradigm. *Journal of Human Development and Capabilities, 21*(1), 1–16.

Laruffa, F., McGann, M., & Murphy, M. (2022). Enabling participation income for an eco-social state. *Social Policy and Society, 21*(3), 508–519. https://doi.org/10.1017/S1474746421000750
Le Guin, U. K. (2014). *Ursula K Le Guin's speech at National Book Awards: 'Books aren't just commodities'*. Retrieved May 28th 2022, from https://outlook.office.com/mail/inbox/id/AAQkADg1OGYxMzUzLTlmYzgtNDVjZC1hZGY2LTdkZDAyMTgxNTEzZgAQAGq9rxAltH1Ktag7Od7vmyM%3D
Lucas-Darby, E. T. (2011). The new color is green: Social work practice and service-learning. *Advances in Social Work, 12*, 113–125.
Lynch, D., Lathouras, A., & Forde, C. (2021). Community development and social work teaching and learning in a time of global interruption. *Community Development Journal, 56*(4), 566–586.
Maglajlic, A. R., & Ioakimidis, V. (2021). Social work and the quest for environment justice; a call to action. *The British Journal of Social Work, 51*(8), 2869–2871.
Malier, H. (2019). Greening the poor: The trap of moralization. *British Journal of Sociology, 70*, 1661–1680.
Marx, K. (1867 [2013]). *Das Kapital*. Wordsworth.
Marx, K. (1875). *Critique of the Gotha Programme*. Retrieved May 28th 2022, from https://www.marxists.org/archive/marx/works/1875/gotha/
Mead, L. (1986). *Beyond entitlement*. Free Press.
Mead, L. (1992). *The new politics of poverty*. Basic Books.
Morton, T. (2013). *Philosophy and ecology after the end of the world*. University of Minnesota Press.
Murphy, M. P. (2023). *Creating and ecosocial welfare future*. Policy Press.
Murray, C. (1990). *The emerging British underclass*. Institute of Economic Affairs.
Murray, C. (1994). *Underclass: The crisis deepens*. Institute of Economic Affairs.
Noble, C. (2016). Green social work – The next frontier for action. *Social Alternatives, 35*(4), 14–19.
Nussbaum, M. (1997). Capabilities and human rights. *Fordham Law Review, 66*, 273.
Nussbaum, M. (2007). Human rights and human capabilities: Twentieth anniversary reflections. *Harvard Human Rights Journal, 20*, 21–24.
Philip, D., & Reisch, M. (2015). Rethinking social work's interpretation of 'environmental justice': From local to global. *Social Work Education, 34*(5), 471–483.
Powers, M., Schmitz, C., & Beckwith-Moritz, M. (2019). Preparing social workers for ecosocial work practice and community building. *Journal of Community Practice, 27*(3–4), 446–459.
Rambaree, K., Powers, M., & Smith, R. J. (2019). Ecosocial work and social change in community practice. *Journal of Community Practice, 27*(3–4), 205–212.
Ranta-Tyrkkö, S., & Närhi, K. (2021). Striving to strengthen the ecosocial framework in social work in Finland. *Community Development Journal, 56*(4), 608–625.
Selbourne, D. (1994). *The principle of duty*. Sinclair Stevenson.
Sen, A. (1985). Well-being, agency and freedom: The Dewey lectures 1984. *The Journal of Philosophy, 82*(4), 169–221.
Sen, A. (1987). The standard of living. The Tanner lectures on human values 1985. In G. Hawthorn (Ed.), *The standard of living* (1st ed., pp. 1–38). Cambridge University Press.
Sen, A. (1999). *Development as freedom*. Knopf.
Sen, A. (2004). *Inequality reexamined*. Oxford University Press.
Sen, A. (2009). *The idea of justice*. Allen Lane.
Sen, A. (2013). The ends and means of sustainability. *Journal of Human Development and Capabilities, 14*(1), 6–20.
Soss, J., Fording, R. C., & Schram, S. F. (2011). *Disciplining the poor: Neoliberal paternalism and the persistent power of race*. University of Chicago Press.
Teixeira, S., & Krings, A. (2015). Sustainable social work: An environmental justice framework for social work education. *Social Work Education, 34*(5), 513–527.
United Nations. (2021). *Conference of parties 26 report*. Retrieved May 28th 2022, from https://unfccc.int/documents/460955
Walker, G. (2012). *Environmental justice: Concepts, evidence and politics*. Routledge.

Whelan, J. (2021a). Specters of Goffman: Impression management in the Irish welfare space. *Journal of Applied Social Science, 15*(1), 47–65. https://doi.org/10.1177/1936724420983578

Whelan, J. (2021b). We have our dignity, yeah? Scrutiny under suspicion: Experiences of welfare conditionality in the Irish social protection system. *Social Policy & Administration, 55*, 34–50. https://doi.org/10.1111/spol.12610

Whelan, J. (2021c). Work and thrive or claim and skive: Experiencing the 'toxic symbiosis' of worklessness and welfare recipiency in Ireland. *Irish Journal of Sociology, 29*(1), 3–31. https://doi.org/10.1177/0791603520957203

Whelan, J. (2022a). From dissent to authoritarianism: What role for social work in confronting the climate crisis? *Aotearoa New Zealand Social Work, 34*(3), 21–33. https://anzswjournal.nz/anzsw/article/view/927

Whelan, J. (2022b). *Hidden voices: Lived experiences in the Irish welfare space*. Policy Press.

Whelan, J., & Flynn, S. (2024). Who's right? What rights? How? Rights debates in Irish social work: a call for nuance. *Critical and Radical Social Work, 12*(2), 274–280.

Chapter 10: Rethinking Transnational Social Work Pedagogy for Climate Change, Migration, and Crisis Preparedness

Janet Carter Anand, Anish K. R., Manish K. Jha, Bipin Jojo, and Satu Ranta-Tyrkkö

Introduction

Human societies are currently troubled with multiple interrelated crises on a global scale. The climate crisis alone, effectively warming the Earth, is an epochal threat with irreversible ecological, economic, and social consequences (UN Climate Action, n.d.), the risks getting greater and adaptation and mitigation more taxing, the warmer the planet gets (Pelttari, 2021). Already, at the time of writing this chapter in 2023, news of unprecedentedly high temperatures around the globe has been daily, as have been estimates on the consequences for human and nonhuman health and ecosystems, including forest fires, drying up of aquatic systems, and unusual rain patterns elsewhere.

The climate system is integral to the Earth's biosphere processes and thus intertwines with other detrimental planetary processes that are caused by human activity. Steffen et al. (2015) identify these as stratospheric ozone depletion, biodiversity loss and extinctions, chemical pollution, acidification of oceans, consumptive freshwater use and consequent changes in the global hydrological cycle, land system change, atmospheric aerosol loading, and nitrogen and phosphorous flows to the

J. C. Anand (✉)
Department of Social Sciences and Business Studies, University of Eastern Finland, Kuopio, Finland
e-mail: janet.anand@uef.fi

Anish K. R
Rajagiri College of Social Sciences, Kochi, Kerala, India

M. K. Jha · B. Jojo
Tata Institute of Social Sciences (TISS), Mumbai, Maharashtra, India

S. Ranta-Tyrkkö
Department of Social Sciences and Philosophy, University of Jyväskylä, Jyväskylä, Finland

C. Forde et al. (eds.), *Teaching and Learning in Ecosocial Work*,
https://doi.org/10.1007/978-3-031-58708-5_14

biosphere and oceans. The further these planetary boundaries are stretched, the greater the risk of “abrupt and irreversible changes” (ibid.; Stockholm Resilience Centre, n.d.). Exceeding the planetary boundaries endangers and aggravates both human and nonhuman well-being and speaks to the very foundations of ecosocial work. In the present moment in social work, other crises and emergencies compete constantly for the attention and resources available.

In this chapter we contribute to the development of ecosocial work with a focus on climate change and forced migration as interconnected issues that have significant social, economic, and environmental implications for minority and indigenous groups in the authors’ own countries India and Finland and globally. Our unique starting point was to focus on the development of crisis knowledge and preparedness in social work (Redondo-Sama et al., 2020). We further believe that ongoing crises are symptomatic of a broader systemic or civilizational crisis, of which the development of capitalism into its current neoliberal stage (Moore, 2016) has been a central driver. Climate crisis involves many consequences for social work, such as a push factor for both internal and cross-border migrations and specifically the teaching and learning of these themes in the context of an intensive course in a transnational (Indo-Finnish) setting.

Climate Change and Curriculum Reform

Dedicated curricula on climate change and crisis preparedness remain peripheral to many European university-based social work education programs (Ellis, 2020). European social work education often prioritizes current agency requirements and national social policy expectations and political priorities but may lack global and futuristic perspectives on emerging roles and possibilities for the profession, as highlighted by Lena Dominelli (2012) and Satu Ranta-Tyrkkö (2017). This is even though the IFSW and IASSW global standards in social work education (Sewpaul & Jones, 2004), together with the global agenda (IFSW/ICSW, 2012), emphasize the future environmental and humanitarian priorities in social work education. Gradually, the necessity to address the climate and other systemic crises such as forced migration and to develop crisis preparedness and an overall relevant curriculum that structurally connects social work with the broader green or ecosocial sustainability transition is being highlighted in the discussions on environmental, ecological, green, and ecosocial social work (Matthies et al., 2001).

Multiple studies have proposed frameworks for academic use to strategically embed environmental content into social work curricula (Besthorn, 2003; Jones, 2013; Melekis & Woodhouse, 2015). Powers et al. (2019) identify how social work educators are often confronted by a conflict between the wish to transform the curriculum and embrace an ecosocial lens and the centrality of the ecological environment on the one hand and the realities of local demands and practical constraints on the social work curriculum only allowing the integration of ecosocial work into the existing program structures on the other. Both transformative and integrative methods are legitimate models for promoting curriculum change. The example given in this chapter represents how European social work academics adopted an integration of ecosocial work knowledge for the modification of the social work curriculum. The illustration also highlights how the Indian context provides more opportunities to embrace the infusion of ecosocial work comprehensively across the curriculum.

Many obstacles stand in the way of much needed changes to the European curriculum. Often university level professional programs are overpopulated with (as such justified) courses on established skills and competences, which makes introducing new courses and content into degree programs limited to a certain number of credits challenging. Together with normal organizational inertia (Filho et al., 2021), there may be resistance to change, such as to including politically nuanced courses that address structural aspects; these may be seen as sacrificing the received wisdom coming from the conventional courses. In such situations the political, economic, and social aspects of the climate crisis are often overlooked due to the reluctance of the education fraternity to provide them with the center stage in contemporary discourses. Instead, climate change is often subsumed under the notions of sustainability education or environmental literacy (Eilam, 2022), with limited scope to engage with crisis and preparedness in detail.

For much of the past, the profession's understanding of person-in-environment has been limited to examining only the social, economic, and built environments (Besthorn, 2003; Matthies et al., 2001). Yet, the connections between human health and ecosystem health, as well as social and environmental injustice and environmental ruin, are indisputable (Besthorn, 2003; Muldoon, 2006). Ironically, social work that "attempts to improve social conditions may be lost if society itself lacks clean air, drinkable water and adequate food" (ibid., p. 2), all of which are adversely impacted due to climate change-induced exposure pathways (WHO, 2021).

The intertwined existence of the global environmental crisis and the social crisis demands close attention and requires modifications in the current social work curriculum. The curriculum needs to provide basic understanding about the processes forming key environmental challenges from major pollutants and carbon emission sources to the political-economic structures of unhindered resource extraction pushed by the dominant development paradigm and discourse. This chapter explores an international team of social work teachers' approach to the development of a transnational curriculum exploring the issue of climate change and migration from a crisis preparedness perspective, offering students renewed professional priorities and opportunities.

Description of the Indo-Finnish Curriculum

A Winter School on Climate Change, Migration and Crisis Preparedness in Social Work was taught in January 2023 in India in the megacity of Mumbai, with an extension in Kochi, Southern India. The intensive course was part of a broader Team

Finland Knowledge (TFK, 2023) higher education mobility project "Sharing Best Practices in Crisis Preparedness in Social Work," funded by the Finnish National Agency for Education. The project partners were the University of Jyväskylä (JYU), University of Eastern Finland (UEF), Tata Institute of Social Sciences (TISS), and Rajagiri College of Social Sciences (RCSS). The partnership was founded on the complementary expertise of the four institutions in the field of social work and the long-term cooperation of the key academics in each university and thus mutual confidence and trust, which is crucial for sustaining projects like this.

The themes of climate change, migration, and crisis preparedness profiled high on the research and educational agenda of each of the participating institutions. Both JYU and UEF share an interest in global social work and crisis preparedness and want to improve the social work curriculum with jointly developed new teaching content with TISS and RCSS, institutions known for innovative and socially responsive curriculum. The main motivation for the project thus was complementary strengths building in this area, shared partnerships in India, and mutual interest in developing social work education so that it meets current global challenges. The project aimed to develop mutually relevant curriculum, pedagogy, and research activities between the member institutions in social work and social policy, including new online teaching content.

The project team was committed to the principles of reciprocity and mutuality. The grant support enabled physical student and staff mobility between the Finnish and the Indian institutions. Several rounds of online zoom meetings helped in planning and organizing for webinars, development of curriculum, identification of resource persons, methodology of teaching, field visits, and community interface. A popular messaging platform, WhatsApp, was actively used for group communication between faculty members, students, and other scholars from Finland and India. This was followed by visits from faculty, research scholars, and students to engage in teaching-learning in summer and winter schools at JYU and TISS-RCSS, respectively. The outcomes included jointly planning for future online teaching modules and webinars for Master- and PhD-level students in social work in both Finland and India, a Summer (2022) and Winter School (2023), internships for both Indian and Finnish social work students and co-created research (joint publications) and recognized themes for future research projects. In the following sections, we elaborate how the winter school module, its curriculum and pedagogy, engaged with ecosocial work, humanitarian challenges, and complex realities of migration.

The transnational winter school format, involving the four institutions and their expertise along with tailored mobility activities, served to develop and deliver social work curricula on the current global challenges of climate change, migration, and crisis preparedness and learn together from the multifaceted topic both as teachers and students. The winter school took place face-to-face for 1 week in January 2023, consisting of lectures, PhD student panel discussion, field visits in Mumbai and Kochi, and a group assignment. In the group assignment, students worked in four- to five-person teams with at least one participant from each participating university. Their task was to identify the responsibilities of social workers in mitigating the effects of climate change and forced migration and provide examples of the roles social workers play and the impediments that they may encounter in promoting crisis preparedness in their day-to-day work. Moreover, they were asked to (1) outline the values, knowledge, and skills that social workers will be required to carry out this role in the future and (2) reflect upon how the experience of transnational exchange has influenced their understanding of social work practice and what changes can/should be made to the social work curriculum. The form of presentation (10–20 min) could be either a short video of 10–20 min, or a short drama, song, game etc., but not an ordinary seminar presentation. Each group was encouraged to interpret the questions and tasks from a unique perspective. Active participation during this week granted the students 2 ECTS credits and strengthened their understanding of the theme.

Beforehand, students were provided access to the online learning Moodle platform with course information and a selection of scientific articles addressing the course themes from various angles. The broad themes for classroom engagement included climate crisis and migration, natural disasters and humanitarian response, and implications for indigenous communities. The course also engaged with the Anthropocene and history of crisis, experiences of trauma and processes of recovery, resilience of communities, and specific case studies of the situations of refugees, migrants, and stateless people. The inputs, discussion, and classroom participation transcended to the field realities through visits to affected families and the agencies working on climate change, community, and livelihood issues. An important aspect of the teaching-learning was active, enthusiastic, and engaged participation of scholars and students. Their shared knowledge and lived experiences brought forth diverse and enriching dimensions of the topics at stake.

Crisis Preparedness and Pedagogical Principles

The approach to teaching adopted a pedagogical framework designed to provide a structured and systematic approach or model to crisis preparedness that guides the planning, development, and implementation of educational practices and strategies. Course presentations were delivered from multidisciplinary perspectives including science, economics, and social sciences and from the interconnectedness of Sustainable Development Goals (UN DESA, 2022). Inclusion of both local and global perspectives on issues encouraged students to understand the global impacts of climate change and its effect on local communities and indigenous cultures. Students were encouraged to research and understand the specific climate issues in their own region. Among others, strategies for mitigation and adaptation include renewable energy, carbon reduction, sustainable agriculture, and effective social and economic policies to combat climate change. Ethical aspects of climate change, such as its disproportionate impact on marginalized communities, were stressed. The transnational nature of the course helped to engage students in practical activities and community-based projects related to climate change.

The inclusion of guest speakers, experts in the field, and field trips facilitated this experience together with the relatively informal nature of the teacher-student engagement. Students were encouraged to think critically about climate policy and their position to advocate for structural change at local, national, and international policy levels. Overall, solidarity was a concept enhanced by this cross-border program, which fostered a sense of unity, cooperation, and support among social workers in times of need, hardship, or crisis. This framework not only aimed to educate students about the science of climate change but also empowered them to act and be responsible citizens.

Reimagining Climate Change and Crisis Preparedness

The dual role of crisis preparedness and responding to urgent needs is an essential aspect of European social policy and by default social work practice (Redondo-Sama et al., 2020), and as such it provides a viable framework for the teaching and learning of global crises. Crisis preparedness does not detract from the reality that social workers are mainly engaged in frontline crisis management, and it offers broader opportunities for workers to play crucial roles in supporting individuals and communities in facilitating their recovery, resilience, and prevention. Although social work programs may include a consideration of environmentalism as an ideology or as a social movement agenda, there are few examples of courses devoted specifically to linking the social and ecological in theory and practice (Jones, 2012). Some dimensions of crisis preparedness are routinely taught in the social work curriculum including risk assessment and planning, building resilience, consciousness-raising, crisis intervention and support, collaboration and coordination, advocacy, and policy development (Sutton & Tierney, 2006; Howard et al., 2018; Rapeli, 2018). In countries like India, disaster preparedness and humanitarian response sometimes find space in teaching-learning within social work programs. However, structuring and analyzing social work specifically through the lens of crisis preparedness has been relatively random and thematically scattered, focus being more commonly at specific dimensions and themes of crisis work as part of service processes (Rapeli, 2018) than crisis preparedness as an overall framework. However, academic literature on the scopes of crisis preparedness and responses has increased rapidly, and these cover a vast array of issues from natural disasters, such as tsunamis, earthquakes, storms, forest fires, droughts, and floods (Jha, 2015), to violent conflicts, refugee situations (Jha & Pande, 2022), ecological, and health challenges. Among others, they point to the necessity of paying attention to structural, class, and gender aspects of disaster response (Jha, 2015), as well as to reimagine (Panagiotaros et al., 2022) and reconceptualize social work in times of crisis.

Especially the COVID-19 pandemic has brought along a surge of scientific articles, often in special issues and edited volumes (Harrikari et al., 2023), that analyze the pandemic and the experience and insights gained, addressing both the systemic and epidemic features of the pandemic but also loss and grief (Johns et al., 2020;

Miller & Lee, 2020), biopolitical nationalism (Kloet et al., 2020), importance of keeping vigilant and hopeful on human rights, and staying resilient, compassionate, and ethical (ISW editorials vol 63, journals 4 and 5 respectively), and justice in the context of the pandemic. Also the climate crisis and the need to engage with its mitigation and adaptation efforts in social work have been addressed (Alston, 2015) from various perspectives, including, for example, calls to engage with deep ecology and ecofeminism in social work practice (Klemmer & McNamara, 2020), make elderly care services responsive to extreme weather events such as heat waves (Teixeira & Krings, 2015; Rapeli & Mussalo-Rauhamaa, 2022) and pay attention to mental health impacts of climate change (Hayes et al., 2018). On several aspects, these contributions either overlap or downright position themselves as part of the broader sectors crossing endeavor of ecological, or alternatively ecosocial, environmental, or green sustainability transition, which also this book represents.

To avoid merely reactionary social work responses to the climate and other crises, crisis preparedness requires both theoretical and practical exercises. To start with, it is crucial to understand the complexity and interrelatedness of the planetary problems at stake. When social work seeks to develop various constructive approaches, along with hands-on practical work, it is important to pay attention to the concepts and discourses used in different economic, political, and cultural contexts. As Paterson (2021) notes, the strategies used to frame and construct problems like the climate crisis influence how and what kind of responses are imagined in the first place and thereafter justified and produced. The crises of our time require a rethink not only of preparedness but how to address and teach them. Social workers require a new and refined set of knowledge and skills to carry out their own work and promote the responsible and balanced use of natural resources, the protection and conservation of ecosystems, and the promotion of social and economic development that is environmentally friendly, socially equitable, and economically viable (Teixeira & Krings, 2015).

The climate crisis together with its social, ecological, and economic ramifications not only puts the achievements and goals of social work, such as dignity, worth, and caring and respectful treatment of everyone, in danger. It also aggravates poverty, inequality, and vulnerability for large sections of society. The evolving crises cause polarization in and between societies (Blühdorn & Welsh, 2007). Therefore, social issues associated with climate change are critical for the future relevance of the social work profession given global trends, such as large-scale forced migration anticipated to take place during the coming decades (Breen et al., 2023).

The refiguring of crisis and crisis preparedness in social work therefore presents both challenges and an opportunity for social work education. Arguably transformative learning theory (Mezirow, 2018) provides an important way forward for social work education in the face of the growing awareness of the need for change. In its emphasis on critical reflection, critical self-reflection, and comparative analysis, it suggests an approach that links the personal with the political—pursuing the goals of critical theory through ideology critique but joining these to individual experience. It also emphasizes the centrality of dialogue and rational discourse through a commitment to engage openly with others in assessing the validity of alternative pathways, including, in this case, an expanded ecological consciousness. More important for social work, it also manifests an action orientation, seeing learning as truly occurring when it translates into change both at the level of individual behavior and through political action for social transformation.

Transnational Teaching and Glocalized Curriculum Development

Currently transnational pedagogy opportunities are supported by higher education policies promoting internationalization, including advances in e-learning (Healy & Thomas, 2020; Schrooten et al., 2023). The advantages of international exchanges include, among others, chances for experiential and holistic (also extracurricular) learning in another culture and with and from people with diverse backgrounds. Yet the critique of transnational exchange programs relates to the very causes underlying the climate crisis, for the (air) travel required for the exchanges to take place is a source of climate emissions. Overall, international exchanges can be accused of being just one of the ways the world's privileged exploit of common climatic and other resources at the long-term expense of others. While we acknowledge these kinds of complex structural inequalities and biases present in education globally, we see them symptomatic of the systemic crises in which we authors are also embedded, having no simple solutions.

Educating social work students for global practice involves an understanding of both local and global perspectives. The concept of "glocal" describes the interplay between global and local dynamics that social workers commonly experience in their day-to-day practice (Lyons & May-Chahal, 2017; Harrikari & Rauhala, 2018; Livholts, 2023) essential to global mindedness (Das & Anand, 2023). Transnational exchange practices provide students with unique opportunities to develop their own understanding of the issue of climate change, mass migration, and different national contexts. Transnational teaching is part of the mainstream enterprise of most universities worldwide and can take various forms of student and staff exchange and mobility (Anand et al., 2014). Teaching across borders involves new combinations of environments, often changing roles and relationships between educators and learners (Kim & Slapac, 2015). It encourages the participants to engage with diverse socio-economic and cultural realities and gain knowledge and skills in the process.

Transnational exchanges are often viewed as involving empowering pedagogy (O'Dowd, 2020), referring to educational practices through which teachers and students create an environment of collaborative empowerment and critical inquiry of cultural identities of one's own and others. Critical pedagogy interrogates differences in terms of how problems are socially constructed based on the prevailing power dynamics (McLaren, 1995) such as diverse positionalities, identities, and subjectivities of different groups of people in relation to a larger social, cultural, and historical context and questions. Critical pedagogy developed through transnational collaboration relied on understanding how "curricular focus on socio political topics, such as climate change," has been constructed and resisted. It facilitates the capacity to work toward changing the mindset and approach toward issues and intervention. The consciousness that "critical pedagogy cannot be created solely by the teacher, but instead must be co-intentional (Freire, 2015; Taylor & Hikida, 2020), or collaboratively constructed" was the guiding principle and approach.

Student Evaluation of Learning and Teaching

The winter school program was evaluated using data from the final day formal debrief session which was recorded and transcribed involving both students and academy ($N = 49$) and a post-course web portal survey student ($N = 16$) in April 2023. The results of the student survey confirmed that the winter school program at Tata Institute of Social Sciences (TISS) and Rajagiri College of Social Sciences (RCSS) was a memorable experience, featuring presentations, field trips, and discussions on climate change, migration, and disaster preparedness. Students gained valuable professional knowledge and made new friends. The sessions on climate justice and adaptive innovation, challenges of interstate migration, and visiting the beneficiaries of the Towards Advocacy Networking and Developmental Action (TANDA), a field action project of TISS, were memorable. TANDA works toward creating self-reliant Nomadic Tribes (NT) and Denotified Tribes (DNTs) and other marginalized communities, ensuring the realization of their constitutional and entitlement rights as citizens of India. The intensive academic program in Kochi provided an opportunity for intercultural learning and understanding the intercontinental climate crisis.

Student Learning on Climate Change The course enhanced students' understanding of the theory and diverse perspectives of climate change and the effects of inequality on ingenuous and poor people. The program helped demonstrate examples and compare the effects of climate change in India and Finland in ways that students had not seen in their previous studies or day-to-day practice. Most importantly it highlighted the possible roles of social work in addressing the issues. The presentations illuminated the importance of SDGs in today's world and the critical relevance of local knowledge in an effective response and basis of future studies and the implications for indigenous and minority groups and communities:

> The sessions I attended helped me understand how climate change is a complex phenomenon that requires a complex solution. It is also a global concern that necessitates cooperation and collaboration among nations across the world. At the same time, it impacts different communities differently because the larger narratives of climate change tend to exclude the poor and the marginalized as well as indigenous knowledge systems. The solution to the climate crisis needs to be informed by the voices that have been excluded from the conversations until recently. (Finnish Student)

> The Finnish context of Climate Change was entirely new for me and their coping mechanism which is state centric approach was quite new learning. (Indian student)

Student Learning on Forced Migration and Crisis Students were provided with new insights as to how people are affected by migration and what makes them become migrants. The field visit at Mumbai slum communities was particularly valued by students as it increased their understanding as to why people migrate, the factors precipitating forced migration, and how the multiple stages and process of migration affect all aspects of people's lives and not just physical surroundings. The discussion and the presentations helped the Finnish students to broaden their views on the loss of ethnicity and lack of inclusivity experienced by many migrants, how internal work-based migration in India functions, and how it affects people. The course thus exposed students to new realities and by doing so challenged their assumptions and stereotypes:

> I knew people migrated and one reason is climate change, but I didn't know that migration happens mostly inside the country, and my knowledge about the migration was limited very much to the European migration and war refugees. (Finnish Student)

> In summary, the course helped the students to build both a global and localized understanding of migration. The program provided an opportunity to learn about the various factors that influenced migration patterns, including climate change, ranging from the underlying social, economic, and political push and pull factors of guest workers in Kerala to the refugee migrant crisis in Finland.

Student Learning on Crisis Preparedness The program promoted an appreciation of the concept of crisis preparedness including the limitations, possibilities, and examples of capacity building for preparedness undertaken by social workers in both countries. Finnish students suggested that crisis preparedness is a decidedly underdeveloped topic in Finnish social work education and appreciated the new insights provided by exposure to the Indian experience:

> Their (Indian Social Workers) appreciation gave me a deeper and wider understanding of crisis preparedness: it is not just waiting for a possible crisis but creating a society that is stable and equal enough to handle unexpected situations. Some students want more information on the intersectionality of crisis work and about the women's position and religious violence in India and how THAT is a crisis. (Finnish Student)

The discussions, presentations, and field visit enabled students to understand the application of models of crisis preparedness and the use of culturally sensitive practices in dealing with crises, also reinforcing the possibilities of mutual and reciprocal learning from colleagues from the global north and south. The general conclusion was that global south countries like India are much more vulnerable to varied environmental crises. Finnish students felt that the exchange gave them new ways to think about and work differently in response to the crisis:

> I also became aware of the scope for social work in crisis preparedness for social workers. I learnt about the nature of it in a global context and different roles social workers could play in it. (Finnish Student)

Linkages between climate change and crisis preparedness were illustrated in discussions and presented to students with comparative differences in disaster management in India and Finland highlighted. Both Finnish and Indian students felt that they gained new knowledge on risk preparedness in social work including the issue of intersectionality, such as aging, impact on social and economic institutions, and delivery of services to those most vulnerable. Overall, the social work students learnt about the importance of crisis preparedness and how to deliver adequate crisis services to the vulnerable population.

Student Evaluation of Teaching Practices The selection of topics and guest speaker's resource persons was greatly appreciated by most students. The faculty members coming from varied disciplines and the presentations relying on multiple sources were found to be cross-cultural and different from the home-based curriculum for both Finnish and Indian students. Field excursions were particularly acknowledged to play a critical role in giving students a sense of reality and the political, social, and religious contexts of social work of which students from different country contexts were largely unaware. The program was particularly effective in promoting an understanding of cultural differences and how social workers need to engage with the diversity of any context. Students also became aware of the cultural differences of conducting education, i.e., the use of time, planning, and opportunities for flexibility, reflecting the different culture as well as the distinct history, structure, and delivery of social work both across and within counties. Both Indian and Finnish students appreciated spending time in Mumbai, Maharashtra, and Kochi, Kerala, given the added experience of regional and national variations in culture and social work practice and context:

> It was good to have winter school in two locations, particularly for the Finnish students who got the chance to explore at least two regions of India, being a country of diverse culture and traditions. As for the Indian students, it allowed them to explore how social work education and practice is organized in a different region. (Indian Student)

Students were quick to point out the opportunities provided for personal and professional networking and its relevance in social work practice and education. Many felt that the best qualities of course were relational, opening opportunities to meet people from different cities and countries. The openness of communication, reciprocity, and mutuality in learning was felt by most students as a form of solidarity:

> There is no way of learning more enjoyable than learning from other people! Cultural differences and very different ways of life are not always easy to understand from far away, but they become so much more understandable and easier to accept when experienced by getting to know each other. That leads you to be more open to ideas and expertise from other cultures and see their value. (Indian Student)

The experience gave students the opportunity to compare the differences and similarities. For Finnish students, insights into indigenous practices and knowledge of their respective societies were particularly insightful. Some striking differences and similarities were interesting to know from others. Basically, the mix of four universities allowed for an enriching exchange of ideas and social work education practices:

> I learned about the Sami Indigenous Community and their struggle with Livelihood and Climate through this winter School Program which is now very useful in my International Social Work and Indigenous People Class… The knowledge presented in this program helped me understand the role of government and social workers in Finland. I could also know the differences and similarities between social work practices in India and Finland. (Indian Student)

The group assignment received positive reviews from students as it effectively facilitated the Indian-Finnish interaction. The participants contributed and interacted without any boundaries in both the planning and display. Field visits were also critical learning tools and provided students with practical viewpoints to the theory provided in the lectures:

> All the components of the course were different from the usual classroom experience and so valuable and monumental in my academic journey. The insights received through lectures and panel discussions were not read before. The field visits and the group assignment were a great experiential learning as it showed how classroom learnings can be implemented effectively on ground. (Finnish Student)

Student experiences of cultural immersion and transformative learning, a final debriefing session among students and teachers at the end of Indian winter summer schools, were filmed and analyzed from the short documentary produced. The recordings reinforced many of the themes already summarized from the post-school survey. Findings from the pedagogical experience or cultural immersion and transnational exchange focused on specific learning outcomes for both Finnish and Indian students. Many of the students found the formal debrief session emotional, but it did help both students and academics make links between the experiences and transformative processes involved. While the Finnish students found their stay, exposure, harsh ground realities, and inequalities overwhelming, they felt emotional about the hospitality, care, and support provided by peers from India. The spontaneity of mutual connection and support belied the anxiety and apprehension of students prior to starting their journey and the program. The formal debrief was held on the last day when students and academics gathered at a venue in the environmental center in Kerala at the end of the program.

Despite different perspectives and views being shared, there was a sense of solidarity in the discussion, which alluded to many interesting points. Students discussed that the main source of new knowledge on climate change and crisis preparedness for students was quite often knowledge sharing or learning from each other, commonly referred to as peer-to-peer learning. Finnish social work students tend to be complacent in the accepted exceptionalism of the Nordic welfare state in providing for the needs of citizens. They had an opportunity to witness the self-reliance and the power of collective action in a context where neither the state and public services nor civil society plays a major role in the very limited provision of resources and assistance. Whereas students from India were exposed to descriptions of the Finnish system where social workers are integral to the operationalization of the welfare state. Both Indian and Finnish students responded with many questions relating to the roles, assumptions, and values behind social work in Finland and India to enhance their understanding of positioning of the profession in general.

Students discussed the need for critical reflection, involving the deconstruction and reconstruction of the positioning of social work. The Finnish students talked about having to accept different cultures and realities. This process involved historical explanations and alternative understanding to help make sense of questions as to how and why differences and divergences occurred and why political and social contradictions as to positioning of social work between civil society and the state exist (Das & Anand, 2014). The importance of cultural and religious beliefs and indigenous knowledge was also acknowledged by both Finnish and Indian students as critical to the learning journey. This offered new insights to Finnish students given the homogeneous and secular nature of Finnish society (Sinnemäki et al., 2019). Of course, this was dependent on individual students' openness to new or opposing perspectives and an openness to bend the boundaries as students experienced transnational learning.

The exchange assumed that while social work is a global profession, its practices and interventions are highly contextualized. The use of cultural immersion in the project was, however, not to encourage students to abandon what they have been taught but rather to critique, modify, and build upon what they have learnt prior to and during the exchange. The learning and teaching process must be ethical and retain respect for local or regional context, welfare system, and legislative mandate.

As transformative learning often takes place in a social context through dialogue and discourse with others, teachers play a critical role in discussion by providing opportunities for dialogue that extends beyond the classroom session to the social interactions and experiences. To achieve this dynamic, both students and teachers must be open to engaging in conversations and debates with each other and respecting others' different viewpoints. This attitude and responsiveness allow for the examination, deconstruction, and reconstruction or reevaluation of one's own perspectives (Anand & Das, 2019). In the winter school, students also noted the importance of reciprocal relationships, solidarity, and empathy. Transformative learning involves actively listening to others, seeking to understand their viewpoints, and considering how one's own beliefs and actions impact others. This represents informal, social, and even serendipitous learning opportunities which can also be introduced and experienced in home curriculum.

Yet the experience of travel also fundamentally changes the power between students and teachers, allowing for more spontaneous and open discussion. The use of a messaging platform was instrumental in facilitating group communication, relating and joint problem-solving to deal with the many practical issues involved. Ultimately the process described involves active engagement and personal agency. Students were encouraged to take responsibility for their own learning, actively seeking out new information, resources, and experiences vis-à-vis sustainable development, climate protection and crisis preparedness. The students felt that they benefited from exposure to concrete examples of issues of climate change and recognized the interrelation in the ecosystem and hence the acknowledgement for the fair and sustainable use of resources.

Exoticism and stereotyping of social work in other countries are counter to critical learning. Exchange programs are often criticized for being superficial and representing a form of academic tourism (Hammersley, 2014) reinforcing stereotypes and binary thinking. An important role of the academics and teachers involved was thus to maintain a degree of objectivity in the process of deconstructing experiences and personal assumptions. Learning outcomes that promote the unrealistic exoticism of social work practice in another country in comparison to one's own, or the reinforcement of negative stereotypes on the superiority or inferiority of one country's welfare system over the another (i.e., between global north and south systems) were challenged by the teachers. Through presentations in different sessions, teachers from India and Finland engaged with the specific context of welfare and social protection that witnessed the evolution of the differential welfare regime. The received binary wisdom around superiority of global north and limits of global south regime was challenged and discussed at length. Students were not encouraged to disengage or avoid critical issues as to the positioning and role of social work in either country but were consistently challenged by faculty to engage in critical issues, debates, and discussions. Teachers encouraged students to deepen their understanding of the effects of globalization and glocalization within their local context and field of practice. This was done through facilitated class participation, mixed group assignments, continuous feedback, and students' presentation.

Teaching Faculty Experiences and Observations

The emphasis on critical reflection and critical self-reflection and comparative analysis appeared to promote transformative learning. For example, linking the personal and the central and rational discourse through a commitment to engaging openly with others provided space and scope for diverse and alternative viewpoints. Students appeared to appreciate critical inquiry and deconstruction of the taken-for-granted subject positions (Dunlop, 1999). The course was observed to encourage students to develop a new understanding of themselves and others, as well as intercultural awareness. The experience enabled students to engage with critical comparative analysis differences and similarities as well as indigenous and culture-specific practices and knowledge of their respective societies.

In summary students' response to the program included initial resistance, overidentification, reproduction of stereotypes, and "real learning" by critical reflection (Anand & Das, 2019). For example, some students had difficulty responding to new learning or changes to existing assumptions. While students did report feeling overwhelmed at times, very few if any students reported that the experience did not encourage them to consider new insights as to the role of social work or to challenge existing definitions and approaches to new or challenging perspectives arising from the exchange. Many students, however, felt that they needed more time and opportunities for discussion to process new information. The impact of the exchange may also result in a position where the student fails to critique the limitations and tensions of new social work practices and exoticizes the practice of the host country. The academic staff thus have a critical role in countering overly idealistic views on social work practice in one country compared to another. Students require considerable knowledge to be able to critique the legal, historical, ethical, and structural contexts of social work practice in other countries. The reaction to being exposed to a new environment and challenges can lead to a position where prejudices and negative comparisons between countries are reinforced. Again, academics need to encourage critical reflection to avoid prejudices. Perhaps the desired position in achieving transformative and reflective learning is that new information is considered, critiqued, contextualized, and integrated for practical application.

Concluding Remarks

Overall, the project program aimed to achieve learning and teaching practices that are critical, reflective, and ethical. Moreover, the winter school sought to compensate for the lack of global and futuristic perspectives within the existing curriculum. The utilization of transnational exchange provided an experiential and holistic (also extracurricular) learning environment in another culture. The educational content exposed students to the global complexities of climate change, migration, and crisis preparedness and their different manifestations, interfaces, as well as response strategies, in the global south and north. The refiguring of crisis and crisis preparedness in social work education represents both challenges and an opportunity for social work education.

Bibliography

Alston, M. (2015). *Women and climate change in Bangladesh*. Routledge.

Anand, J. C., & Das, C. (2019). *Global mindedness in international social work practice*. Bloomsbury Publishing.

Anand, J. C., Das, C., Campbell, A., Davidson, G., Dill, K., Duffy, J., Hayes, J., Montgomery, L., & Pinkerton, J. (2014). *Internationalising the social work curriculum: A working toolkit for higher education facilities*. Higher Education Academy. https://www.academia.edu/70696195/Internationalising_the_Social_Work_Curriculum_A_Working_Toolkit_for_Higher_Education_Facilities

Anish, K., Borrmann, S., Nguyen-Meyer, N., Zhao, Y., Moen, H. B., & Liu, W. (2023). The art of not-knowing: Valuing diversity and intercultural competences in international social work education. *International Social Work, 66*(4), 991–1004. https://doi.org/10.1177/00208728211065708

Besthorn, F. H. (2003). Radical ecologisms: Insights for educating social workers in ecological activism and social justice. *Critical Social Work, 4*(1), 66–106.

BIOS. (2021). *Sivistys ekologisen jälleenrakennuksen maailmassa*. https://bios.fi/sivistys-ekologisen-jalleenrakennuksen-maailmassa/

Blühdorn, I., & Welsh, I. (2007). Eco-politics beyond the paradigm of sustainability: A conceptual framework and research agenda. *Environmental Politics, 16*(2), 185–205.

Braidotti, R. (2019). A theoretical framework for the critical posthumanities. *Theory, Culture & Society, 36*(6), 31–61.

Breen, K., Greig, M., & Wu, H. (2023). Learning green social work in global disaster contexts: A case study approach. *Social Sciences, 12*(5), 288.

Christie, M., Carey, M., Robertson, A., & Grainger, P. (2015). Putting transformative learning theory into practice. *Australian Journal of Adult Learning, 55*(1), 9–30.

Das, C., & Anand, J. C. (2014). Strategies for critical reflection in international contexts for social work students. *International Social Work, 57*(2), 109–120.

Das, C., & Anand, J. C. (2023). Global Mindedness in Social Work for Anti-oppressive Solidarity. In: *The Palgrave Handbook of Global Social Problems. Palgrave Macmillan.* https://doi.org/10.1007/978-3-030-68127-2_121-1

Dominelli, L. (2012). *Green social work: From environmental crises to environmental justice*. Polity.

Dunlop, R. (1999). Beyond dualism: Toward a dialogic negotiation of difference. *Canadian Journal of Education/Revue canadienne de l'education, 57–69.*

Dunlop, J. M., & Fawcett, G. (2008). Technology-based approaches to social work and social justice. *Journal of Policy Practice, 7*(2–3), 140–154.

Eilam, E. (2022). Climate change education: The problem with walking away from disciplines. *Studies in Science Education, 58*(2), 231–264. https://doi.org/10.1080/03057267.2021.2011589

Ellis, L. (2020). *Sustainable social work: A response to the climate emergency from social work education and practice in Aotearoa, New Zealand: A thesis presented in partial fulfilment of the requirements for the degree of Doctor of Philosophy in Social Work at Massey University, Albany, New Zealand* (Doctoral dissertation, Massey University).

Filho, L. W., Sima, M., & Sharifi, A. (2021). Handling climate change education at universities: An overview. *Environmental Sciences Europe, 33*, 109. https://doi.org/10.1186/s12302-021-00552-5

Freire, P. (2015). *Pedagogy of indignation*. Routledge.

Hammersley, L. A. (2014). Volunteer tourism: Building effective relationships of understanding. *Journal of Sustainable Tourism, 22(6), 855–873.*

Harrikari, T., & Rauhala, P. L. (2018). *Towards glocal social work in the era of compressed modernity*. Routledge.

Harrikari, T., Mooney, J., Adusumalli, M., McFadden, P., & Leppiaho, T. (Eds.). (2023). *Social work during COVID-19: Glocal perspectives and implications for the future of social work.* Routledge. https://doi.org/10.4324/9781003374374

Hayes, K., Blashki, G., & Wiseman, J. (2018). Climate change and mental health: Risks, impacts and priority actions. *International Journal of Mental Health Systems, 12*, 28. https://doi.org/10.1186/s13033-018-0210-6

Healy, K. (2022). *Social work theories in context: Creating frameworks for practice*. Bloomsbury Publishing.

Healy, L. M., & Thomas, R. L. (2020). *International social work: Professional action in an interdependent world*. Oxford University Press.

Howard, A., Agllias, K., Bevis, M., & Blakemore, T. (2018). How social isolation affects disaster preparedness and response in Australia: Implications for social work. *Australian Social Work, 71*(4), 392–404.

International Association of Schools of Social Work, International Council on Social Welfare and International Federation of Social Work. (2012). *Global Agenda for Social work and Social Development. Geneva: IFSW.*

Jha, M. K. (2015). Liquid disaster and frigid response: Disaster and social exclusion. *International Social Work, 58*(5), 704–716.

Jha, M. K., & Pande, S. S. (2022). Production of knowledge and methodologies in conflict induced displacement and forced migration. In N. Chowdhory & P. Banerjee (Eds.), *Gender, migration and identity in India* (pp. 17–42). Palgrave Macmillan.

Johns, L., Blackburn, P., & McAuliffe, D. (2020). COVID-19, prolonged grief disorder and the role of social work. *International Social Work, 63*(5), 660–664.

Jones, P. (2012). Transforming the curriculum: Social work education and ecological consciousness. In *Environmental social work (pp. 213–230). Routledge.*

Jones, P. (2013). Responding to the ecological crisis: Transformative pathways for social work education. *Journal of Social Work Education, 46*(1), 67–84.

Jones, D. N., & Truell, R. (2012). The global agenda for social work and social development: A place to link together and be effective in a globalized world. *International Social Work, 55*(4), 454–472.

Kim, S., & Slapac, A. (2015). Culturally responsive, transformative pedagogy in the transnational era: Critical perspectives. *Educational Studies, 51*(1), 17–27.

Klemmer, C. L., & McNamara, K. A. (2020). Deep ecology and ecofeminism: Social work to address global environmental crisis. *Affilia, 35*(4), 503–515.

Kloet, J. D., Lin, J., & Chow, Y. F. (2020). "We are doing better": Biopolitical nationalism and the COVID-19 virus in East Asia. *European Journal of Cultural Studies, 23*(4), 635–640.

Livholts, M. B. (2023). *The body politics of glocal social work: Essays on the post-anthropocentric condition* (1st ed.). Routledge. https://doi.org/10.4324/9781003193593

Lyons, K., & May-Chahal, C. (2017). Glocal social work. *European Journal of Social Work, 20*(1), 1–4.

Matthies, A. L., Närhi, K., & Ward, D. (Eds.). (2001). *The eco-social approach in social work.* SoPhi 58, University of Jyväskylä.

McLaren, P. (1995). *Critical pedagogy and predatory culture: Oppositional politics in a postmodern era.* Psychology Press.

Melekis, K., & Woodhouse, V. (2015). Transforming social work curricula: Institutional supports for promoting sustainability. *Social Work Education, 34*(5), 573–585.

Mezirow, J. (2018). Transformative learning theory. In *Contemporary theories of learning* (pp. 114–128). Routledge.

Miller, V. J., & Lee, H. (2020). Social work values in action during COVID-19. *Journal of Gerontological Social Work, 63*(6–7), 565–569.

Moore, J. W. (Ed.). (2016). *Anthropocene or capitalocene?: Nature, history, and the crisis of capitalism.* PM Press.

Muldoon, A. (2006). Environmental efforts: The next challenge for social work. *Critical Social Work, 7*(2), 1–7.

Närhi, K. (2004). *The eco-social approach in social work and the challenges to the expertise of social work* (Jyväskylä Studies in Education, Psychology and Social Research) (Vol. 243). University of Jyväskylä.

O'Dowd, R. (2020). A transnational model of virtual exchange for global citizenship education. *Language Teaching, 53*(4), 477–490.

Panagiotaros, C. V., Boddy, J., Gray, T., & Ife, J. (2022). (Re-)imagining social work in the Anthropocene. *The British Journal of Social Work, 52*(8), 4778–4794.

Paterson, M. (2021). The end of fossil fuel age? Discourse politics and climate change political economy. *New Political Economy, 26*(6), 923–936.

Pelttari, M. (2021). *Lämpenevä maa: Ilmastolukutaidon käsikirja.* Vastapaino.

Powers, M., Schmitz, C., & Beckwith Moritz, M. (2019). Preparing social workers for ecosocial work practice and community building. *Journal of Community Practice, 27*(3–4), 446–459.

Ranta-Tyrkkö, S. (2017). Sosiaalityön tulevaisuuden etiikka epävarmuuden ja ympäristöriskien maailmassa. In R. Enroos, M. Mäntysaari, & S. Ranta-Tyrkkö (Eds.), *Mielekäs tutkimus. Näkökulmia sosiaalityön tutkimuksen missioihin* (pp. 113–138). Tampere University Press. http://urn.fi/URN:ISBN:978-952-03-0606-9

Ranta-Tyrkkö, S., & Talentian ammattieettinen lautakunta. (2023). *Sosiaalialan ilmastokäsikirja.* Sosiaalialan korkeakoulutettujen ammattijärjestö Talentia. https://www.talentia.fi/wp-content/uploads/2023/08/talentia_ilmasto-opas_2023_final.pdf

Rapeli, M. (2018). Social capital in social work disaster preparedness plans: The case of Finland. *International Social Work, 61*(6), 1054–1066.

Rapeli, M., & Mussalo-Rauhamaa, H. (2022). Intensive and residential elderly care services responding to heat wave – Case Finland. *Nordic Social Work Research*, 1–12. https://doi.org/10.1080/2156857X.2022.2047767

Redondo-Sama, G., Matulic, V., Munté-Pascual, A., & de Vicente, I. (2020). Social work during the COVID-19 crisis: Responding to urgent social needs. *Sustainability, 12*(20), 8595. https://doi.org/10.3390/su12208595

Schrooten, M., Evans, F., Kloppenburg, R., Csoba, J., & Kinos, S. (2023). An argument for transnational learning in social work programmes: Experiences from the Erasmus+ "Urban Diversities" course. *Social Work Education, 2023*, 1–20.

Sewpaul, V., & Jones, D. (2004). Global standards for social work education and training. *Social Work Education, 23*(5), 493–513.

Sinnemäki, K., Nelson, R. H., Portman, A., & Tilli, J. (2019). The legacy of Lutheranism in a secular Nordic society: An introduction. In K. Sinnemäki, A. Portman, J. Tilli, & R. Nelson (Eds.), *On the legacy of Lutheranism in Finland: Societal perspectives* (Studia Fennica Historica) (Vol. 25, pp. 9–36). Suomalaisen Kirjallisuuden Seura.

Steffen, W., Broadgate, W., Deutsch, L., Gaffney, O., & Ludwig, C. (2015). The trajectory of the Anthropocene: The great acceleration. *The Anthropocene Review, 2*(1), 81–98.

Stockholm Resilience Centre. (n.d.). *The nine planetary boundaries*. Retrieved July 19, 2023 https://www.stockholmresilience.org/research/planetary-boundaries/the-nine-planetary-boundaries.html

Sutton, J., & Tierney, K. (2006). Disaster preparedness: Concepts, guidance, and research. *Colorado: University of Colorado, 3*(1), 3–12.

Taylor, L. A., & Hikida, M. (2020). Unpacking everyday critical pedagogy: Languaging critique and dialogue. *Literacy Research: Theory, Method, and Practice, 69*(1), 266–284.

Team Finland Knowledge. (2023). https://okm.fi/tfk-verkosto

Teixeira, S., & Krings, A. (2015). Sustainable social work: An environmental justice framework for social work education. *Social Work Education, 34*(5), 513–527.

UN Climate Action. (n.d.). *What is climate change?* Retrieved July 19, 2023 https://www.un.org/en/climatechange/what-is-climate-change

UN DESA. (2022, July). *The Sustainable Development Goals report 2022*. UN DESA. https://unstats.un.org/sdgs/report/2022/

WHO. (2021, October 30). *Climate change and health*. https://www.who.int/news-room/fact-sheets/detail/climate-change-and-health#:~:text=Climate%20change%20affects%20the%20social,malaria%2C%20diarrhoea%20and%20heat%20stress

Part III
Connecting Teaching, Learning and Practice in Ecosocial Work

Chapter 11: Becoming Environ*mental* Writers! The Futures of Reshaping Knowledge and Learning in Ecosocial and Post-Anthropocentric Social Work

Mona B. Livholts

Introduction

During the last decade, international social work organisations have strengthened the commitment for social work to promote education about "environmental justice" (Global Standards of Social Work Education, 2020), "environmental sustainability" (Global Agenda, 2012, 2020), and "listening to and learning from Indigenous peoples around the world" (Global Definition of Social Work, 2014). These commitments are also connected with the fulfilment of the UN's Agenda 2030 of the SDG goals comprising economic, social, and environmental sustainability (2015–2030). However, despite the valuable contribution from the emergent literature on environmental, ecosocial, green, and post-anthropocentric scholarship (e.g. Matthies et al., 2001; Coates et al., 2006; Gray et al., 2013; Dominelli, 2012; Matthies & Närhi, 2019; Rambaree, 2020; Bozalek & Pease, 2021), the question of how knowledge and learning are shaped by writing as a methodological tool to promote critical and creative knowledge and learning has remained marginalised (Livholts, 2022a, b, 2023).

In this chapter, I propose that directions that engage with environmental perspectives in social work, such as ecosocial and post-anthropocentric social work, need to include writing as a methodological tool to reshape their futures (Livholts, 2023; see also Livholts & Södergran, 2024). Laurel Richardson (1997, p. 1) asked the following questions: "How do the specific circumstances in which we write affect what we write? How does what we write affect who we become?" From this perspective, I understand writing as a creative activity which needs to be included as part of the research design when we enter into research and learning. Creative writing is a practice where one can learn both about oneself and the topic and the shaping of the

M. B. Livholts (✉)
Department of Social Sciences, University of Helsinki, Helsinki, Finland
e-mail: mona.livholts@helsinki.fi

C. Forde et al. (eds.), *Teaching and Learning in Ecosocial Work*,
https://doi.org/10.1007/978-3-031-58708-5_15

text can function as an interactive dialogic space that forms relationships to readers. In social work, creative writing and art-based methods have been developed as alternative practices to knowledge and learning. These practices bridge theory and practice, promoting situated knowledges, critical reflexivity on power and intersectionality, as well as creative approaches when creating data with research participants. Some examples are the personal essay, memoirs, the academic novella, poetry, and photography, all of which contribute forms of knowledge that allow diverse forms of expressivity, voices, and storytelling (cf. Livholts, 2001/2011; Witkin, 2014; Eliassi, 2012; Chambon, 2005). More recently, social work research has argued that posthuman approaches that strive to better understand knowledge and learning through the pandemic, climate crisis, and more-than-human entanglements, require alternatives to mainstream textual forms which makes it possible for emotions and feelings of risk and uncertainty to be expressed. Mayor and Pollack (2022, p. 383; see also Livholts, 2021) demonstrate how creative writing during the COVID-19 pandemic visualises "how all levels of social work are complicit with or can work to disrupt systems of white supremacy, hetero/sexism, colonialism, and classism". Livholts (2023, p. 112) writes post-anthropocentric essays "to shape writing towards creative, critical and dialogical encounters with the writing self, with human and non-human forms, and environ*mentalities*".

The overall aim of this chapter is to invite readers to familiarise themselves with theories and practices of environ*mental* writing inspired by posthuman philosophy, knowledge, and learning in ecosocial and post-anthropocentric social work. I propose that creative modes of writing can be used as a transformative force towards eco-social consciousness and resilience. By thinking with posthuman philosophies and using creative life writing genres such as memory work, diaries, poetry, and photography, the possibilities of creating new knowledge about our interconnection with more-than-human existence can radically reshape knowledge and learning (see also Livholts, 2023; Livholts & Södergran, 2024). How do we shape our academic texts and image worlds when we create knowledge about ecosocial and post-anthropocentric transformations in social work? How can the conceptual framing and practices of environmental writing contribute to a creative reshaping that leads towards situated and decolonised knowledge and learning? My chapter includes illustrations and scenes thematically related to childhood memories from a farm, exhaustion as a way of rewording social problems and water as a pathway for social justice. These illustrations are inspired by my recent book (Livholts, 2023), an article on the pandemic (Livholts, 2021) and a chronicle (Livholts, 2022a). The chapter includes brief guidelines for readers who wish to make use of one or several of the genres discussed there for their own environmental writing.

As I initially mentioned, we have seen the emergence of a rich body of literature that uses diverse definitions, such as "environmental", "ecosocial", "green", "glocal", or "post-anthropocentric" social work. Due to the focus of this chapter on writing in eco-social and post-anthropocentric social work, it mainly draws on literature related to these directions and does not engage in a discussion on differences and interconnections, possibilities, and limitations between diverse directions. However, I suggest that environmental writing, as it is composed as an umbrella concept for genre-transgressive writing in this chapter, can be used as a methodological tool to promote creative life writing methodologies more broadly across diverse directions.

In the next section, I present the philosophies and theoretical ideas that are important as a departure to guide our understanding of environmental writing. This is followed by three sections with examples and illustrations of thematically organised topics and uses of diverse writing genres. The chapter concludes with a section that brings out the main features of environmental writing in ecosocial and post-anthropocentric social work.

Philosophical and Theoretical Framings for Environmental Writing

Inspired by posthuman philosophies in post-anthropocentric social work, I use creative writing to situate processes of knowledge and learning. The Anthropocene is a rather recent conceptualisation created through the combination of "Anthropos" meaning human and "cene" new or recent referring to a geological time-period when humans made an impact on the environment on Earth (e.g. Crutzen & Stoermer, 2000; Crutzen, 2002). Post-anthropocentric means "after" the Anthropocene and critiques the centralisation of human life and experiences. A few years ago, the first book on post-anthropocentric social work, which builds on critical posthuman and new materialist perspectives was published by Vivienne Bozalek and Bob Pease (2021). Bozalek and Pease (2021) describe how critical posthuman philosophy provides a theoretical departure for how social work can become responsible for better understanding of humans' responsibility for climate destruction and the environment and specifically the interconnection between humans and the natural world. It is an understanding that rejects the view of humans as the dominant species and uses posthuman thought to challenge dichotomies such as "nature/culture, material/discursive, object/subject, human/animal, man/woman and North/South" (Bozalek & Pease, 2021, p. 1). Posthumanism builds on the argument that intersecting relations such as ethnicity, gender, class, heteronormativity, disability, and age are intimately intertwined with all life forms such as animals, water, forests, and nature. This does not mean a devaluation of humans but visualises in what ways humans in diverse life situations live in close relation not only to each other but also other living and non-living subjects. In my recent book (Livholts, 2013; see also Livholts & Södergran, 2024), I use *assemblages* as an open-ended conceptualisation, to learn from the viewpoint that bodies have multiple forms, such as bodies of water, cities, institutions, and so on (Puar, 2012, p. 57). As a way of creating knowledge that is attentive to intersectional relations of both human and non-human embodiment, I make use of feminist, Indigenous, and postcolonial writing methodologies and draw on my own work with situated life writing genres, such as diaries, letters, photography, and poetry, to rethink and promote renewed practices of what I in this chapter refer to as environ*mental* writing to offer an alternative to mainstream prose in social work (cf., Livholts, 2012, 2017, 2018, 2019b).

A recurring question in the literature of environmental, ecosocial, and post-anthropocentric social work is the transformation of what is understood by a person in a situation and the holistic perspective. Närhi and Matthies (2018, p. 495) argue for a broadened perspective of the person in the environment "to understand how human and nature are interconnected and what that means in social work practice". The framework of this model should integrate global, local, and ecosocial theories and be used as "a practical action model". Rambaree (2020, p. 558) writes that social work practice "is developed with a holistic perspective that focuses on multiple, reverberating transactions between people and their social as well as biophysical environment". Rambaree (2020, p. 569) argues for pedagogical transformation through which students "read and write in relation to the awakening of their consciousness about their social reality and developing transformative pathways for responding to SD [social development] issues, concerns, policies and practices". In post-anthropocentric social work, the reframing of knowledge and learning is guided by critical posthuman theory to promote situated and material embodied processes of knowledge. In social work, Bell (2021, p. 65) writes that posthumanism is helpful as a transformative philosophy to achieve renewed understanding of holistic practices that "situate(s) humans as interdependent and interconnected with all other living things and the material world" and visualise that "humans are not necessarily the only creators of knowledge".

Environ*mental* writing then becomes a practice that intertwines the environing (surroundings) as situational, material, and ecological structures as well as the process of becoming writers through embodied agencies of multispecies and mentalities (acting, thinking). As Frichot (2019, pp. 30–31) argues with reference to Agrawal (2005, p. xiv), "environmental subjects come to think and act in new ways in relation to the environment". Oakley et al. (2018) point out that environmental writing is not new but that the "new nature writing" is linked with art activism to promote creativity and imagination. This allows a better understanding of climate destruction and at the same time enables "a sense of the future, an ability to imagine another, less destructive way of being" (p. 2). Indigenous scholarship has contributed to a critical rethinking of the importance of bringing renewed forms of writing and storytelling into academic texts on the environment and ecologies. Gloria Anzaldúa (1987) illustrates such intimate interconnections through writings on humans/animals/plants/stones/seas, which also have consequences for the responsibilities of caring:

> […] no one is an island and has never been. Every person, animal, plant, stone is interconnected in a life-and-death symbiosis. We are each responsible for what is happening down the street, south of the border or across the sea. (Anzaldúa, 1987, p. xxviii)

It is important to decolonise writing and be guided by a critical and creative perspective on writing that situates knowledge in order to limit its knowledge claim and promote accountability (Livholts, 2019b, 2022b). Ulmer (2018) emphasises the importance of thinking and discusses how thinking *with* and thinking differently are necessary to challenge mainstream forms of knowledge and allow multiple subjects to be present. Thinking differently also calls for doing research differently. Guttorm et al. (2021, p. 114) demonstrate how decolonising research storying means changing perceptions about authoring and co-authoring an academic text: "to hear the Earth, to feel the Moon, to think like the forest, and to write with these ontologically different epistemologies, where do the non-words get translated into words in our writing process". This draws attention to how environmental writing includes more-than-human actors as co-authors in our texts. I have been inspired by the writings of Anna Tsing (2015), who makes use of the interaction between renewed forms of ethnographic writing in the Anthropocene and includes sketch drawings of animals and plants in her texts.

In the forthcoming sections, I draw on examples from my earlier publications (Livholts, 2019b, 2021, 2022a, 2023) to illustrate how I have used memory work, diary writing, poetry, and photography thematically related to childhood memories from a farm, exhaustion as an alternative to social problems during the pandemic, and "writing water" for social justice. I include brief guidelines to inspire readers' own uses of these diverse and overlapping genres and motivate how they can promote a pedagogy of learning. I conclude with reflections on the futures of becoming environmental writers in ecosocial and post-anthropocentric social work.

Situating Bodies of Knowledge and Learning

Picking Stones from the Field Before Sowing Barley

Every time I pass a ploughed field or a field with barley (see Fig. 1 below) in rural space on a warm summer day, I imagine how I used to walk behind the tractor with my sisters each year, to pick up stones from the earth to prepare for the sowing. It was hot in the sun, and we wore tall rubber boots. I tried to keep the side of my face towards the tractor to avoid breathing in the dust that swirled up from the earth. We were trained to choose the right-sized stones; not too large, not too small, but middle sized, stones we were strong enough to pick and carry with us in our buckets, allowing the field to breathe and yet regenerate the growth of the earth. There was a particular rhythm in the movements when leaning forward, picking the stones, moving the bucket, and a clicking sound, "clock", "clock", "clock", could be heard as another stone was dropped into the bucket.

The scene that introduces this section is collected from my book (Livholts, 2023, p. 82) based on a memory from the seasonal work in the farm where I grew up in a rural part of the mid-Sweden area. It was in this rural landscape that my gendered childhood was shaped in a patriarchal family setting, a place where whiteness was taken for granted, an unnamed invisible privilege for my family. I worked with and cared for the animals: horses, pigs, sheep and our dog. I was often in conflict with the dominant values concerning animals, who were inevitably subordinate and not emotionally valued. I sometimes sneaked out early in the morning to check the traps my stepfather had set in the forest. I pleaded that the farm animals, who were my closest friends, should be spared from slaughter. I felt loss and grief for long periods after their death. All memories from this childhood bring with them embodied

Fig. 1 Barley. (Source: Wikimedia commons. Retrieved June 14, 2022 https://commons.wikimedia.org/wiki/File:Hordeum_vulgare_(6-row_barley,_Gerste)_dry_specimens_(2019-08-07).jpg)

sensory perceptions of the landscape: the heat when walking behind the tractor, the sound of stones and the animals, the smell when cleaning the stables, and the relationship to animals.

It was an odd feeling of awakening when I realised that some of the discomfort and alienation I felt in academia was caused by the way my embodied experiences and emotions of growing up in a rural environment were not acknowledged as part of Western academic knowledge. The sensory perceptions of how my thinking and knowledge had been formed by the landscape, the forest, the sound of the wind, the companionship of the large birch tree next to the gravel road each morning when I waited for the school bus; the snowfall each winter that covered the landscape and brought silence, the crunching sound of my horse eating, an encounter with an elk in the forest, and how I learned how to swim in the cold water of a stream. The social work paradigm that (silently) required specific reading choices, definitions of social problem orientation, and theoretical departures consequently leaves out the question about the importance of how knowledge is *shaped* through interconnections with the *living environment* and how knowledge is shaped by *the way it is written*.

Using the Memory Work Method to Trace Embodied Experiences

In the late 1990s, when I was halfway into my doctoral studies, I came across the method of memory work, and I began to write a parallel memory text thematically related to the articles I wrote. This form of writing short memory episodes opened up another way of expressing experiences and attending to the richness of memories and emotions involved in situations across time and space (Livholts, 2019b, pp. 27–31). The method was developed in Germany by Frigga Haug (1987) and Anette Kuhn (2002, 2010), whose memory work related to film and photography. This methodology first emerged as a collective form and has later been renamed collective biography (e.g. Davies & Gannon, 2006). The process of doing collective memory work includes technologies of writing, talking, reading, listening, and seeing, which interconnect specific moments of memories into a flow of interlinked stories (Davies & Gannon, 2006, p. 7). At the heart of memory work is to attend to situations, fragments, and scenes in our everyday life experiences because they provide important knowledge about relations of power that are often not visualised. Haug (1987, 2008) has argued against narrative coherence and a linear understanding of life experiences and has recommended writing in the third person, thematically focusing on the body, emotions, gendered relations of power, and knowledge production. Working with memories as fragments of lived experiences is a powerful methodological tool (Livholts, 2019b; Haug, 1987) due to the way it visualises relations of power in specific situations of lived experiences. This way of working with details, actualising emotions, and listening to the sound of the memory when it is read aloud allows for different senses to be actualised in the learning process. Baikie (2020, p. 42) writes about decolonisation through Indigenous memory work promoted by a "need to routinely navigate in-between the worldviews of the colonizer and the colonized in both research and practice". Since Eurocentric systems of colonisation have colonised land and resources, as well as the mind and the memory, it is necessary to develop alternative critical and creative practices and ways of creating knowledge. Baikie (2020, p. 52) demonstrates how the process of writing, reading, and listening to memories provides a possibility to decolonise storytelling practices. The re-evaluation of knowledge as shared, as well as creating safe spaces for storytelling, includes changing the terminology of science from expressions such as "owning data" to "caretaker of data".

Guidelines for your own practice

Think of and choose one of the first memories that come to your mind related to experiences from *the environment of your childhood*. Write down the memory on your computer or on paper. Limit the writing to a maximum of half a page. Pay special attention to detail and to emotions. You are encouraged to use the third person (he, she, they, or your own or an invented name). Choose which language you write with (your mother tongue(s), other chosen language).

For further analytical reflections, you are encouraged to read the memory aloud (record it or read to a colleague or friend) and use active listening. How did you perceive the memory when you listened to it? Did it express what you thought about when you wrote it? Were there particular details, words, or impressions from this memory that were particularly interesting to spur further writing or reading? What are your thoughts about how this memory writing exercise can contribute to learning in social work?

In this chapter I promote memory work to recognise and visualise the importance of fragments of memories in our lived realities where our personal selves co-exist with other living species and non-human agents in the environment. I have found this method very useful as a way of becoming attentive to how working with memories (writing, reading, listening, rewriting) provides opportunities to engage critically in understanding embodied and emotional personal experiences and their meaning in relation to both human and non-human existence in the environment. By working with memories and attending to details that we may not see as important at first, multifaceted experiences of one's situated self can emerge. Memory work can be used in teaching as a way of doing individual preparation related to a theme, as well as in group work to share, read, and listen and rewrite memories. It can also be used as a point of departure when planning to write an essay or a research project to learn more about one's own relationship to a topic. Whenever I use memory work in teaching sessions collectively, I remind participants of the ethics of not revealing biographic information about other people and not choosing a memory that is too sensitive to share with others. It is important to create equal and safe spaces for sharing and working with memories.

Exhaustion as an Alternative Wording to Social Problems

13 June 2022, Writing with Exhaustion

It is a warm and sunny day in Helsinki. The plant on the front side of my house has grown so fast during the last week that it has transformed the front side of the house from a brick wall to a surface covered with leaves in an intense green colour. As I struggle to compose the writing of this chapter, I think back on the keynote "Exhaustion, Environmental Writing and Structural Change" that I presented in Jyväskylä 17th of February 2022 in the Social Work Research Days. The theme of the conference was "Environments and Structures in Social Work", and I started the talk with a brief introduction about the paths of research and teaching that led me to create the title. Exhaustion is a philosophical and practice-based concept that I came across some years ago when I was invited to do a review of the philosopher and architect Helene Frichot's (2019) *book Creative Ecologies. Theorizing the Practice of Architecture. As you can see, the title revealed nothing about the methodology of exhaustion that runs like a thread throughout the book. It was during this reading that I found that the concept of exhaustion was a much better word to use in social work than "social problems". Ever since I came to social work, I have felt a discomfort with the notion of using social problems, which mainly refers to marginalisation without taking into account the problem of privilege.*

Exhaustion conveys the meaning that humans are intimately intertwined with their mental, physical, ecological, and cognitive environmental worlds. Exhaustion challenges the endpoint of extinction. It can be used as creative resistance when we have used up all other possibilities. As Frichot (2019, *p. 71): writes:*

> *Exhaustion is more profound than tiredness; its structure of feeling is rather about an anxious restlessness. Exhaustion is about wakefulness and the distance of sleep. It is insomnia. Exhausted, awake, not tired, a constant wondering plagues you, as you ask yourself: Have I exhausted all that is possible? Have I done everything I can to ameliorate my local environment-world? It is an exhausting challenge.*

Every time I talk to people, whether it is a lecture, a seminar, or in conversation with a person, the response is that exhaustion is "the new normal" of our time, that it is a word that better describes the way we perceive the current condition of multiple crisis, of environmental destruction, the health crisis of the pandemic, the ongoing war after Russia's invasion of Ukraine, and the food crisis.

I recently read the World Inequality Report (2022), *which is a document about exhaustive inequalities from a world perspective. It shows how wealth and income are unequally divided to the extent that "the richest 10% of the global population currently takes 52% of global income, whereas the poorest half of the population earns 8.5% of it" (p. 10). Historical patterns of colonisation and imperialism continue to uphold unequal life conditions, and inequalities have significantly increased not only between but also within countries. I believe that exhaustion offers a possibility to think with all these aspects of the used-upness of resources and as an alternative concept for social problems. I contend that it is important to transform the language of social problems in social work and to be aware that its legacy is a discipline that has been complicit with oppressive acts. "How does it feel to be a problem?" as the African American sociologist and civil rights activist Du Bois (*1903, *p. 3) once wrote. This is the unasked but constantly present question that even though it is not said aloud, it is always present in the "double consciousness" of someone who is compelled to see themselves through the eyes of others.*

The reading of Frichot's book and other texts on exhaustion (e.g. Barth, 1984; Deleuze, 1988) inspired me to create a photo-poetic diary during the pandemic (Livholts, 2021). In the following section, I share scenes extracted from my diary notes in this study between 4 April and 4 June 2021. My writing was guided by the question: How can the language of exhaustion offer an alternative language for social work beyond social problems and vulnerability?

Tracing Exhaustion During the Pandemic

Tiredness/trötthet/väsymys
My heart beats pounding in my ears
The chorus of a whistling/susande/viheltävä/ sound
Restlessness, insomnia, dreams, shivers
Dots with red legs, dancing in the darkness/mörkret/pimeys
Spring trees/vårträd/kevätpuita
Arms and legs of bark skin
Stretching out, touched by the wind
Rooted in the earth/jorden/maa

My skin/min hud/ihoni
The largest organ of the body
Breathing/andas/hengittää
Connecting with all other bodies/kroppar/ruumiit (Livholts, 2021, p. 55)

4 April 2020: Closer to Things

The last few weeks have gradually infused a different rhythm into my life. I spend more time with myself, I read more, I avoid going to the grocery store if it is not necessary, and I keep my distance with people. I watch TV more often to keep updated with news about the virus. Every morning I talk to my 78-year-old mother, who lives in isolation. The epidemic has brought an increased awareness to my daily life, an uncertainty, sometimes a sense of failure. I use the camera to photograph my surroundings. I feel a renewed attachment to windows, buildings, furniture, trees, flowers and parks (See Fig. 2 a and above). It is as if social distancing between people brings me closer to things; makes me attentive to all forms of life, materials and architecture. I take a photo of the window in the room where I most often work at home. I have always loved this window, but I now see it as more than a window; it is an extension of the flat intertwined with the large trees outside, with the ivy that climbs and embraces the tree trunks. (Fig. 2b; Livholts, 2019b, *p. 57)*

11 May 2020: Reflexive Bridges, Traces of rust, and Exhausted Things

This morning I photographed shadows under the bridge. It felt like I too was made of water, stone and mud. The sunbeams broke through the water movements, created shadow patterns on the on the concrete archway under the bridge. I used this moment as a reflexive bridge (Fig. 3). I continued my walk and unexpectedly saw traces of rust on the bridge by the waterfront. I felt the smell of mud, of something rotting, and passed a container filled with steel garbage. Besides it was a row of shopping carts. At first, I just passed, but when I came back I took a photo. I understood then that traces of rust came from the bikes and shopping carts that had been brought up from the bottom of the lake (Fig. 4). I thought there was something hopeful about the scene. I thought it should be made into a permanent exhibition; an invitation for people to think and rethink their relationships to things. (Livholts, 2021, *p. 59)*

1 June 2020: "I Can't breathe"

Protests have been going on for almost a week across cities all over the United States and many other countries, as a response to the death of George Floyd, an African American man who died from police brutality May 2020. The death of George Floyd visualises violence and death in a context of colonial history. The words he said before dying—"I can't breathe"—have become an expression in the many protests against racism, and a call for social justice and equality. I think about how the pandemic, in a similar way to other disasters, has the most serious consequences for people who are racialised. Sweden is no exception. In Stockholm, there is an over-representation of deaths in suburban areas where many people with migrant backgrounds live. Colonialism and racism have exhausted minority and Indigenous people for generations. Who is seen as worthy and unworthy is intimately intertwined with embodiment, education, resources, nation and community. (Livholts, 2021, *p. 61)*

Fig. 2 (**a**) Spring trees and (**b**) Bark skin. (Source: Photographed by the author)

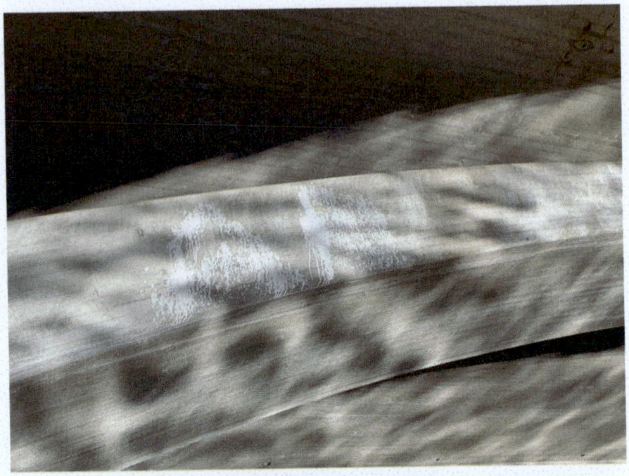

Fig. 3 The reflexive bridge. (Source: Photographed by the author)

Fig. 4 Exhausted things. (Source: Photographed by the author)

Writing a Diary and Sculpting Life

Writing a diary in the form of dated notes from everyday life has a long history along with other life writing genres such as letters and photography (Plummer, 2001). People write diaries in many different forms, and Lejeune (2009, p. 173) has described diary writing as a formative document that "sculpts life as it happens and takes up the challenge of time". I find that this way of expressing the shape of diary writing shows the intimate connection between the human body, the textual body, and the spatial and material surroundings. Lejeune (2009, pp. 194–196; cf., Livholts, 2019b, pp. 31–35) suggests that diary writing has four functions: (1) to *express* oneself, to release emotions, thoughts, and to communicate. To make use of writing is an act of expression and separation that connects the body (hands, fingers) with the pen, or the keyboard, into a form of meaning-making through words; (2) to *reflect*, allowing the writer to develop thoughts and analytical reflections from the everyday world and to write about crises, (3) to *freeze time*, which has been described as a way of creating "a memory out of paper, to create archives out of lived experience, accumulate traces, prevent forgetting (…)"; and (4) to take *pleasure* in writing through "a flow of energy that courses through the practice of writing" (Lejeune, 2009, p. 195).

Diaries have mainly been used in historical, anthropological, feminist, and critical race studies and more recently in human geography and social work, often combined with other methods (Kenten, 2010; Latham, 2020; Livholts, 2019b). Contemporary diaries take multiple forms, such as notebook diaries, preexisting diaries, online diaries, written, and video and photo diaries; they can be time- or events-based and intrapersonal or interpersonal (Latham, 2020). Hyers et al. (2006) emphasise the advantages of diary writing in documenting unequal treatment and marginalisation in everyday life, which is not often included in oral telling.

Guidelines for your own practice

Write a diary entry for at least 2 or 3 days each week over a period of 2–3 weeks on the theme of "Exhaustion". If you wish, you can make an entry every day or on multiple occasions for a shorter or longer period. Mark your diary entrances by *dating* them on each occasion. Include photographs or other collected images with your written diary if you think they add illustrative value. Write in *a freely chosen language*, your mother tongue and/or English. If you wish, you can create headings for each diary entry.

For further analytical reflections, you are encouraged to read your diary after the period of writing the diary is over, and write a further diary note based on reflections of this reading.

If this diary practice is used in the context of social work education or research, the diary notes/images can be used as a basis for writing a one- to two-page analytical reflection guided by the main purpose of your course/teaching or your doctoral project.

Using a diary for the purpose of becoming an environmental writer is an excellent way to document events, thoughts, and memories from the flow of everyday life. There is no one way of writing a diary but many. Some notes can be short, others longer; the style of language can shift, it can be hand-written or computer-written, and it is also possible to make sound recordings of a diary note. The diary is a communicative act with the self, with other humans, as well as with the natural world. My own diaries have had many different forms over the years with periods of writing on loose pieces of paper, my computer, or a notebook, and I feel that these diaries have offered a writing space to document, shape, and sculpt life as it is lived in relation to many different environments.

Writing *with* Water for Social Justice in Social Work

Letter of the Sea

It is mid-March and the time of the ice break (Fig. 5 below) in Helsinki when I begin to write this letter. It is a time when I often walk to the seafront; listening to the clashing of ice blocks; the seeping sound of the ice-slush; the slowness of water movements. I perceive water during these moments as a source of spiritual recovery. This sea also has meaning for me on a personal level as it concerns Swedish-Finnish connectivity and (be)longing. At the same time, it is a political and economic space for fishing, shipping, trading and travelling.

Am I floating	*I rörelse,*
in the sun beams,	*min kropp,*
of the spring sea,	*mina tankar,*
the surface of ice and open water?	*I solljuset,*
Am I floating	*flyter jag,*
towards the open sea,	*mot öppet hav,*
slow water movements,	*lyssnar till,*
the heartbeat of ice-blocks,	*isblockens hjärtslag;*
asking me to listen?	*I rörelse,*
Am I floating	*min kropp,*
in this moment,	*mina tankar,*
of intense sunlight,	*I solljuset,*
with the sea of the Gulf of Bothnia,	*långsamma vågrörelser,*
connecting and separating	*isblockens hjärtslag,*
Finland and Sweden?	*sammanlänkar och åtskiljer,*
	Finland och Sverige.

When I was a child, I learned to swim in the cold water of a stream. The forest around it and the small beach where I used to warm my feet on sunny days were part of a landscape that shaped my ways of knowing. Learning to swim was also about learning how to float, and floating was an art because it required relaxation, breathing, and trust. There were also some uncomfortable things at the beach. There were leeches in the water. The changing room had no locks, and boys would come barging into the room where I was struggling to remove my wet bathing suit. As I remember it, we were all white children on this small beach. While my gendered

Fig. 5 Thinking with the Ice-Thaw. (Source: Photographed by the author)

body was exposed to sexual harassment, my white body was a privilege I took for granted.

Several years ago, I wrote the academic novella *Writing Water* when I wanted to think *with* water in order to better understand the disciplinary boundaries and possible transgressions in social work (Livholts, 2013). I wrote about water as part of our embodied selves and the Earth and how water can symbolise the complexity of understanding social change through memories and stories. Water provides the possibility of seeing and listening to life beyond the human self, to raise awareness of different histories and everyday realities from perspectives of social injustice, exclusion and inclusion, and fear and safety. It can express personal, political, economic, ecological, and social layers and promote changing structures of social injustice in social work (see also MacShane and Moynihan, this volume). In this book, I continue to write stories of water as geopolitical spaces of knowledge. One example is how colonialism and invasion make themselves visible in the entanglements between bodies, such as excluding black and privileging white bodies when travelling and migrating across water, swimming, prohibition, fear, and safety at the beach are concerned. These thoughts have been visualised by the South African poet Koleka Putuma in her poem "Water": https://www.youtube.com/watch?v=8dfq3C8gNrE.

At the same time as Putuma uses poetry to engage in personal, political, and social movements and relations of power through water, these movements and relations become a site for creative resistance and social change. As I write the last words of this text, the ice blocks have melted, and I listen to the waves of the open sea. I hope that you who read my words are inspired to take up life writing genres such as letters, diaries, poetry, and photography, to "write water" as a situated, place-based, embodied, and more-than-human practice in social work. What are your memories and stories of water? How can they be written and used to promote empowerment and agency for social change and social justice? (Livholts, 2022a, n.p.)

The Photo-Poetics of Writing

This illustrative section with poetry, photography, and memories is collected from my blogpost for Sosiaalityön tutkimuksen seura/Föreningen för forskning i socialt arbete/The organisation for research in social work (Livholts, 2022a, n.p.). I wrote this blog on the theme of water as a continuation of my previous writing with water (Livholts, 2013). The human body is made up mostly of water. So is the Earth. Flooding, heavy rainfall, and tears are part of the natural world where human embodiments are interconnected. Photo-poetics is a rich inspiration for environmental writing. Perhaps this combination of textual and visual composition that I have used in this chapter in the section on exhaustion promotes some of the most vivid expressions of the interconnections and even permeability of the human body and the natural world. I would go as far as to say that binaries are always transgressed with images and poetry. Patricia Leavy (2015, p. 97) writes about how poetry offers a language for "a third space", "a third voice". In an article by Szto et al. (2005), the links between poetry and photography in qualitative research are discussed and illustrated in relation to the concept of the person-in-environment perspective in social work. They (Szto et al., 2005

, pp. 148–149) pose the question: "What is photography in the context of research and practice?" and illustrate it through a set of expressions that bind together the diverse genres I employ in this chapter:*it is writing with light*
photographs are the eyes with a memory
it is a one to one correspondence to the world
language characterised by compression, image and metaphor
the distillation of the essence of being, through language

The value of using photography in environmental writing is that it can visualise (and question) the power of humans to portray themselves as often in the centre. Jasmine Ulmer (2021, p. 239) has provided insight into such processes in her photo essay on the "Anthropocenic gaze". She challenges the way that humans take up most of the space within their own photographs (classically through selfies) and instead presents a series of photographs in which humans have been zoomed out and the landscape, the place, plants, and water take up the space instead. As Ulmer writes: "photography can help us put ourselves into perspective". Ulmer (2018) emphasises how using diverse genres, attending to the details, scenes, and experiences of everyday life promotes a slower pace of writing. This can help to increase the attention paid to specific aspects that have caught one's interest or provoked the reaction of the writer/photographer. This slower way of working means practising reading, listening, and responding to events and the environment in the local community. The importance of reinventing ways of understanding embodiment in the process of becoming environmental writers also promotes rethinking and shaping renewed ways of understanding diversity and intersectionality. Tiainen et al.'s (2020, p. 50) inspiration from new materialism and more-than-human agency through *the middle* is a possibility that challenges structures of oppression by producing "open-ended relationalities happening across social, material, discursive, human and more-than-human areas of activity". I interpret middling as a methodological move in which the material, the discursive, and the organic come together.

Guidelines for your own practice

In line with the uses of poetry and photography in this section and the other sections in this chapter, you are encouraged to create your own use of poetry and photography thematically related to "Water". Imagine how the multiple forms of water can be an inspiration to shape your writing.

In your analytical reflections, you are encouraged to pay attention to the sensory perceptions and shaping of knowledge by using poetry and images. Listen to Koleka Putuma's poem, and listen to your own poetry and/or look at an image of water. Think about the political and power aspects related to water and how they are relevant for knowledge and learning in social work.

Poetry is a very valuable genre that allows for mediating experiences that would otherwise often remain unarticulated. Its emotional, compressive, and rhythmic expressivity allows for gaps and in-betweenness to be voiced and communicated. Over the years my own writing of poetry has often been co-produced with photographing, and as you see in this chapter, I often use photography when I write a diary. Photography is useful for visualising ways of seeing, and images have their own way of telling stories. At the same time, the context of the photography, who/what is portrayed and who is seeing, is a complex relationship. I often use photography to teach me to become more attentive to details, to engage in a better understanding of the entanglements between human and more-than-human agency, and to capture unexpected moments of seeing. The consciousness of the power of photographic images to privilege specific viewpoints can be taken as a departure for consciousness-raising and critical practice.

Becoming Environ*mental* Writers!

18–19 June 2022, Creative Resistance, Imagination and Writing to Reshape the Future

The rain is intense, and I had to hold hard onto my umbrella in the strong wind when I went out to buy food this evening. I'm working with this chapter, reading and rereading the text repeatedly, editing and deleting and adding to the guidelines. I want to say something more about the future, but I know the future is already here. I put "writing to reshape the future" in the entry title because I find that some days are intolerable, even if I have resources and should "feel safe". The environmental situation is pressing; ecologically, economically, socially and mentally: "How do we creatively resist our present situation when circumstances become intolerable?" asks Frichot (2019, p. 190). The multiple crises of the pandemic, the climate, violence, and death due to illness, war, racism and sexism, exploitation of workers and the food crisis. Anna Tsing (2015, p. 22) argues for the need to reshape the imaginaries of progress as well as rethink the practices of how we write the future: "As long as we imagine that humans are made through progress, nonhumans are stuck within this imaginative framework too." I believe that this is what we risk as researchers and teachers in social work, that we are stuck in fast science production and haven't taken the time to rethink, reimagine and rewrite our work away from the fast science of knowledge production. I read Isabelle Stengers's (2018, p. 106) book where she argues that another science is possible. She refers to a past a few years ago when theses' were written from the perspective of future generations as a critique of "the unsustainable character of what we call development", but now:

"[…] we … realise that the future is coming towards us at full speed."

These words are important. They are written with such clarity, stated briefly, straight to the point. When the future is hurtling towards us at full speed, we who have positions in academia need to "imagine how we will answer to those who are not here, but who nevertheless already exist" writes Stengers (2018, p. 106). We need to make time to find ways of rewording and rethinking knowledges from the diversity of embodied and situated locations in the academic world and beyond. How can we use writing to situate our knowledges, to situate and resituate ourselves together with human and more-than-human co-authors who have so far often remained in the margins of eco-social and post-anthropocentric social work? In this chapter, I proposed that by replacing what is commonly regarded as "social problems" by making use of exhaustion, it is possible to create interconnections through and beyond words to materialities where both humans and more-than-human embodiments and life forms are interdependent. The multiple and overlapping genres of creative life writing open for creative and critical explorations of how we think we know what we know, how we can also rethink and unthink the dominance of the written form as an expression for fast science.

It is now morning. The rain has stopped, and the sky is blue. I take a break and walk to the sea. I listen to the waves. The ice blocks are no longer here, but they still remain lively in the sensory perception of my memory world; I can hear the sound of their movements and think and write to the sound that occurs when they clash into each other and separate, the sound of ice-slush during the ice thaw, the heartbeats of the Gulf of Bothnia.

28–29 May 2023, The Futures of Creating Spaces for Environmental Writers

It is the end of May when I write this last diary entry. The chapter is almost finished, but I have been asked to say something about what could be the future challenges for applying arts and creative writing in social work education, including new technologies. My first thought is that notions of "the future" have been radically transformed, and in the title of this chapter I use the word futures to signal that there are many different futures, depending on where we are situated as people and in our knowledge. I think this is an important starting point for co-learning and co-writing practices in the social work classroom. Becoming environmental writers potentially changes the space of the classroom because it broadens the concept of equality to include more-than-human life forms and non-living existences as part of the learning process.

I wish to promote trust, creativity and imagination as lodestars when working with art-based methods and creative writing in social work education. After many years as a leader of writing workshops and using creative writing in research, I often encounter students and colleagues who do not trust what they think is important, who are held back in their creative ability to think anew and do research. They do not study social work in a creative and imaginative way because the norms for how to work and study in academia are restricted. Imagination is necessary if we are to achieve environmental justice, a goal that is not being fulfilled today. Without imagination and hope for a better future for all living creatures and the natural world, research-based knowledge in ecosocial and post-anthropocentric social work will be difficult to achieve.

In digitalised societies it is important to take into consideration how sites and expressivity concerning writing tend to blur boundaries between the material and the discursive, and between the physical and digital words and sounds of writing. During the last year, my writing and learning practices have taken on more digitalised forms, both in the way I write and am inspired by the writing of others. I often record rather than write my diary notes. Visiting the same place and noting the differences of the conditions in that place is a valuable way of learning. As I record my diary entries, my voice blends with the sound of water movements and my thoughts, emotions, memories, ways of seeing and learning are entangled in a rhythm that is different each time due to the water. Sometimes when specific weather conditions occur like the ice-thaw, I photograph and record the sound of waves and listen to them afterwards. I use them as a soundscape when I create some of my lectures, and they inspire my thinking about achieving environmental justice through more-than-human agency. At the same time, I also learn from recordings like Koleka Putuma's poetry on water in the context of South African inequality, which I discussed in an earlier section of this chapter. What I wish to say is that as the art-based practices of creative writing shift between genres and become environmental, they challenge the idea of writing as something that is limited to words. The philosophical and theoretical departures that promote writing in the post-Anthropocene offer perspectives to understand human and more-than-human entanglements better. The way writing is combined with practices of reading, rewriting, photographing and recording more-than-human existence, makes material existence, sound and movements from the natural world co-creators in writing. Voices of animals and water movements can be heard and listened to. Images of seeing a view of a landscape encourage us to see again. Becoming environmental writers can contribute to knowledge and learning that are inclusive to more-than-human agency. I hope that this chapter has inspired students, teachers and researchers to create their own ways of interpreting the richness of posthuman philosophy, using diverse genres, expressions and practices.

Bibliography

Agrawal, A. (2005). *Environmentality: Technologies and government and the making of subjects*. Duke University Press.
Anzaldúa, G. (1987). *Borderlands/La Frontera: The new mestiza* (4th ed.). aunt lute books.
Baikie, G. (2020). Indigenist and decolonizing memory work research method. *Journal of Indigenous Social Development, 9*(1), 41–59.
Barth, J. (1984). The literature of exhaustion. In *The Friday book: Essays and other non-fiction* (pp. 62–76). John Hopkins University Press.
Bell, K. (2021). A philosophy of social work beyond the Anthropocene. In V. Bozalek & B. Pease (Eds.), *Post-anthropocentric social work. Critical posthuman and new materialist perspectives* (pp. 58–67). Routledge.
Bozalek, V., & Pease, B. (2021). *Post-anthropocentric social work. Critical posthuman and new materialist perspectives*. Routledge.
Chambon, A. (2005). Social work practices of art. *Critical Social Work, 6*(1), 1–11.
Chancel, L., Piketty, T., Saez, E., & Zuchman, G. (2022). *The world inequality report*. Retrieved July 18, 2024, from https://wir2022.wid.world/www-site/uploads/2022/01/Summary_WorldInequalityReport2022_English.pdf
Coates, J., Gray, M., & Hetherington, T. (2006). An "ecospiritual" perspective: Finally a place for indigenous approaches. *British Journal of Social Work, 36*(3), 381–399.
Crutzen, P. J. (2002). Geology of mankind. *Nature, 415*, 23.
Crutzen, P. J., & Stoermer, E. F. (2000). The "Anthropocene". *Global Change Newsletter, 41*, 17.
Davies, B. &., Gannon, S. (Eds.) (2006). *Doing collective biography: Investigating the production of subjectivity. Open University Press.*
Deleuze, G. (1988). *Spinoza. Practical philosophy. City Lights.*
Deleuze, G. (1995). The exhausted. *SubStance, 24*(3), 3–28.
Dominelli, L. (2012). *Green social work: From environmental crises to environmental justice*. Polity Press.
Dominelli, L. (2018). *The Routledge handbook of green social work*. Routledge.
Du Bois, W. E. B. (1903/1989). *The souls of black folk*. Penguin Books.
Eliassi, B. (2012). Political terrains of writing belonging, memory and homeland. In M. Livholts (Ed.), *Emergent writing methodologies in feminist studies* (pp. 83–97). Routledge.
Frichot, H. (2019). *Creative ecologies: Theorizing the practice of architecture*. Bloomsbury.
Guttorm, H., Kantonen, L., Kramvig, B. & Pyhälä, A. (2021). Decolonized research-storying: Bringing indigenous ontologies and care into the practices of research writing. In Virtanen, P. K., Keskitalo, P., & Olsen, T. (Eds.), *Indigenous Research Methodologies in Sámi and Global Contexts (113–143). Brill Sense.*
Gray, M., Coates, T., & Hetherington, T. (2013). *Environmental social work*. Routledge.
Haug, F. (2008) Memory-work as a method of social science research. A detailed rendering of the memory work method. In Hyle, A. E., Ewing, M.S., Montgomery, D., &., Kaufman, J. S. (Eds.) *Dissecting the mundane. International Perspectives on memory work (pp. 21–41). University Press of America.*
Haug, F. & Others (1987). *Female sexualisation. A collective work of memory. London: Verso.*
Hyers, L., Swim, J., & Mallett, R. (2006). The personal is political: Using daily diaries to examine everyday prejudice-related experiences. In S. N. Hesse-Biber & P. L. Leavy (Eds.), *The practice of qualitative research* (2nd ed., pp. 313–335). Sage.

Kaun, A. (2010). Open-ended online diaries: Capturing life as it is narrated. *International Journal of Qualitative Methods, 9*(2), 133–148.

Kenten, C. (2010). Narrating oneself: Reflections on the use of solicited diaries with diary interviews. *Forum Qualitative Sozialforschung/Forum: Qualitative Social Research, 11(2).*

Kuhn, A. (2002). *Family secrets: Acts of memory and imagination. Verso.*

Kuhn, A. (2010). Memory texts and memory work: Performances of memory in and with visual media. *Memory Studies, 3(4), 298–313.*

Latham, A. (2020). Using diaries to study urban worlds. In A. Ward (Ed.), *Researching the city* (pp. 104–124). Sage.

Leavy, P. (2015). *Method meets art: Art based research practice*. The Guilford Press.

Lejeune, P. (2009). *On diary*. University of Hawaii Press.

Livholts, M. B. (2001/2011). *'Women', welfare, textual politics and critique. An invitation to a thinkingwriting methodology in the study of welfare. Dissertation, Lambert Academic Publishing.*

Livholts, M. B. (2012). *Emergent writing methodologies in feminist studies*. Routledge.

Livholts, M. B. (2013). Writing water. An untimely academic novella. In L. Stanley (Ed.), *Documents of life revisited: Narrative and biographical methodology for a 21st century critical humanism* (pp. 177–192). Routledge.

Livholts, M. B. (2017). "What we learn how to see": A politics of location and situated writing in glocalised social work. In M. Livholts & L. Bryant (Eds.), *Social work in a glocalised world* (pp. 89–105). Routledge.

Livholts, M. B. (2018). Narrative writing as art based method. *Synnyt: Taidekasvatuksen Tiedonala, 21, 9–30.* https://wiki.aalto.fi/download/attachments/145123907/Livholts_OK.pdf?version=1&modificationDate=1547579585070&api=v2

Livholts, M. B. (2019a). A situated reading diary of exhaustion as a creative methodology: Mona Livholts reads creative ecologies written by Hélène Frichot. *Reading Writing Quarterly*. https://site-writing.co.uk/wp-content/uploads/2019/12/Mona-Livholts-reads-Creative-Ecologies-written-by-Hél-ene-Frichot.pdf

Livholts, M. B. (2019b). *Situated writing as theory and method: The untimely academic novella*. Routledge.

Livholts, M. B. (2021). Exhaustion and possibility: The Wor(l)dlyness of social work in (G)local environment worlds during a pandemic. *Qualitative Social Work, 20*(1–2), 54–62. https://doi.org/10.1177/1473325020973314

Livholts, M. B, (2022a). *Writing water for social justice in social work. [Swedish: Att skriva vatten för social rättvisa i socialt arbete]*. Sosiaalityön Tutkimuksen Seura. Retrieved March 19, 2023, from http://www.sosiaalityontutkimuksenseura.fi/Ajankohtaista?id=177

Livholts, M. B. (2022b). De-colonising writing in post-anthropocentric social work. *SocialDialogue, 26*, 17–26.

Livholts, M. B. (2023). *The body politics of glocal social work. Essays on the post-anthropocentric condition*. Routledge.

Livholts, M. B., & Södergran, F. (2024). The futures of writing *with* posthuman feminism in social work. In C. Noble, L. Harms Smith, L. Rasool, & G. Muñoz-Arce (Eds.), *Routledge international handbook of feminisms and social work*. Routledge.

Matthies, A-L., & Nähri, K. (2019). *The Ecosocial Transition of Societies: The Contribution of Social Work and Social Policy. Routledge.*

Matthies, A.-L., Närhi, K., & Ward, D. (Eds.). (2001). *The eco-social approach in social work*. Sophi.

Mayor, C., & Pollack, S. (2022). Creative writing and decolonizing intersectional feminist critical reflexivity: Challenging neoliberal, gendered, white, colonial practice norms in the COVID-19 pandemic. *Affilia: Feminist Inquiry in Social Work, 37*(3), 382–395. https://doi.org/10.1177/08861099211066338

McCulliss, D. (2013). Poetic inquiry and multidisciplinary qualitative research. *Journal of Poetry Therapy, 26*(2), 83–114.

Närhi, K., & Matthies, A.-L. (2018). The ecosocial approach in social work as a framework for structural social work. *International Social Work, 61*(4), 490–502.
Oakley, K., Ward, J., & Christie, I. (2018). Engaging the imagination: "New nature writing", collective politics and the environmental crisis. *Environmental Values, 27*(6), 687–705.
Plummer, K. (2001). *Documents of life 2: An invitation to critical humanism*. Sage.
Puar, J. K. (2012). "I would rather be a cyborg than a goddess": Becoming-intersectional in assemblage theory. *Philosophia, 2*(1), 49–66.
Putuma, K. (2016). *Water*. Retrieved March 19, 2023, from https://www.youtube.com/watch?v=8dfq3C8GNrE
Rambaree, K. (2020). Environmental social work: Implications for accelerating the implementation of sustainable development in social work curricula. *International Journal of Sustainability in Higher Education, 21*(3), 557–574.
Rambaree, K., Sjöberg, S., & Turunen, P. (2019). Ecosocial change and community resilience: The case of "Bönan" in glocal transition. *Journal of Community Practice, 27*(3–4), 231–248.
Richardson, L. (1997). *Fields of play. Constructing an academic life*. New Rutgers University Press.
Spooner, M. (2011). *A clinic for the exhausted* (Doctoral dissertation, RMIT University).
Stengers, I. (2018). *Another science is possible: A manifesto for slow science*. Polity Press.
Szto, P., Furman, R., & Langer, C. (2005). Poetry and photography: An exploration into expressive/creative qualitative research. *Qualitative Social Work, 4*(2), 135–156.
The Global Agenda for Social Work. (2012). Retrieved November 2, 2023, from https://www.iassw-aiets.org/global-agenda/
The Global Definition of Social Work. (2014). Retrieved November 2, 2023, from https://www.iassw-aiets.org/global-definition-of-social-work-review-of-the-global-definition/
The Global Standards for Social Work Education and Training. (2004/2020). Retrieved November 2, 2023, from https://www.iassw-aiets.org/global-standards-for-social-work-education-and-training-archive/
The New Global Agenda for Social Work. (2020). Retrieved November 2, 2023, from https://www.iassw-aiets.org/global-agenda/
Tiainen, M., Leppänen, T., Kontturi, K., & Mehrabi, T. (2020). Making middles matter: Intersecting intersectionality with new materialisms. *NORA – Nordic Journal of Feminist and Gender Research, 28*(3), 211–223.
Tsing, A. (2015). *The mushroom at the end of the world: On the possibility of life in capitalist ruins*. Princeton University Press.
Ulmer, J. (2018). Minor gestures: Slow writing and everyday photography. *Qualitative Research in Psychology, 15*(2–3), 317–322.
Ulmer, J. B. (2021). Refocusing the Anthropocenic gaze: A photo essay. *Journal of Posthumanism, 1*(2), 235–243.
United Nations. *Sustainable Development Goals: 2015–2030*. Retrieved November 2, 2023, from https://sdgs.un.org/goals
Witkin, S. L. (Ed.). (2014). *Narrating social work through autoethnography*. Columbia University Press.

Case Study 4: Sustainability Practices in Portugal: A Reflection on Ecosocial Practice

Daniela Freitas

As a social worker with 10 years of professional experience in community intervention and as a recent researcher of ecosocial practice in Portugal, I believe that social work has a fundamental mission in promoting sustainability. Social work should include the environmental dimension in the core of its intervention. This inclusion in practical terms can be achieved through innovative, creative, and integrative community intervention models and projects, with strategies for solving/mitigating social and environmental problems. In the design and planning of the intervention projects, the different sectors of the community (citizens, groups, and public and private institutions) should be invited to look holistically at the problems and together seek resolution strategies for them.

In Portugal, in 2020, the Healthy Neighborhoods Program was created and financed by the Portuguese state, with the aim of promoting community intervention projects designed, planned, and presented by associations, collectivities, non-governmental organisations, civic movements, and residents' organisations. Social workers and other professionals in the social, medical, architecture, civil engineering, landscape, arts, and multimedia fields work in these public and private organisations. More specifically, in public organisations, in addition to social workers, professionals from education (teachers), architecture, landscaping, urban planning, construction, and health professionals such as doctors, nurses, physiotherapists, etc. were involved in the programme. Private organisations such as non-governmental organisations and foundations involved social workers, sociologists, psychologists, social entrepreneurs, trainers, visual and musical artists, as well as technology and multimedia professionals. This programme was created with the aim of being a participatory instrument that boosts the partnership between local institutions and the population, with a view to improving the quality of life of communities,

D. Freitas (✉)
ISCTE University Institute of Lisbon, Lisbon, Portugal
e-mail: daniela_oliveira_freitas@iscte-iul.pt

C. Forde et al. (eds.), *Teaching and Learning in Ecosocial Work*,
https://doi.org/10.1007/978-3-031-58708-5_16

including the environment. Across the continental region of Portugal,[1] 246 projects were funded and cofinanced out of 752 applications eligible for funding. As a result, 135 projects were cofinanced, and 111 projects were fully funded by the Healthy Neighborhoods Program.

To know the reality of community intervention in Portugal and doing a PhD on ecosocial intervention, I started the doctoral research with a qualitative study, with content analysis of the projects financed by the Healthy Neighborhoods Program. Data from the 246 applications of the funded projects were analysed. Since it is impossible to describe all the projects, I present two projects that proposed actions to respond to social and environmental problems simultaneously, revealing a potential ecosocial practice.

Promoting Environmental Awareness with Young People

One of the projects took place in a coastal town, in the district of Lisbon, involving two local youth associations and two parish councils. The project aimed to introduce awareness and actions of circular economy and empowerment of young people in a neighbourhood, involving mainly young NEETs (*Not in Employment, Education or Training*) in the construction of skateboards, using recycled materials. They are young people who usually use skateboards to get around and as a hobby. The creation of new skateboards with recycled materials was intended for their own use but also for sale to the public (physically and online), thus constituting paid employment for a small number of young people. The target audience directly involved in this project was young people, especially NEETs, but it is important to mention that its indirect target audience was their families and the whole community, since one of the objectives of the project focused on the insertion of environmental and circular economy practices that responded to local employment and entrepreneurship needs.

As there was already a skatepark in the community, widely used by young people, the idea of creating and selling skateboards made from recycled materials connected with an activity of choice for them. In fact, benefits of this intervention included the involvement of NEET young people in an activity of their interest but also the development of artistic, multimedia, and economic, youth, and local entrepreneurship skills. At the same time, this project responded to environmental problems such as excess waste and the need for recycling and reuse of materials, raising awareness and involving young people and the whole community in the collection, recycling, and reuse of recycled materials, as well as in the promotion of cleaning actions of local spaces and beaches. It was a social intervention project with an

[1] The programme was only launched in continental Portugal, excluding the islands. The islands have their own governments.

impact on the environment and on the lives of these young people in the short term. One of the benefits of this project has been to create business and paid employment opportunities that can last over time. Another benefit is the environmental awareness and behaviour change in the whole community in favour of the environment.

Caring for People with Dementia Without Forgetting the Environment

The second project took place in a city in the interior of Portugal, in the district of Santarém, where the population is older. The project was promoted by the Alzheimer's Association in Portugal and local social institutions for the elderly, where social workers work. The main objective was to improve the quality of life of people with dementia and their caregivers, through the provision of specialised services at home (provided by clinical professionals who work in partner public health institutions) and access to sensory/cognitive stimulation materials, created with recycled materials, by unemployed citizens of the community. Some examples of the materials created are exercises and games for cognitive and sensory stimulation and for maintaining motor skills, such as puzzles and geometric figures with different textures and which emit different sounds when moved.

The target audience involved in this project was unemployed community members as well as people with dementia (as beneficiaries and users of the materials produced) and their caregivers. Throughout the project, caregivers received training on cognitive and sensory stimulation of people with dementia and were involved in the evaluation of the materials produced and proposing improvements.

The idea of creating sensory and cognitive stimulation materials for people with dementia, based on the recycling and rehabilitation of recycled materials, emerges from the growing context of people with dementia (Alzheimer Europe, 2019) and the need to introduce environmental and recycling awareness in the community. This project focused on providing specialised support to a vulnerable public (people with dementia and their caregivers), through the involvement of unemployed people (who were paid), thus contributing to the improvement of the quality of life of the community in general.

In addition to the explicit benefits of this project, both for people with dementia and their caregivers and for unemployed people and the environment, this project made possible the creation and subsequent commercialisation of these types of materials for the different stages of dementia, becoming a successful example of innovation and local entrepreneurship with endorsement from the Alzheimer's Association of Portugal.

Reflections on the Projects

Involving different target audiences, both projects developed actions for specific social problems, incorporating environmental awareness, especially of the emerging need to recycle and reuse materials. They were developed by local entities in partnership and funded 100 per cent by the Healthy Neighbourhoods Program, in the amount of €50,000 and €49,545, respectively. Funding for Healthy Neighbourhoods 2020 projects ended in June 2023, but the Portuguese government has already announced that it will open a new application for the programme. The individual results of these two projects have not been released, but a total of 407 jobs have been created, 284 of which will continue after the funding ends, through self-funding.

I consider that the projects presented are practices that promote sustainability in Portugal, as they incorporate a concern with environmental issues and with the need for communities to think and adopt more responsible behaviours towards nature and the environment. Although they have an environmental focus, these projects essentially seek to respond to social problems. Both have an orientation towards the mitigation of specific social problems, through actions that include and benefit the environment. In the first example, the aim is to promote the development of skills and employment opportunities of young NEETs, and in the second example, the aim is to promote an improvement in the quality of life of people with dementia and their caregivers. In both, the environmental intervention objective appears as a strategy, which together with other actions aims to mitigate social problems. The environmental dimension is present in these two projects, through environmental awareness actions, cleaning public spaces, and encouraging recycling and reuse. These projects demonstrate recognition of the link between the social and environmental systems in solving community problems.

Training Social Workers for Ecosocial Practice

Ecosocial practice presupposes the inclusion of the environmental dimension in the core of social work intervention. These projects demonstrate environmental concerns, and I believe that with training, social workers will play a key role in promoting strategies and responses focused on mitigating environmental problems. For this, a change in the perspective of social workers is needed. Some already have an integrative view of the environmental dimension in their interventions, but this is not enough. The climate crisis and its devastating consequences demand more from social work. For this change to take place, it is essential that the training of social workers incorporates specific content of ecosocial practice, including content on values, rights, and global challenges, not only social but also environmental and economic, addressing issues related to ecology, sustainability, the circular economy, education, and environmental justice. As we can see in the second project, the right to dignified ageing and improving the quality of care for people with dementia was crossed and linked with environmental awareness and sensitisation. In the training of social workers, it is also important to introduce intervention models that recognise the interdependence between the individual, community, and environment systems, such as the ecosocial model (Belchior-Rocha, 2018). The training of social workers should be guided by the promotion of reflexivity, critical thinking, and immersive experiences. The use of immersive methodologies promotes participation, reflection (interdisciplinary, individual, and class), and, consequently, learning through experimentation and sharing. The possibility for students to plan, develop, and participate in community projects where they have to propose answers to existing social and environmental challenges encourages greater and more in-depth learning of social reality and its interdependent relationship with the environment. With immersive experiences, students will be able to understand how environmental problems produce and accentuate social problems or, on the contrary, how environmental issues can enhance social development (the environment at the service of society in a sustainable way, promoting employment through example) but also understand how environmental situations can be resolved through community/social activation. These immersive experiences provide students with opportunities to participate in social intervention projects, in their different phases (community consultation/diagnosis, design and development of actions/strategies, and monitoring/evaluation) with other professionals, citizens, and local institutions, preparing them for ecosocial practice.

With training in ecosocial practice, social workers will have the values, knowledge, and skills to respond to humanity's greatest challenge, acting locally and globally in favour of fairer and more sustainable societies and environments.

Bibliography

Alzheimer Europe. (2019). *Dementia in Europe yearbook 2019: Estimating the prevalence of dementia in Europe*. Alzheimer Europe. ISBN 978-99959-995-9-9. Accessed on 20 Oct 2023.

Belchior-Rocha, H. (2018). Social work practices and the ecological sustainability of socially vulnerable communities. *Sustainability, 10*(5). https://doi.org/10.3390/su10051312

Chapter 12: Shifting the Human-Nature Relationship in Outdoor and Nature-Based Social Work Programs: Contributions to Teaching and Learning Relational Ecosocial Work

Emmanuelle Larocque

Introduction

Contemporary human-Nature[1] relations are identified as critical relational drivers of the socio-ecological and biodiversity crises (Artmann, 2023). Conversely, developing and sustaining harmonious and partnership-based interactions between people and local ecosystems can positively impact human and planetary well-being (Seymour, 2016) while also facilitating transformative social change and environmental stewardship (Lugg, 2007). Despite the centrality of human-Nature connectedness in moving towards sustainability and socio-ecological equity (Artmann, 2023), the field of social work has paid little empirical attention to the relational aspects of social-ecological change processes. Yet, the human-Nature relationship is an area of research that is of interest to teaching and learning ecosocial work because it (1) represents an important but understudied avenue to promote holistic and integrative approaches to well-being (Boetto, 2017); (2) contributes to shifting the dominant, anthropocentric (human-centered) perspective to one that is based on (re) connection, reciprocity, and reconciliation to "Earth others" (Willis, 2011); and (3) symbolizes a crucial step in working towards a just, equitable "social-ecological transition" (SET) (Audet et al., 2021), a concept which refers to a complex, multilevel, and transdisciplinary change process that fundamentally aims to drive social

[1] While I recognize that the term "Nature" is based on Western dualistic ideology, the terminology will be used herein to refer to the natural world and to all living organisms that enliven ecosystems. The words "Nature" and "Forest" are capitalized to account for the "personhood" of nonhuman living beings and to position humans within this broader and interconnected web of life (Pereira & Bina, 2020).

E. Larocque (✉)
School of Social Work, Université du Quebec en Outaouais, Gatineau, Canada
e-mail: emmanuelle.larocque@uqo.ca

C. Forde et al. (eds.), *Teaching and Learning in Ecosocial Work*,
https://doi.org/10.1007/978-3-031-58708-5_17

and ecological innovations to respond to the global environmental crisis (Huntjens, 2021). This chapter aims to fill a gap in literature by providing theoretical, methodological, and practice-based strategies to articulate Nature connectedness as an integral dimension of ecosocial work.

Presented as a case study, this chapter discusses an outdoor and Nature-based program (entitled *Feet to Earth*) designed for young adults (18–30 years old) and implemented in a Forest setting in Eastern Ontario (Canada) as part of a collaborative intervention research (Stassart et al., 2008). With the hope of forging new pathways for teaching and learning ecosocial work, the specific objective herein is to critically reflect on the steps undertaken to develop the ecosocial component of the program and to discuss the strategies pursued to generate transformative, resonant (Rosa, 2019), and embodied (Wacquant, 2015) human-Nature relations in Nature-based intervention settings.

The chapter is divided into four sections. The first part presents the theoretical underpinnings of the research. The method section follows and highlights the strategies carried out to center on human-Nature relations during the program. I then present the program's objectives, protocol, and main practice frameworks (outdoor and Nature-based therapy, narrative therapy, yoga therapy) in relation to how they were applied to create conditions for the emergence of embodied and resonant experiences *with/in* Nature. To enrich this section, data collected during the program are presented for illustrative purposes. Finally, key implications for ecosocial work teaching and learning are laid out and include (1) multiplying experiential learnings to foster reflexivity and connectedness, (2) emphasizing collective action to move beyond individual and moralistic approaches, (3) unpacking and re-storying human-Nature relationships in social work settings, and (4) recognizing the centrality of the body in teaching and learning ecosocial work.[2]

Positionality

The subjective thinking and learning processes presented in this chapter highlight *possibilities* for transformative change which are framed within a specific geosocial context and standpoint. As a white francophone woman, mother, ecological activist, social worker, and professor who lives on Algonquin Anishinaabe unceded territory (in Eastern Ontario, Canada), I recognize that my social positionality is not value-free and impacts my understandings of human-Nature relations and of alternative worldviews, such as ecocentrism. As such, this nonlinear journey of *becoming* an ecosocial worker involves a constant reorientation of our ways of *being*, *knowing*,

[2] While various terminologies are found in literature (e.g., green social work, environmental social work), the term "ecosocial work" is used to refer to a radically transformative and ecocentric approach, which aims to transcend Western sociocultural and colonial perspectives of Nature and the environment (Boetto et al., 2022).

and *doing* (Boetto, 2017) rather than an end product. To reduce sociocultural biases, I adopted a self-reflective approach and, through an empirical process, complemented my own situated standpoint with a "plurality of knowledge" (Pereira & Bina, 2020, p. 315) to co-develop the program. Ultimately, I hope this chapter will spark collective reflections on ways to nurture and nourish harmonious human-Nature relations through ecosocial work teaching and learning.

Theoretical Groundings

Following Boetto's transformative ecosocial work model (2017) which incorporates ontological (*being*), epistemological (*thinking*), and methodological (*doing*) dimensions of practice, ecosocial work can be broadly defined as a critical, multi-level, and transformative practice approach committed to challenging modernist and colonial worldviews (Boetto et al., 2022). Recognizing the interdependencies between humans and all living systems, this radically relational approach to ecosocial work involves shifting towards a decolonial[3] (Boetto, 2017), holistic, and decentralized practice framework (Coates & Gray, 2019) across all levels and areas of social work (Rambaree et al., 2019).

Despite the intention to promote planetary health, which includes a sustainable approach to restore and regenerate biodiversity and the health of ecosystems (Dominelli, 2012), ecosocial work scholarship has mainly focused on the impacts of climate-related issues and power relations on *human* populations. As such, the rights and well-being of "Earth-others" (Willis, 2011) are generally left aside in existing literature. This lack of studies on human-Nature relations is surprising since it consists as a key element to "confront and resolve the fundamental epistemological and ontological divide that stands in the way of transformative change" (Pereira & Bina, 2020, p. 311). Hence, addressing this fractured relationship represents a fundamental proactive response to global sustainability challenges (Besthorn, 2000). This chapter contributes to strengthening the relational discourse in social work by focusing on how human-Nature relationships can evolve and be nourished through outdoor and Nature-based therapy (Gargano, 2022). To highlight the role of affective, discursive, and embodied experiences in the expansion of human-Nature relations beyond anthropocentric and utilitarian understandings, it seems most appropriate to adopt a bifocal theoretical lens of resonance (Rosa, 2019) and carnal sociology (Wacquant, 2015).

[3] Decolonization is recognized "as a long-term process involving the bureaucratic, cultural, linguistic and psychological divesting of colonial power" (Smith, 1999, p. 98).

(Re)Connecting to Nature Through Experiences of Resonance

In line with transformative ecosocial work's critical endeavors (Boetto, 2017), Rosa's comprehensive theory of resonance (2019) is useful to problematize the social-ecological and biodiversity crises as *relational* crises, embedded in the types of relationships promoted and sustained by modernist societal systems. Indeed, growth- and acceleration-oriented Western societies, rooted in industrialism and capitalism, have dramatically sped up all facets of social life (Rosa, 2009). This significant increase in the pace of everyday life activities, combined with technological hyper-connectivity, is affecting our collective ability to focus, connect (to self, others, and Nature), and critically reflect (Hari, 2022). These modern politics of speed (Rosa, 2019) are having unprecedented effects on human and planetary well-being and are also changing the way humans are relating to time, space, and place (Rosa, 2019), therefore resulting in a "crippling disjuncture between the individual, society, and Nature, and the modern competition over natural resources" (Fuchs, 2020, p. 334). By encouraging hierarchal and utilitarian relations, modernity's project has simultaneously widened the divide between humans and Earth-others, where human-Nature interactions have progressively become "mute, non-affective and thus non-responsive" (Artmann, 2023, p. 3). In this context, contemporary problems, such as the macro-scale climate crisis or the micro-level ailments that infuse everyday life (e.g., depression, burnout, and the increasingly diffused label of ecoanxiety), can therefore be understood as by-products of life-temporal acceleration, a process which triggers and generates disconnection and alienation to self, others, and Nature (Rosa, 2019). The concept of social acceleration is therefore relevant to critically examine the "hegemony of speed" (Rosa, 2009, p. 27) and engage in counter-narratives which embrace embodied, mindful, meaningful, and Nature-centered interactions.

To overcome this disjuncture, Rosa proposes the concept of "resonance" (Rosa, 2019, p. 185) as an alternative, vibrant, and connected way of engaging with the world. Constructed as a process and based on theories of empathy, resonance refers to an "anthropological ability that connects human beings with others, with the world and its objects" (Fuchs, 2020, p. 254) and entails a receptive, responsive, affective, embodied, and active form of being-into-the-world (Rosa, 2019). Resonance therefore involves a bidirectional movement of affection (being touched or moved by something) and emotion (the response given to establish a connection) (Rosa, 2019). At the practical level, building resonant relationships involves creating space for the development of deep, meaningful, intuitive, and embodied connections with the world (self, others, and Nature). Specific intervention tools will be discussed in the section on the main practice frameworks but to spark curiosity; these can include (1) promoting slow and reflective practices (e.g., meditation, mindfulness, breathing techniques) to foster attunement to our senses and "prepare" the body to feel and adopt Nature's rhythms (Larocque, 2022), (2) deconstructing dominant ideas about Nature through narrative approaches to enable a re-storying process (Nicholas, 2021), and (3) taking the time to learn about the natural world by combining outdoor and Nature-based therapy with ecopedagogy and environmental ethics (Payne & Wattchow, 2009). Because resonance is not solely a cognitive process but also an affective, sensory, and embodied experience, the body plays a central role in cultivating resonant experiences *with/in* Nature and is therefore viewed as an instrument *of* and anchor point *to* experiences of resonance (Plard, 2016).

Carnal Sociology: Engaging the Body in Human-Nature Relations

Whereas the body is a vital conduit for sensory experiences and for the development of attunement and connection to Nature (Larocque, 2022), carnal sociology (Wacquant, 2015) complements Rosa's theory by casting light on the immersive and embodied dimensions of resonance. This social ontology constructs the social being as a "sensate, suffering, skilled, sedimented, and situated corporeal creature" (Wacquant, 2015, p. 2) and focuses on the lived experience and embodiment of individuals within social contexts (Quidu, 2022). Carnal sociology, which considers the body as a legitimate site of knowledge and mediator of social interactions and experiences, is especially useful to capture the singularity and nuances of resonant experiences as they occur *with/in* Nature. Thus, attending to the "active, perceiving, and sensuous corporeality of the body with other bodies (human and more-than-human)" (Payne and Wattchow, 2009, p. 16) calls for a temporal adjustment to "attune" to body, space, and time. This can be accomplished by, for example, taking the time to slow down intervention processes through bodily and immersive practices (yoga, mindful forest walks or bike rides, journaling). Finally, carnal sociology leverages "carnal comprehension by deepening [the researcher's] social and symbolic insertion into the universe he studies" (Wacquant, 2015, p. 4) and is therefore most appropriate to support fieldwork, such as participant observation, which considers researchers' experiences as empirical data (Quidu, 2022).

In sum, two theoretical frames are mobilized within this research to examine and rethink dominant ways of relating to Nature. Rosa's theory of resonance offers an entry point to question life-temporal acceleration which organizes everyday social life and generates disconnections. It also provides a critical and ethical framework to cultivate alternative visions of human-Nature interactions by ascribing Nature with "relational qualities" and pursuing partnerships based on mutual responsiveness (Rosa, 2019), reciprocity, and coexistence (Artmann, 2023). Carnal sociology complements this approach by considering the embodied and sensorial dimensions of transformative experiences of resonance, as they are lived *by* and *through* the body.

The Research Process: A Collaborative, Transformative Approach to Ecosocial Program Development

An intervention research methodology (Stassart et al., 2008), a derivative of action research, was applied to promote purposeful and transformative action, favor a place-based approach, and bring forth a range of voices to co-design the protocol for the outdoor and Nature-based program (named *Feet to Earth*). The research was conducted in three phases, including (1) the preparation phase (literature review on the concept of "social-ecological transition" and outreach to community partners), (2) data collection through consultations with youth eco-activists, and (3) co-development and implementation of *Feet to Earth*.

Phase 1: Situating the Program Within a Social-Ecological Transition Perspective

To prepare for the fieldwork and to find ways to position the program as a strategy to engage with the "social-ecological transition" (hereafter, SET) (Audet et al., 2021), a sociohistorical review of literature on this concept was conducted to study its emergence, evolution, and appropriations within scientific scholarship. Constructed as a response to the limits of growth and to the accelerating social and environmental crisis, the SET refers to the systemic shift needed to construct sustainable, equitable, and resilient economic and societal models (Huntjens, 2021). The SET is therefore viewed as a process and involves a broad range of significant and interconnected changes (at the economic, political, environmental, and sociocultural levels) which aim to address local and global socio-ecological issues through various innovation strategies (from surface-level technocentric solutions to deeply transformative innovations) (Larocque, 2023a). The concept of SET is relevant for ecosocial work because it calls for a systemic, integrative, and transdisciplinary approach to social change, an approach which resonates with the macrosocial and paradigmatic transformation advocated by ecosocial work (Boetto, 2017). Further, instead of focusing on reactive solutions, as can be observed in conventional social work responses to climate change (Krings et al., 2020), the SET provides a solution-focused approach to address climate issues and ecological injustices. Adopting a proactive lens is therefore useful to contribute to minimizing socio-ecological vulnerabilities and injustices by addressing the fundamental *relational drivers* of the climate crisis (such as relationships to Nature). If ecosocial work is committed to engaging with deeply transformative practices, we must first broaden our understanding of environmental justice to shift towards an ecological justice approach (Rambaree et al., 2019) which includes Earth-others in the social justice project. Specifically, whereas "environmental justice" is rooted in anthropocentrism because it focuses solely on human experiences, "ecological justice, on the other hand articulates a broader, ecosocial vision of justice – justice for the ecosystem as a whole with the understanding that humans are only one part of the ecosystem" (Rambaree et al., 2019, p. 2). Focusing on human-Nature relationships therefore represents an attempt to embody and promote ecological justice in ecosocial work teaching and learning.

Considering the importance of knowledge plurality to understand how the SET is being imagined and played out in everyday life, this phase of the research also involved reaching out to various social agents (community partners, local eco-leaders) to better understand the geosocial specificities and ecological needs of the community in which the program was implemented (a francophone rural community in Eastern Ontario, Canada).

Phase 2: Enhancing the Ecosocial Component of the Program Through Youth Eco-Activists' Experiential Knowledge

The second phase of the intervention research involved consulting youth eco-activists to better understand how this unique group of young people defined the SET. This choice was motivated by three main reasons. First, this unique population holds both insider and practice-based knowledge (as youth experiencing the multi-dimensional effects of the climate crisis and engaged in climate action), which is essential to broaden current, mainly technocentric conceptions of the SET (Audet et al., 2021). From a critical perspective, youth are also recognized for their willingness and capacity to disrupt hegemonic ways of knowing (Barraclough et al., 2021). I thus hypothesized that youth eco-activists' experiential knowledge would complement existing practice-based ecosocial literature by bringing forth new types of ideas to enhance the ecosocial component of the outdoor and Nature-based program. Finally, constructing a youth-inclusive research agenda consisted in a strategy to engage with knowledge plurality and represents a political choice, as it reinstates importance of foregrounding disenfranchised voices (such as young people's) in decision-making processes (Schusler et al., 2019).

After receiving ethics certification, data was collected through semi-structured interviews (September–December 2020) with ten youth eco-activists from Canada (Québec, Ontario, British Columbia) and a focus group (conducted with five of those same youth) and analyzed through thematic analysis (Paillé & Muchielli, 2012). While empirical results are detailed elsewhere (Larocque, 2023a), it is worth mentioning that most eco-activists endorsed a people-focused narrative when discussing intricate elements of the SET. However, nuances in their discourses were observed and seemed to be especially influenced by their geosocial location. As such, participants from rural areas showed greater concern for biodiversity issues while social justice issues were at the top of urban activist's priorities. Also, contrary to scientific literature, which mainly frames the SET around economic and political issues, youth narratives revealed a broader, socially oriented conceptualization, framing the SET around three main processes: (1) intersectional climate justice, (2) intergenerational allyship, and (3) place-based connections (Larocque, 2023a).

To move these findings from abstractions to applicable ecosocial practices, youth eco-activists shared concrete examples of transformative experiences or interactions that shifted the way they related to the world. For example, one activist mentioned the importance of taking an intergenerational approach to ecosocial work, where involvement of parents or of community members in Nature-based programs may facilitate intergenerational allyship, break down communication barriers between generations, and extend ecosocial engagement to everyday life (instead of limiting this engagement to the social work context). All activists were keen on experiential approaches and hands-on learning, as immersive approaches helped them acquire new skills, gain a deeper understanding of climate-related issues and solutions, and enhanced their sense of agency and empowerment. Another aspect that was mentioned by most activists was the vital role of the arts, of poetry, stories, and metaphors in their personal change process, especially as it related to deepening their understandings of other's experiences. Although not all recommendations were included in the program due to time constraints or foreseeable ethical issues,[4] youth voices and ideas were prioritized when designing the program.

Phase 3: Program Implementation: Zooming in on the Human-Nature Relationship

Following this data collection phase with eco-activists, other actors from the community were consulted to collaboratively develop the program, including community partners (n = 3) (a community health center, a sport and recreation social enterprise, and the local department of forestry), local "eco-leaders" (n = 6), and program participants (n = 5). Two research questions guided the intervention phase of the study: (1) How can an ecosocial outdoor and Nature-based program be a lever for transformative social-ecological change? (2) What mechanisms or processes enable alternative (nondominant) ways of relating to Nature and well-being?

[4]One temporal and ethical issue concerns youth eco-activists' recommendation to include Indigenous voices in the program, an idea they approached cautiously due foreseeable tokenizing or cultural appropriation issues. After consulting with colleagues and to avoid taking a "check-box approach," it was decided that an essential first step would be to find ways to build authentic and intentional allyship instead of rushing into collaborations that may reproduce epistemic injustices or perpetuate neocolonial ideology (Fellner, 2018).

Recruitment

Recruitment was conducted through social media and by reaching out to local youth-oriented social organizations and aimed to obtain a small purposive sample (Frechette et al., 2021). Group size was limited due to public health recommendations related to the global COVID-19 pandemic. Eligibility criteria required that participants be (1) aged 18–30, (2) physically capable of practicing outdoor sports (mountain biking, hiking), (3) able to communicate in French, and (4) live in the rural francophone region of Prescott-Russell (Ontario, Canada). An intake interview was conducted with participants to explain the research objectives and review the consent form.

A total of five young people, aged between 19 and 29 (she/her = 3; he/him = 2) and living in various villages of Prescott-Russell participated in the program. Because the program was implemented during the COVID-19 global pandemic (in Fall 2021), the number of participants was limited with respect to public health recommendations. While the intention was to obtain a diversified sample, participants' sociodemographic and cultural backgrounds turned out to be quite homogenous. Four participants held postsecondary education diplomas—all were employed full- or part-time and identified with a middle-class socioeconomic status. All had a preexisting affinity with sports and/or the outdoors.

Data Collection and Analysis

To capture data in the most "natural" ways possible, questions were asked mainly during the welcome and closing circle held at the beginning and end of each session, during reflective activities or through organic and targeted interviews. All sessions were audio-recorded by attaching a recording device to my backpack. To uncover the rich landscape of experiences and meaning-making processes, other complementary data-gathering strategies were used, including (1) participant observation (Quidu, 2022), a posture which invites a full, embodied immersion into both the research and practice fields; (2) a research journal, used as an observation and critical reflective tool (Frechette et al., 2021), and (3) focus group, conducted 1 month post-intervention to inquire about participants' transformative moments, key takeaways, and recommendations for future ecosocial work programs.

Audio recordings were transcribed verbatim and imported into Nvivo, a qualitative data analysis software. All data was analyzed through thematic data analysis (Paillé & Muchielli, 2012), including the notes produced in my journal. Now that the three main phases of the intervention research have been laid out, the following section focuses on the specificities of *Feet to Earth*, as it was developed to foster human-Nature connections.

Program Overview: Cultivating Human-Nature Connectedness

The *Feet to Earth* program took place in the Larose Forest, the largest intact forested site in Eastern Ontario (roughly 26,000 acres). Set amidst farms and growing villages, the Forest is made up of 30% wetlands, natural Forest ecosystems, and a network of managed pine plantations. It is home to over 540 plant species and 122 birds and 22 mammal species, including a significant moose population (Hanrahan, 2006). Reflecting the colonial history of the region, the primary Forest was deforested by settlers in the nineteenth century for agricultural purposes, but the sandy soils proved to be infertile. The land was gradually reforested, mostly with red and white pine and white spruce. Today, the Forest combines recreation, logging, and wilderness, and can be easily accessed through various entry points near villages.

Drawing from data collected in the previous phases of the research (literature review and consultations with youth eco-activists), the program *Feet to Earth* was designed according to three main interrelated experiential pillars. As depicted in Table 1, these pillars each answered to specific objectives and were framed according to intertwining practice frameworks (outdoor and Nature-based therapy, yoga therapy, and narrative therapy).

As detailed in the program protocol (Table 2), *Feet to Earth* included 8 weekly sessions (3 h each) and concluded with a group-led collective action (August 2022). Reflecting the holistic orientation of the program as well as my expertise in yoga

Table 1 Three main pillars, corresponding practice frameworks and objectives for an ecosocial outdoor and Nature-based intervention program

Component	Therapeutic/Reflective ⟺	Sport/Movement ⟺	Eco-pedagogical
Practice/ conceptual frameworks	Outdoor and Nature-based therapy, Narrative therapy, Yoga therapy	Outdoor and Nature-based therapy, Narrative Therapy, Yoga therapy	Environmental Education, Social-Ecological Transition Studies
Objectives ↓ **Interrelated to each component and practice framework**	• Through holistic approaches, enhance global well-being of participants and facilitate the development of ecological identity. • Unpack and deconstruct preconceived ideas about relationship to Self, Other and Nature, and well-being. • Engage participants in a re-storying process. • Emphasize slow pedagogy/therapeutic practices to create space for reflective and embodied experiences.	• Engage in outdoor, movement-based activities to connect with the body, Self, Others and Nature. • Focus on the development of social bonds, self-confidence and problem-solving skills through sport and outdoor physical activity. • Through sport and outdoor activities, foster opportunities to experience the Forest differently, linking these learnings with everyday life.	• Enable place-based connections by interacting with the Forest, local territory, and community. • Collaborate with various eco-leaders and community partners to foster connections, intergenerational allyship and mentorship opportunities. • Engage in a "giving back to Nature" process through a group-led collective action.

Table 2 Program protocol: thematic element, main theme, sub-themes and activities related to each pillars

Element	Theme and subthemes	Reflective/ therapeutic	Sport/movement	Ecopedagogical
Session 1: ether (space)	*Theme*: awakening and renewal *Subtheme*: relationship to nature; noticing natural cycles	*Meditation*: body in nature awareness exercise *The companion tree exercise* *Tree of life*: roots—documenting my understory (exploring the roots of my relationship to nature and well-being)	Mountain bike basics (coaching session) and group ride (6 km)	Ethics of the land Linking human and nature cycle Establishment of a collaborative contract
Session 2: air	*Theme*: letting go and (re)connecting *Subthemes*: movement, transition, adaptation, go with the flow	*Meditation*: awakening the five senses, setting an intention *The companion tree exercise* *Tree of life*: trunk personal strengths and abilities and the qualities which enable you to foster a respectful relationship with nature	Mountain bike coaching session and group ride (12 km)	Discussion on meanings attributed to "connection to nature" (what it means to participants; what images come to mind)
Session 3: Earth	*Themes*: getting to know the Earth that sustains us and learning to sustain the Earth in return *Subthemes*: reciprocity, interrelationship, food security, local food systems, regenerative farming	*Breath work*: alternate nostril breathing with focus on exchange of breath or life force energy between the body and the Earth *Yoga*: standing postures for anchoring to the Earth, enhancing stability and security *Tree of life*: branches (right side)—dreams and aspirations for self and community	Mountain bike ride on gravel roads in the forest (10km) to a local alpaca and permaculture farm	*Permaculture workshop with local expert*: learn about the ethical principles of permaculture; learning to notice and accept retroaction from nature Workshop and interactions with alpacas

(continued)

Table 2 (continued)

Element	Theme and subthemes	Reflective/ therapeutic	Sport/movement	Ecopedagogical
Session 4: Earth	*Themes*: forest as teacher *Subthemes*: observation, respect, forest as place and as teacher; *Communitas*, receptivity, wonderment	*Mindfulness exercise*: walking with full awareness to notice, observe, and feel *Closing meditation* on the honorable harvest (Kimmerer, 2013) *Tree of life*: branches—dreams and aspirations to cultivate your friendship to nature (left side branches)	Guided hike (6km): experiencing slow pedagogy principles	*Discovery hike with a Larose Forest forestry expert*: learn about the social history of the forest to better understand it's current state and the communities' relationship to it, tree and soil identification, leave no trace ethics, land stewardship
Session 5: fire	*Themes*: finding your voice to find your path *Subthemes*: rhythm, confidence, self-expression, spirituality, creativity	*Tree of life*: leaves—preparing the path to move toward my dreams and aspirations *Rhythm awareness meditation*	Short hike Yoga practice based on the fire element (included a meditation on noticing the rhythms of the body and the rhythms of nature)	*Drumming circle* (facilitated by a musician and key local figure in community garden initiatives) Linking the arts with connectivity to nature and to climate action Collective vegan meal organized by participant
Session 6: water	*Themes*: resilience *Subthemes*: resilience, sensitivity, empathy, fluidity	*Guided meditation*: the gifts of the forest *Tree of life*: flowers—the gifts you received from the forest Fruits—the gifts you want to give to the forest	Mountain bike ride in the forest (12 km) to get to a lake for the Land Art Activity	Session co-facilitated by a youth activist (Phase 1 of research) *Land Art Activity*: creating our collective tree of life

(continued)

Table 2 (continued)

Element	Theme and subthemes	Reflective/ therapeutic	Sport/movement	Ecopedagogical
Session 7: ether (space)	*Themes*: sensory awakening *Subthemes*: feeling the body in space and time, silence, awareness	*Mindfulness activity*: riding in silence in a nocturnal context and meditation on the senses *Tree of life*: the storms of life (obstacles to the development of our connection to nature or aspirations for health)	Twilight mountain bike ride in the forest (12 km)	*Stargazing activity* (offered by participant) but cancelled due to rain Group discussion on the collective action (putting reciprocity into action)
Session 8: the five elements	*Themes*: integration of learnings and putting reciprocity in action *Subthemes*: reciprocity, gratitude, heart opening	*Tree of life*: the forest of life—harvesting the fruits of the program—recapitulation of key learnings, focus on the processes, the discoveries	Slackline activity (facilitated by one participant) Heart meditation (gratitude) and yoga session	*Collective action*: paint wooded signs (recycled wood) to indicate new pollinator habitats in the forest (co-animated by a local elderly artist involved in arts-based eco-initiatives)

therapy,[5] the sessions were built thematically around the five elements (Earth, Water, Fire, Air, Ether), and subthemes were carefully chosen to cultivate a sense of interconnection to Nature (Larocque, 2022).

The protocol was kept purposefully flexible so as to remain receptive to (1) the moment-to-moment changing natural landscape and cycles (i.e., a full moon; the start of autumn, the first snowfall, etc.) and to (2) participants' standpoints and personal change process. Subthemes were chosen to reflect the changes occurring in Nature and to respond to participants' interests. Without making this a goal, each participant enriched the program through unique contributions. For example, one participant (Virginie, a young entrepreneur)[6] organized a collective vegan meal, and another (Alexander, a university student) took on the role of co-facilitator by introducing the group to slacklining, an activity he presented as meditative and useful to build self-awareness.

[5]Yoga therapy involves a professional certification approved by the International Association of Yoga Therapists (IAYT) and aims to "bring balance to the body and mind, through an experiential understanding of the primary intention of yoga: awakening of spirit, our essential nature" (IAYT, Lepage, 2022).

[6]A pseudonym is used to preserve participants' anonymity.

While the program was co-facilitated by myself (researcher-practitioner) and by another social worker who specialized in narrative therapy and in sport and Nature-based therapy, collaborations with various eco-leaders were key to diversify knowledge and standpoints during eco-educational activities and to extend the program into the community. Ranging from a youth eco-activist who co-facilitated the water element session (Session 6) to a musician and community gardens expert who led a drumming circle activity (Session 5), these collaborations proved to be invaluable in showing the various means by which transition innovations can take form (e.g., arts, gardening, music, land stewardship) and to move key concepts (e.g., social-ecological transition, reciprocity) from abstractions to tangible, accessible practices. For example, following a bike outing to a permaculture farm (session 3), Frédérique, a physiotherapist, realized: "it doesn't have to be large scale to make a difference, I can do it [permaculture] on my own, without having to change my whole life and quit my job." Collaborations also multiplied opportunities for experiential learnings and created space for participants to connect *with* and to feel moved *by* community members who embodied a reciprocal understanding of the human-Nature relationship, as per Marika, a young mother and nurse, who expressed feeling inspired by the forestry technician (Session 4) who shared her knowledge of the Forest "with such passion in her eyes."

As a strategy to build attachment to place and experiment with reciprocity – the act of giving back to Nature (Kimmerer, 2013) – a youth-led collective action was incorporated as the final objective of the program. In consultation with the department of forestry, the group opted to create wooded signs to indicate new pollinator habitats in the Forest. The specific objective, determined by participants, was to (1) raise awareness of these habitats and promote their safekeeping and (2) encourage "Forest-visitors" to slow their pace to notice and appreciate local ecosystems. The signs were created with the assistance of a local elderly artist renowned for her ecological methods (Session 8) and installed by participants in August 2022, at which point the group participated in removing invasive plant species and in seeding the habitat with native wildflowers to help with ecological restoration and promote biodiversity. This experiential process was deemed essential by participants to understand *how* to shift away from individualism and value "selflessness," says Virginie but also to learn *how* to put reciprocity in practice, therefore highlighting "the importance of learning by doing," as expressed by Alexander.

Frames for Practice

The program protocol (Table 1) was developed at the nexus of outdoor and Nature-based therapy, narrative therapy, and yoga therapy, approaches that are linked by their experiential learning goals and emphasis on embodied knowledge. The chosen approaches complemented each other as they each fulfilled a specific role in answering to program objectives and reflected my own and the co-facilitator's professional knowledge base and scope of practice. This section aims to explain these approaches in relation to how they were applied for ecosocial and transformative purposes.

Outdoor and Nature-Based Therapy

The richly sensuous, natural context of outdoor and Nature-based therapies makes it appealing to study relationship to Nature and embodied experiences of resonance. Rooted in experiential education, outdoor and Nature-based programs can involve diverse activities (from contemplative interventions to challenging ventures) aimed at psychosocial development (Gargano, 2022). Carried out in contexts ranging from urban forests to deep wilderness, this approach intertwines with environmental education (Sauvé, 1996) and combines therapy and ecopedagogy to foster learnings *with/in/from* Nature. Beyond their well-documented health and psychosocial benefits, these Nature-based modalities are increasingly being used by social workers (Gargano, 2022) and have the potential to deepen the connection with "Earth-others" as well as foster ecological identify development (Willis, 2011), processes which can lead to transformative social change (Lugg, 2007). However, because the field of outdoor and Nature-based therapies has only recently started expressing concern for its role in the social-ecological transition, dominant discourse and contemporary social constructions of Nature continue to influence interventions as well as the types of human-Nature interactions being promoted (Sauvé, 1996).

Rethinking Outdoor and Nature-Based Modalities Through an Ecosocial Lens

Considering that outdoor sport and leisure activities can have detrimental effects on the environment and even foster utilitarian and consumerist mindsets (Mounet, 2007), one question that remained at the forefront of the program design process was *how* can ecosocial work programs contribute to the well-being of participants while also being of service to the well-being of the Forest and of local ecosystems? Thus, to minimize negative ecological impacts, it seemed crucial to instill practices which aimed to develop ecological identity and a place-based ethics of care. Several strategies were implemented. The first included integrating rituals to help create a sense of familiarity with the Forest *as place* and as a *shared living, interactive, sacred space*. For example, upon arrival, participants were given time to settle into their own bodies and Nature's rhythms. Each welcome circle included land acknowledgment and a message of gratitude to the Forest. Body and Nature attunement was facilitated by guided meditations and/or breathing practices incorporated in the welcome routine.

Because sport does not intrinsically possess ecosocial proprieties, all physical and movement-based activities were framed to provide opportunities to critically reflect on interactions with the Forest or to learn about/from the Forest. To extend learnings and therapeutic objectives beyond individual skills, physical activity was incorporated into the program as a tool to experience the Forest as "an ecocentric place of patient observation" (Payne and Wattchow, 2009, p. 23). Indeed, outdoor sport is documented as a possible sphere of resonance and gateway to connect with the body, place, and space (Plard, 2016). For example, a twilight bike ride (Session 7) was established to encourage participants to pay attention to their senses and notice their bodies moving in space and time as they silently rode the dark trails. Experiencing the Forest at night was useful to deconstruct ideas around fears and worries attributed to stepping outside of one's comfort zone (e.g., fear of the Forest at night/of the unknown/of darkness). This activity also provided a context to learn about the nocturnal sounds of the Forest and, as per Alexander's account, experience "the natural state of all living beings when they're in an 'animal' context." Participants talked of their "instincts" kicking in, where they naturally got out of their heads and into their bodies, experiencing the world differently, and about getting a sense of what it might be like to rewild themselves, a process that facilitates the development of a coexistence mindset (Bekoff, 2014).

Integrating eco-learning goals to outdoor physical activity was another way to use sport as a mediator to enhance ecological awareness. For example, learning about the colonial history of the Forest and becoming familiar with the ecosystems that enliven "recreational" natural spaces were key to excavate preconceived ideas and assumptions about the "purposes" of a Forest and critically examine our individual and collective relationship to this specific Forest. This reflective process proved to be eye opening for participants but also for myself, as researcher-practitioner experiencing the Forest through my own mind/body. Through journaling, and as a regular "Forest-visitor," I realized that I experienced the Forest primarily through a utilitarian, "Forest as commodity" viewpoint, even if I felt attached to this local, natural space. As illustrated by a journal excerpt, small but significant shifts occurred within the process of (re)discovering the Forest:

> I'm used to passing through this section of the Forest on my mountain bike, but it's the first time that I linger here and truly take the time to soak in what I observe, hear, and feel. At first view, this part of the Forest seems dead. Many trees are fallen and rotting. Upon closer look, the liveliness becomes apparent, where I can first smell and then see the abundance of lichens and mushrooms. It's so alive. (Session 3)

These shifts were indeed key to critically examine my own assumptions and notice epistemological paradoxes inherent to combining "recreational" and ecosocial practices. Reflecting on how outdoor sports can be integrated as an ecosocial development tool in outdoor and Nature-based programs represents an interesting and understudied avenue for future research.

Narrative Therapy

As a second practice framework adapted for ecosocial purposes, narrative therapy was key to deconstructing dominant assumptions about human-Nature relations. Based on the fundamental premise that realities are socially constructed, organized, and maintained through stories (White & Epston, 1990), narrative therapy purports that the stories we believe about ourselves and the world are anchored in discourse.

In this sense, language shapes experience and the meaning we ascribe to our lives and relationships and provides the discursive space to (re)negotiate relationships (e.g., human-Nature relations) and construct new meanings for problematic situations (e.g., the ecological crisis). Whether applied to individuals, groups, or communities, narrative therapy aims to generate new possibilities through (1) deconstruction of dominant stories that limit possibilities for change and (2) re-storying processes that help create preferred storylines and identities. From a macrosocial perspective, narrative therapy shifts attention to the sociocultural, political, and contextual elements (power structures, privileges, oppressions, geosocial location) that shape experiences and in which dominant stories are rooted (Denborough, 2008). Re-storying relationship (to Nature, for example) involves "thickening" the storyline around alternative ways of relating to the world. A variety of narrative techniques can support this shift, from narrative questions to art-based techniques, to collective projects and relational language (Bird, 2004). In line with the temporal conditions of resonance, narrative therapy involves lingering on processes rather than on end-goals, or what Brinkmann (2021, p. 573) refers to as "patiency," meaning "a more receptive, contemplative, and 'dwelling' mode of being in the world."

Unpacking and Re-storying Relationship to Nature Through Narrative Ideas

Narrative therapy was mainly applied by adapting the Tree of Life methodology (Denborough, 2008), a reflective tool that aimed to examine participants' relationship to Nature and well-being. As detailed in Table 1, each section of the Tree of Life (root, trunk, branches, etc.) was attributed to narrative questions, asked at the end of each session. Efforts were made to formulate questions according to participants' emerging narratives. For example, when Nature was described as a "friendship to nurture" (Virginie) (Session 2), this relational metaphor, which nuanced the "Nature as commodity" discourse, gave us (the co-facilitator and I) precious discursive material to build on for the duration of the program and formulate questions that would "thicken" the storyline around the "Nature as friend" narrative (e.g., How do you hope to develop and maintain your friendship to Nature? What abilities or knowledges would Nature most appreciate about you?). Participants documented their reflections in their journal and shared their thoughts (this was optional) during the closing circle to support a collective deconstruction and re-storying process. Because narrative therapy focuses on discursive aspects of the change process, complementing this approach with holistic and embodied practices was essential to account for the carnal dimension of experiences.

Yoga Therapy

Yoga therapy consists of the third practice framework adapted for the program. As a holistic mind-body-spirit discipline based on the ancient sciences of yoga and Ayurveda, a healing science rooted in Indian Vedic culture, yoga therapy is composed of various techniques (breath work, meditation, relaxation, postures, chanting, mindfulness) that aim to provide integrative and adaptive care (Wang & Tebb, 2018). Whereas yoga is increasingly being applied in social work settings, it is mainly used as an individualistic practice anchored in normative, responsibilizing health discourse (Larocque & Moreau, 2020). As such, the philosophical and spiritual underpinnings of yoga are often lost in clinical contexts, which poses an epistemological dilemma for ecosocial work, which operates from a holistic and decolonial perspective (Larocque, 2023b).

Reflecting on these issues, the goal was to incorporate basic philosophical principles (e.g., interdependence, unity) into yoga practices to generate mindful awareness of body, space, and time. At the philosophical level, environmental ethics can be found in yogic philosophy, as in Patanjali's Eight Limbs of Yoga, which serves as an ethical "road map for self-development, relationships, and approaches to life" (Wang & Tebb, 2018, p. 302). Yoga philosophy is congruent with the fundamental premise of ecosocial work and ecological justice because it also considers that all sentient beings matter equally (Besthorn, 2000). In terms of human-Nature resonance, somatic techniques and focused attention exercises increase the sense of connection to one's body (Wang & Tebb, 2018) and help bring awareness to the vibrations of the natural world (Larocque, 2022) through multisensory and embodied experiences. Evidence suggests that connecting to the body and the sense in Nature settings enables Nature-connectedness and can help dissolve the boundaries between the self and the natural world (Rickard & White, 2021). In line with the concept of "patiency" (Brinkmann, 2021, p. 573), which entails committing to a slowing down process, yoga not only provides the means to engage with Nature via the senses but also provides tools to cultivate stillness, a bodily and mental state that is conducive to resonant experiences (Rosa, 2019).

Embodying Ecosocial Values Through Yogic Philosophy

Aside from cultivating space for an embodied practice, principles of yogic philosophy were integrated to challenge the fundamental assumptions of disconnection and separation (mind-body/human-Nature) that permeate the social work field (Larocque, 2023b). With the objective of creating conditions that created space for participants to interact differently with Nature, sessions were thematically constructed around the five elements (Earth, Water, Fire, Air, Ether). The purpose was to bring focus the physicality and subtilities of these elements, found in Nature and the body, and enable connections with the sensuous and felt qualities of interconnectedness through yoga techniques (mainly breathing exercises and guided meditations). Turning to yogic principles facilitated the integration of key concepts, such as reciprocity, which were applied concretely through a range of therapeutic or eco-educational activities and complemented with yogic techniques. For example, in preparation for the Earth session, which entailed visiting a local permaculture farm, we began with a grounding yoga practice to bring attention to the reciprocal exchange of Earth-body energies. Sensing the body in relation to the Earth was indeed an embodied introduction to permaculture, a practice that foregrounds an ethic of working in partnership, reciprocity, and connection with Nature's natural cycles. Yoga provided the philosophical and embodied tools to physically and energetically relate to the concepts presented. In terms of resonance, participants reported that yoga helped "prepare" their bodies and senses to enhance receptivity to the teachings of Nature (Alexander). Others, such as Marika, shared that yoga was useful to center herself, slow down, and observe, therefore cultivating presence and responsiveness to the natural world.

To summarize, the active ingredients embedded in each of these complementary approaches provide powerful tools to integrate ecosocial values and to nurture alternative (experiential, discursive, and embodied) ways of relating to Nature.

Implications for Ecosocial Work Teaching and Learning

From the theoretical, methodological, and practical processes outlined in this chapter, four main implications for teaching and learning ecosocial work are identified and discussed as possible gateways to develop resonant and embodied human-Nature relations in Nature-based therapy contexts.

1. *Multiplying Experiential Opportunities to Transcend Dichotomic Thinking:* I first mention the value of adopting a participant observation stance to experience the change process *with* participants. In contrast to holding an expert position, this decentered, co-learner stance was useful to disrupt power dynamics and create space for participants to tap into their insider knowledge and feel valued as experts of their own experiences (White & Epston, 1990). This immersive posture also enabled me to experience the epistemological barriers inherent to rethinking dominant relationships to Nature, especially when starting from a modernist and anthropocentric standpoint. Considering the heuristic and transformational value of experiential pedagogies and practices (Rambaree et al., 2019), integrating more immersive approaches in social work curriculums would allow social work actors (including educators, researchers, students, and practitioners) to experience first-hand these epistemological challenges and perhaps, to collectively find solutions to enhance connectedness to Nature and transcend the human-Nature divide. Teaching and learning ecosocial work should therefore focus more strongly on the relational roots of the socio-ecological crisis and multiply opportunities for students to question the power relations that forge contemporary Nature-human relationships, namely, by developing ecological identity through place-based practices (Willis, 2011).
2. *Shifting from Individual Action to Community Connections:* Experiential learnings through a pluralist knowledge approach (via consultations with youth eco-activists as well as with community partners and local eco-leaders) proved to be a key transformative aspect of the program. On the one hand, consulting with youth eco-activists provided novel insights on ways to make *Feet to Earth* rele-

vant for ecosocial purposes and to refine outdoor and Nature-based therapy in ways that better positions this modality as a socio-ecological innovation strategy. On the other hand, program participants shared that interactions with eco-leaders were most impactful to broaden their perspectives about Nature but also about different ways to deal with personal mental health issues, such as anxiety and depression. One risk of micro-level approaches refers to the social injunctions they may inadvertently enforce. As stated by Rosa (2019, p. 404), the way we interact and make-meaning of the world is not an "individual accomplishment" (p. 404) but rather is anchored in institutionalized social structures. To avoid imposing the weight of ecological responsibility on individuals, incorporating collaborators and deconstructing power relations through narrative ideas were key strategies to displace responsibility towards macrosocial elements. Valuing community connections extended the project of social change to the collective sphere while also fostering interdisciplinary learnings. Including a group-led collective action in ecosocial programs represents another strategy to (re)politicize intervention and shift the focus away from individual responsibility and towards collective awareness. In this study, participants reported feeling empowered by their capacity to engage in the change process, as per 19-year-old Mikael, who stated that he felt full of pride for helping restore habitats where endangered species could flourish. Virginie, on her part, shared feeling a growing sense of attachment to the Forest and was looking forward to returning to the pollinator habitats because "it was because us, our group, we did it together," an action which, she says, "was healing to my heart and soul." From and ecosocial work teaching and learning perspective, social work students must be prepared to face the multi-level injustices and complexities of the socio-ecological crisis. More importantly, they should also be better equipped to find and foster hope in this process of co-building more resilient and just societies. In this regard, it seems crucial to help social work students build ecological identity and leadership through active, place-based pedagogies that generate participation, collective hope, and socio-ecological innovation.

3. *Unpacking and Re-storying Relationship to Nature*: To contribute to transformative change and avoid reinforcing narratives that limit opportunities to develop a philosophy of coexistence, critical self-reflection should precede all interventions conducted in natural settings. Indeed, as more Nature-based programs are being developed by social workers (Gargano, 2022), reflection is needed if the goal is to focus on ecological justice, which aims to increase humans and Earth-others' well-being. It is timely to unpack dominant narratives and reflect on how language and actions can either reinforce and reify dominant discourses (such as the romanticized discourse on "nature as resource for self-care" or the "conquering nature" narrative) or destabilize the status quo to create space for new stories to emerge about relationships to Nature (Willis, 2011). Narrative therapy provides powerful tools to assist this process. Learning about a Forest's social history (especially about how it was impacted by colonization) should be a fundamental objective of every outdoor and Nature-based programs. From an embodied standpoint, immersing themselves in natural spaces where programs

are implemented would allow social work actors to get to know the various ecosystems of local natural spaces through embodied presence and to critically examine their own relationship to Nature. In the context of deconstructing dominant discourse and decolonizing relationship to Nature, advocating for a stronger approach to Indigenous collaborations (Fellner, 2018) and seeking pluriversal guidance on ways to cultivate harmonious, reciprocal relationships to Nature/place (Pereira & Bina, 2020), would be another strategy to unpack dominant ideas about Nature, while also embracing knowledge plurality in this ongoing re-storying and reconciliation process (Fellner, 2018).

4. *Enriching Ecosocial Teaching and Learning Through Embodied Practices:* Considering the near absence of body ontologies in social work research and practice literature and the potentiality of embodied knowledge to foster experiences of resonance (Plard, 2016), the body is a key human dimension to consider in human-Nature relations. Integrating embodied practices are fruitful to (1) "prepare" the body-mind-spirit to enhance receptivity to Nature's teachings, (2) meaningfully engage with the natural world through resonant experiences, and (3) instill a sense of interconnectedness between the self, others, and Nature (Larocque, 2022). Considering the centrality of experiential learning to cultivate a sense of ecological identity and attachment to place (Payne & Wattchow, 2009), I argue that ecosocial work curricula could benefit from increasing embodied interactions with the natural world to help social work students develop the ability to tune in to Nature's rhythms and teachings. Finally, embodied teaching and learning would help rebalance the overly cognitive and disembodied approaches that dominate current social work practices (Larocque & Moreau, 2020).

Limitations

Aside from the size and homogeneity of the group, another limitation of this study relates to the financial cost of outdoor and Nature-based programs, which can entail equipment rental fees and, in this case, fair monetary compensation for community collaborators. Offering accessible and free programs can therefore be challenging. Thus, establishing community partnerships was essential to help (1) cover the costs of certain services (e.g., bike rentals, mountain bike coaching services) and (2) obtain a community innovation grant to cover other expenses (e.g., participants' transportation costs, materials for the collective action). Considering participants' sociodemographics (white, middle-class) despite my efforts to obtain a diversified group, another limitation refers to the broader issue of inclusion and diversity in Nature-based programs. As noted by Ho and Chang (2022), most Nature-oriented recreational activities are geared towards nondisabled, privileged bodies. Therefore, discussions on building inclusivity in Nature-based approaches should be at the top of social workers' list. Finally, because the concept of resonance is anchored in a Western worldview, future studies should pay more attention to gendered, racialized, and classed dimensions of human-Nature experience (Fuchs, 2020) and be further informed by pluralistic, postcolonial, and alternative knowledge systems (Pereira & Bina, 2020).

Conclusion: Reorienting the Teaching and Learning *Foci* Towards Human-Nature Relations

The objective of this chapter was to present the steps undertaken to co-design an ecosocial outdoor and Nature-based program through a collaborative methodology (intervention research). More specifically, it aimed to forge possible pathways to generate transformative, resonant, and embodied human-Nature relations in Nature-based intervention settings.

Presented as a case study, the research process highlighted the value of including plural voices to design and implement the program, which aimed to examine the conditions that foster meaningful and alternative relationships to Nature and well-being. It also described the main theoretical approaches used to frame the program, namely, Rosa's (2019) theory of resonance and Wacquant's (2015) carnal sociology, therefore enriching current ecosocial work literature through sociological analysis. The main practice frameworks (outdoor and Nature-based therapy, narrative therapy, and yoga therapy) were presented in relation to how they were applied for ecosocial purposes and provide notable pathways to explore in future studies.

Significantly, this chapter encourages social work actors to reflect on the types of relationships to Nature produced and promoted by their everyday social work practices. By incorporating ecosocial teaching and learning that aims to create a shift in the way humans relate to the world (Rosa, 2019), the social work profession will surely be better equipped to address the foundational epistemological drivers of the socio-ecological crisis while also contributing to moving towards resonant and reciprocal human-Nature relationships.

Bibliography

Artmann, M. (2023). Human-nature resonance in times of socio-ecological crisis – A relational account for sustainability transformation. *Ecosystems and People, 19*(1), 1–15. https://doi.org/10.1080/26395916.2023.2168760

Audet, R., Butzbach, C., Manon, M., & Folco, J. D. (2021). Transition sociale et écologique: vers un récit citoyen. *À Bâbord, 87*, 17–19. https://id.erudit.org/iderudit/96351ac

Barraclough, A. D., Sakiyama, M., Schultz, L., & Maren, I. E. (2021). Stewards of the future: accompanying the rising tide of young voices by setting youth-inclusive research agendas in sustainability research. *Sustainable Earth, 4*(2), 1–6.

Bekoff, M. (2014). *Rewilding our hearts: Building pathways of compassion and coexistence*. New World Library.

Besthorn, F. H. (2000). Toward a deep-ecological social work: It's environmental, spiritual and political dimensions. *The Spirituality and Social Work Forum, 7*(2), 2–7.

Bird, J. (2004). *Talk that sing: Therapy in a new linguistic key*. Edge Press.

Boetto, H. (2017). A transformative eco-social model: Challenging modernist assumptions in social work. *British Journal of Social Work, 47*(1), 48–67. https://doi.org/10.1093/bjsw/bcw149

Boetto, H., Närhi, K., & Bowles, W. (2022). Creating 'communities of practice' to enhance eco-social work: A comparison between Finland and Australia. *The British Journal of Social Work, 52*(8), 4815–4835. https://doi.org/10.1093/bjsw/bcac092

Brinkmann, S. (2021). Resources and resonance: Notes on patiency as world relation. *Culture and Psychology, 26*(4), 572–576. https://doi.org/10.1177/1354067X211017306

Coates, J., & Gray, M. (2019). How green is social work? Towards an ecocentric turn in social work. In M. Payne & E. Reith-Hall (Eds.), *The Routledge handbook of social work theory* (pp. 171–180). Routledge.

Denborough, D. (2008). *Collective narrative practice: Responding to individuals, groups and communities who have experiences trauma*. Dulwich Centre Publications.

Dominelli, L. (2012). *Green social work: From environmental crises to environmental justice*. Polity.

Fellner, K. D. (2018). Embodying decoloniality: Indigenizing curriculum and pedagogy. *American Journal of Community Psychology, 62*, 283–293. https://doi.org/10.1002/ajcp.12286

Frechette, J., Bitzas, V., Aubry, M., Kilpatrick, K., & Lavoie-Tremblay, M. (2021). Capturing lived experience: Methodological considerations for interpretative phenomenological inquiry. *International Journal of Qualitative Methods, 19*, 1–12. https://doi.org/10.1177/1609406920907254

Fuchs, A. (2020). Resonance: A normative category or figure of uncertainty? On reading Hartmut Rosa with Thomas Mann's The Magic Mountain. *Journal of Political Power, 13*(3), 353–365. https://doi.org/10.1080/2158379X.2020.1828756

Gadais, T., Lacoste, Y., Beaumont, J., Daigle, P., & Bergeron, N. (2021). L'intervention éducative en contexte de plein-air comme objet de recherche et d'enseignement en milieu universitaire au Québec. *Nature et récréation, 10*, 62–73.

Gargano, V. (2022). Les pratiques centrées sur la nature et l'aventure et le travail social: Perspectives disciplinaires et théoriques. *Intervention, 155*, 151–165. https://doi.org/10.7202/1089312ar

Hanrahan, C. (2006). Larose forest: History and ecology. *Ottawa Field-Naturalists' Club*. Retrieved April 19th 2023 from https://ofnc.ca/programs/conservation/larose-forest/laroseforest-historyand-ecology

Hari, J. (2022). *Stolen Focus. Why you can't pay attention – and how to think deeply again*. Crown.

Ho, Y. C. J., & Chang, D. (2022). To whom does this place belong? Whiteness and diversity in outdoor recreation and education. *Annals of Leisure Research, 25*(5), 569–582. https://doi.org/10.1080/11745398.2020.1859389

Huntjens, P. (2021). *Towards a natural social contract. Transformative social ecological innovation for a sustainable, healthy and just society*. Springer.

Integrative Association of Yoga Therapists. (2022). *Contemporary definitions of yoga therapy: Joseph Lepage*. Retrieved May 5th 2023 from www.iayt.org

Kimmerer, R. W. (2013). *Braiding sweetgrass: Indigenous wisdom, and the teachings of plants*. Milkweed Editions.

Krings, A., Victor, B. G., Mathias, J., & Perron, B. E. (2020). Environmental social work in the disciplinary literature, 1991–2015. *International Social Work, 63*(3), 275–290. https://doi.org/10.1177/0020872818788397

Larocque, E., & Moreau, N. (2020). What is our epistemic responsibility? Reflections on social work's orientation in yoga therapy research. *International Journal of Yoga Therapy, 30*(1), 103–109. https://doi.org/10.17761/2020-D-19-00044

Larocque, E. (2022). Le yoga en contexte de plein-air: une pratique écospirituelle au service de la résonance au vivant. *Revue Éducation Relative à l'Environnement – Recherches, Regards et Réflexions, 17*(2), 1–23. https://doi.org/10.4000/ere.8917

Larocque, E. (2023a). Co-envisioning the social-ecological transition through youth eco-activists' narratives: towards a relational approach to ecological justice. *Journal of Community Practice, 31*(2). https://doi.org/10.1080/10705422.2023.2208577

Larocque, E. (2023b). *Pour un travail écosocial centré sur le vivant: contributions à l'élargissement du concept de transition sociale-écologique et au co-développement des rapports incarnés, résonants et réciproques à la Nature-territoire* [Doctoral Thesis, University of Ottawa]. https://doi.org/10.20381/ruor-29387

Lugg, A. (2007). Developing sustainable citizens through outdoor learning: Possibilities for outdoor education in Higher Education. *Journal of Adventure Education and Outdoor Learning, 7*(2), 97–112. https://doi.org/10.1080/14729670701609456

Mounet, J.-P. (2007). Sport de nature, développement durable et controverse environnementale. *Natures, Sciences & Sociétés, 15*(2), 162–166. https://doi.org/10.1051/nss:2007043

Nicholas, E. (2021). Seeing the forest for the trees: Exploring the forest aspect of the Tree of Life process to sustain and nourish socioecological activism. *International Journal of Narrative Therapy & Community Work,* (1), 1–9.

Paillé, P., & Mucchielli, A. (2012). *L'analyse qualitative en sciences humaines et sociales*. Armand Collin.

Parlavecchio, L., Moreau, N., Favier-Ambrosini, B., & Gadais, T. (2021). Principles of actions used by and eductrainor to create social bonds through sport in a psychosocial intervention program. *Journal of Sport for Development, 9*(2), 30–53. https://jsfd.org/2021/08/01/principlesof-action-used-by-an-eductrainer-to-create-socialbonds-through-sport-in-a-psychosocialintervention-program/

Payne, P. G., & Wattchow, B. (2009). Phenomenological deconstruction, slow pedagogy, and the corporeal turn in wild/environmental outdoor education. *Canadian Journal of Environmental Education, 14*, 14–32.

Pereira, L., & Bina, O. (2020). The IPBES conceptual framework: Enhancing the space for plurality of knowledge systems and paradigms. In J. Pereira & A. Saramago (Eds.), *Non human nature in world politics. Frontiers in international relations*. Springer. https://doi.org/10.1007/978-3-030-49496-4_15

Plard, M. (2016). Éléments de réflexion géographique à la faveur d'un cogito corporel: s'explorer, un savoir-faire géographique. *L'information géographique, 2*(2), 114–133. https://doi.org/10.3917/lig.802.0114

Quidu, M. (2022). La participation observante de Loïc Wacquant comme nouvelle « doxa méthodologique » au sein des Sciences sociales du sport ? Arguments pour la réhabilitation d'un pluralisme réflexif. *Loisir et Société, 1*, 218–242. https://doi.org/10.1080/07053436.2022.2091714

Rambaree, K., Powers, M. C. F., & Smith, R. C. (2019). Ecosocial work and social change in community practice. *Journal of Community Practice, 27*(1), 1–8. https://doi.org/10.1080/10705422.2019.1660516

Rickard, S. C., & White, M. P. (2021). Barefoot walking, nature connectedness, and psychological restoration: the importance of stimulating the sense of touch for feeling closer to the natural world. *Landscape Research, 46*(7), 975–991. https://doi.org/10.1080/01426397.2021.1928034

Rosa, H. (2009). *High speed society: Social acceleration, power, and modernity*. The Pennsylvania State University Press.

Rosa, H. (2019). *Resonance: A sociology of our relationship to the world*. Polity Press.

Sauvé, L. (1996). Environmental education and sustainable development: A further appraisal. *Canadian Journal of Environmental Education, 1*, 7–34.

Schusler, T., Krings, A., & Hernandez, M. (2019). Integrating youth participation in ecosocial work: New possibilities to advance environmental and social justice. *Journal of Community Practice, 27*(3-4), 460–475. https://doi.org/10.1080/10705422.2019.1657537

Seymour, V. (2016). The human-nature relationship and its impact on health: A critical review. *Frontiers in Public Health, 18*(4), 260. https://doi.org/10.3389/fpubh.2016.00260

Smith, L. T. (1999). *Decolonizing methodologies: Research and indigenous people*. Zed Books.

Stassart, P. M., Mormont, M., & Jamar, D. (2008). La recherche-intervention pour une transition vers le développement durable. *Économie rurale, 4*(306), 8–22. https://doi.org/10.4000/economierurale.432

Wacquant, L. (2015). Pour une sociologie de chair et de sang. Translated by M. Busset & Cordey, M. *Terrains et travaux, 1*(26), 219–225. https://doi.org/10.3917/tt.026.0239

Wang, D., & Tebb, S. (2018). Yoga as a method to integrate body/mind/spirit with social work clients. In M. Y. Lee, S. Ng, P. P. Y. Leung, & C. I. W. Chan (Eds.), *Integrative body mind-spirit social work: An empirically based approach to assessment and treatment* (pp. 302–317). Oxford Press.

Willis, A. (2011). Re-storying wilderness and adventure therapies: Healing places and selves in an era of environmental crisis. *Journal of Adventure Education and Outdoor Learning, 11*(2), 91–108. https://doi.org/10.1080/14729679.2011.633375

White, M., & Epston, D. (1990). *Narrative means to a therapeutic end.* Norton.

Chapter 13: The Mushroom House: A Socio-economic Learning Initiative to Carry Out Ecosocial Work in the Azores

Eduardo Marques, Elsa Alves, and José María Morán Carrillo

Introduction

In the twenty-first century, we are witnessing a collapsing planet on which people's and ecosystems' well-being is waning because of an "alienated" and predatory economy, dubbed by Löwy (2015) a "Capitalist Catastrophe", as our excessive consumption and destruction of natural resources have a devastating impact on Earth. We, as citizens, researchers, practitioners and educators, have less than a decade to mitigate the effects of global warming's effects, which we are already witnessing. Thus, there is an emergent need for low-cost solutions adapted to communities and capable of addressing the complex sustainability problems we face. In this context, social work should conform to and uphold sustainable principles oriented to degrowth (Latouche & Hapagés, 2011), with "ecosocial work" forming part of the main framework in this transitional process, engaging and including the most vulnerable populations. Närhi (2002, p. 37), quoting Matthies (2001), observes that "the eco-social approach in social work is an umbrella concept that provides a useful vantage-point for studying the interface of ecological and social issues". Even if we adopt different terms to refer to ecosocial work, such as "green social work", "ecological social work" or "environmental social work", the main idea is always the same. Ecosocial work should be able to guide professionals, promoting an active

E. Marques (✉)
CLISSIS – Lusíada Center for Research in Social Work and Social Intervention, University of the Azores, Ponta Delgada, Portugal
e-mail: eduardo.js.marques@uac.pt

E. Alves
School of Health, Polytechnic Institute of Bragança, Portugal, Bragança

J. M. Morán Carrillo
Universidad Pablo de Olavide de Sevilla, Sevilla, Spain

C. Forde et al. (eds.), *Teaching and Learning in Ecosocial Work*,
https://doi.org/10.1007/978-3-031-58708-5_18

pedagogy of hope and co-responsibility through projects. We believe that another world is possible through community participation and inclusive science to find new solutions that can enhance our common well-being and further the achievement of the Sustainable Development Goals.

Regrettably, we are witnessing a surge in inequalities, poverty and environmental destruction. Thus, social work must be increasingly holistic, eclectic, transdisciplinary, critical and present in situ (Hart, 2019). Social work education must adapt its curricula, putting theory into practice and empowering students to co-participate in the designing of solutions and strategies to solve complex societal problems. This was the goal behind the implementation of "the Mushroom House" (Casa do Cogumelo) project (Jaeger & Kopper, 2014; Stephens et al., 2008), and this case study aims to demonstrate the importance of social work internships as a real-world laboratory, putting sustainability into practice through transdisciplinary research, developing new strategies and cocreating knowledge on ecosocial work education (Renn, 2018; Wanner et al., 2018). Using the concept of the circular economy and following a project-based learning (PBL) methodology, it was possible to devise a collaborative learning process that linked the supervisor/teacher to their students and vice versa (González-Domínguez et al., 2020). The project was created as a partnership between the University of Azores (Social Work degree) and a non-governmental organization (NGO), the Kairós Cooperative, located in the Azores (Portugal), which aimed to demonstrate the potential of ecosocial work through the intersecting paths of the social sciences, along with nutrition sciences and social economy organizations, striving for community well-being through the circular economy. Interdisciplinary collaboration encourages the exchange of ideas, fosters creativity and allows for a more comprehensive understanding of how to teach ecosocial work. Interdisciplinarity transcends traditional disciplinary boundaries and encourages the melding of knowledge. It also recognizes that many real-world problems are multifaceted and complex and cannot be fully understood or solved within the confines of a single discipline, like social work. Thus, it promotes a cross-fertilization of ideas, fosters innovation and helps address challenges that require diverse expertise, knowledge and methodologies, through a holistic approach and understanding, to develop adapted and comprehensive solutions (Gardiner, 2020).

Higher education institutions (HEI) have a crucial role to play in the implementation of the development goals entailed by ecosocial work, as its professionals must seek solutions to contemporary challenges. That is, they must participate in societal transformation, facilitating positive shifts: social and environmental and as regards economic sustainability (Chankseliani & McCowan, 2021). Curricular internships integrated into the training of future social workers are an ideal format for educational innovation, experience, research and the cocreation of knowledge (Butler-Warke & Bolger, 2021). Such possibilities allow us to theorize and develop professional skills adapted to local, real-world problems, thereby helping to bolster students' trust and abilities, giving rise to future professionals with profiles as motivated, critical, confident and self-determined people committed to improving their skills and to social and political participation, in general and as ecosocial practitioners, in particular.

The success of this challenge depends on a strategic alliance and cooperation between HEIs and social and solidarity economy enterprises/institutions where students develop their preprofessional skills and practices through dialogue. This helps to empower collaborative learning partnerships, cocreating strategies based on evidence to produce positive solutions to problems. By this way we believe we can create impactful transformations at organizations and communities and knowledge cocreation. We believe that this praxis makes social and environmental justice real through shared social responsibility, whose presence on the agenda of every social actor is a pressing need, whether they are in the public or private spheres. A sustainable society requires responsible and sustainable companies, NGOs and universities (Cañeque, 2006). Today it is imperative that NGOs assume their social responsibility, changing some processes to become greener and more sustainable, thereby demonstrating their real commitment to environmental issues. According to Rambaree (2020, p. 47) "Corporate Social Responsibility (CSR) can be seen entailing activities in the field of social work, with companies getting involved in enhancing well-being and promoting social change and development". Thus, we underscore the importance of cocreated solutions in ecosocial work.

The Mushroom House as an Active Strategy for Ecosocial Work Practices

The *Mushroom House* was conceived as a pedagogical strategy that marshals knowledge and facilitates the development of competencies in students, with the goal of gathering evidence that allows us to think, integrate and critically evaluate the potential of social entrepreneurship and environmental education as ways to implement ecosocial work projects.

Inspired by a practical experience shared within the framework of the ESCUTA Erasmus+ project, this initiative was constituted as a pilot project (*Mushroom House*) between the Cooperativa Kairós[1] and the University of Azores, which, through its implementation, became a practical case study for learning and understanding possibilities for ecosocial work in the Azores. Kairós NGO was founded in 1995 as a Solidarity Economy Initiative Incubation Cooperative. It offers different social services and economic initiatives to respond to local societal problems, namely, "solidarity economy initiatives", "solidarity business initiatives" and "initiatives for social innovation". Kairós embraces a systemic and integrated approach to promoting equal opportunities among young and adult populations facing social exclusion, seeking to increase their employability and social integration skills, autonomy and full exercise of citizenship. Kairós is one of the biggest social organizations on São Miguel Island in the Azores and provides social support services to

[1] https://www.kairos-acores.pt/kairos.php

different groups (children, youth and adults) and the provisioning of social services in education, social inclusion, social enterprises, sports, etc.[1]

This pilot project, which was designed based on new and holistic perspectives, developed a forum for communication between the social sciences and nutrition sciences due to its potential for the local production and consumption of mushrooms, healthy food production on a local scale and support for the social economy. These elements clearly point to its potential as an inclusive tool for social entrepreneurship (intra-entrepreneurship) in the context of social and solidarity economy organizations, allowing for environmental education from an ecosocial work perspective.

The aim of *Mushroom House* is to assess the impact of mushroom production and consumption on health and on the economy, from a circular economy perspective, and to reflect on the potential role of social entrepreneurship-driven social workers in a local community development perspective. During the internship Social Work students are empowered to develop intra-entrepreneurship, understood as "the practice of developing a new project within an existing organization, carried out by the employees or managers, to explore a new opportunity and create economic value" (Pereira et al., 2019, pp. 92). We start with the "Mushroom House" project case study to carry out a reflective and holistic analysis to understand the advantages of an interdisciplinary approach to achieving goals promoting sustainable and community development, in harmony with nature. Aware of the difficulties involved in advancing ecosocial work, we propose to present strategies that bring together the actions and answers of science to meet the fundamental needs of social, environmental and economic challenges. *Mushroom House* is a social innovation, place-based project that allows us to respond to the need to research and produce knowledge of ecosocial work. At the same time, it involves testing out a nature-based solution (NBS) that links social work to circular/local development based on a socially conscious economy. Through an entrepreneurship initiative (mushroom production), we aim to use circular, locally rooted development to bolster social well-being and pursue nature-based solutions (NBS) featuring the kind of healthy food production that is needed for a successful ecosocial transition. The nature-based solution (NBS) concept, introduced in 2015 by the European Union, represents an excellent opportunity for urban planning and governance to create interdisciplinary and systemic solutions that can be used in ecosocial work, promoting social well-being and nature rights and enriching "structural social work with new perspectives and opportunities to follow the principles of sustainable development", according to Närhi (2018, p. 499).

Our quality of life depends on our ability to promote new policies that can stop, mitigate or reverse global warming and the decrease in biodiversity. This calls for the adoption of new lifestyles, as well as a "new" economy and "new" production and consumption practices (including food), evolving social organizations embracing "new" policies and governance aimed at preserving and regenerating the environment.

As Dubey et al. (2020, p. 268) states: "Satisfying food and nutritional security is a mandate for good physical and mental health, well-being, and quality of life"

(Dubey et al., 2020). Aligned with Dubey's statement, the importance of including "nature within social work to reconnect humans with nature for health, well-being, healing, and justice" is also recognized by Slattery et al. (2022, p. 2). These authors assert the need to include an integrative environmental model for social work to assist future practice and education.

As Frantzeskaki et al. (2022) state, social work must be able to understand the importance of NBS, because social workers bridge the historical gap between ecologically oriented solutions and the social justice approach. In this way, ecosocial work reinforces community well-being and helps to further the idea of "living well" (*buen vivir*). This knowledge paradigm of Latin American indigenous groups is understood as "an equilibrium of rights of people and nature, with a 'solidarity economy' emphasising equities, equality and freedoms, social justice, and ecological justice" (Godden et al., 2021, p. 354). It offers an alternative vision to conventional development models and calls for a transformation in how we understand and pursue well-being for individuals and societies.

The development of the Mushroom House project demonstrates the possibilities for teaching ecosocial work and offers a reflective analysis with a holistic approach that aims to demonstrate the value of an interdisciplinary perspective to achieve different Sustainable Development Goals (SDG), as well as to promote social inclusion through entrepreneurship.

Practical Implications of the Project in the Context of the Sustainable Development Goals

A commitment by social workers to the Sustainable Development Goals (SDGs) is assumed in the profession's various regulatory documents. In Portugal, the Deontological Code of Social Workers (2018) states that social and sustainable development is one of the sectors of intervention by social workers. Looking at the SDGs, we see a range of opportunities for the professional practice of social workers, who can, thus, restructure the profession and play a more active role in building a fairer social world. Traditionally, the focus of social work has been more centred on Objective 1 (Eradicating Poverty), Objective 2 (Eradicating Hunger), Objective 3 (Quality Health), Objective 4 (Quality Education) and Objective 5 (Gender Equality). Today, however, we must be able to influence and intervene in the context of Goal 11 (Sustainable Cities and Communities) and Goal 12 (Sustainable Production and Consumption), among others (United Nations, 2016).

The relationships between human communities and nature are, increasingly, part of social work practices and discussions between the academic world and social workers contribute to expanding and explaining the concept and framework of what social work entails in the environmental field. According to Besthorn (2012, p. 3), a "growing core of social workers from around the world have spoken forcefully concerning the importance of incorporating environmental awareness into the profession's theoretical formulations and practice modalities". Due to the COVID-19 pandemic, it is crucial to take an integrative approach to health and well-being. This means that it is vital to be proactive to prevent disease and to promote an economy that respects all living beings, that is, that respects the symbiosis between humans and nature. Based on the "Green Care umbrella", there are some activities—such as gardens, vertical farming, edible neighbourhoods, community composters and others—that can serve as examples of social inclusion, entrepreneurship and health and environmental education, promoting well-being and citizenship in harmony with nature (McBride et al., 2021; ONU, 2018). These are some examples based on the Green Care concept described in the Green Care manual:

> In Green Care activities, nature is used systematically to maintain and promote the well-being of people. The activities are based on nature-assisted working methods that can be used in a forest or garden, on a farm or even in an urban environment. Some working methods are also applicable for indoors. The three basic elements of the Green Care working method are nature, action, and community. Well-being is generated by combining and emphasising these elements in various ways: through the revitalising effects, experiences, and involvement that nature has to offer. The Green Care services are divided into care and empowerment services. (Luke & THL, 2017, p. 7)

We argue that Green Care and NBS can be integrated into ecosocial work through social entrepreneurship and ecotherapy activities, based on the Green Care model and following the subcategories created by McGeeney, which include Social and Therapeutic Horticulture, Animal-Assisted Interventions, Care Farming, Environmental Conservation, Green Exercise, Nature, Arts and Crafts, Wilderness Therapy and Ecotherapy, promoting wellness therapy (McGeeney, 2016).

On a planet whose resources are increasingly limited, but where the population keeps growing, it is necessary to increase the production of food rich in nutrients in a sustainable way. But this concern is not present in the dominant capitalist mode of production, which is based on intensive, monocrop agriculture, sustained by industrial fertilizers and pesticides that leave traces of contamination in all existing forms of life and that aggravate ecological imbalances, with their inevitable damage to human and non-human communities. According to Mehta et al. (2022, p.122), "many countries have announced an alarming situation due to the excessive accumulation of pesticides and fertilizers in the food chain. The presence of agrochemical residues in air, soil, water, and its accumulation in adipose tissues of humans means a threat to our human physiologic ecosystem". Pesticides and chemical residues in food are an invisible threat to human health and an added difficulty in the struggle for food sovereignty. According to Chigangaidze (2022), food insecurity is a pressing humanitarian challenge in a world where almost 795 million people experience chronic hunger and some two billion people suffer from hidden malnutrition, while, at the same time, "poor nutrition and obesity is a serious problem in many countries" (Peláez et al., 2022, p. 11).

We live in a world where hunger has persisted for too long, where around 30% of the food produced is wasted, while a large part of the world's population suffers from poor and unhealthy diets (Willett et al., 2019). To deal with this scenario, it is necessary to emphasize the transformation of foods systems in order to promote justice, health and social well-being, creating spaces for reflection, debate, organization and popular mobilization, which calls for social work as a fundamental agent of social transformation. The mobilization of social work in this context is fully justified, because, according to the summary report put out by the EAT-Lancet Commission (2019, p. 5):

> A radical transformation of the global food system is urgently needed. Without action, the world risks failing to meet the UN Sustainable Development Goals (SDGs) and the Paris Agreement, and today's children will inherit a severely degraded planet where much of the population will increasingly suffer from malnutrition and disease avoidable.

To reverse the current scenario, it is time for action and mobilization by the academic community to put into practice strategies to mobilize students for ecosocial work. In the complex context of the world we live in, new learning emerges to develop community, solidarity, resilience and sustainability. Thus, we have a lot to learn and absorb from mushrooms and other macro fungi, which share cities, forests and the whole planet with us.

Beyond Mushrooms: Health, Culture, Innovation and Economy

Traditionally, mushrooms have constituted more than just an economic asset in rural territories, transcending this role by also forming part of their endogenous heritage, from mushroom picking to the promotion of gastronomy, cooking and tasting workshops and to guided tours, which enhance local entrepreneurship and development (Lino & Carvalho, 2018).

Mushrooms are also an economic asset of great value. Mushroom production has shown growth over the past 20 years. The COVID-19 pandemic had a positive impact on demand, which was reflected in a 6.3% increase in 2020 compared to the average annual growth from 2017 to 2019. The projection for 2028 predicts a 6.74% increase, with an expected production of 24.05 million tons. These are encouraging data for the pursuit of economic, environmental, social and health sustainability at a global level. Production occurs under humid conditions, with little sun exposure, in the spring and autumn, and mushrooms have a short life cycle. These situations create opportunities for the technology-based industry, with canned or frozen commercial options. Mushrooms' multiple uses in food, agriculture, industry and pharmacology favour their production by different actors, both public and private, and in different sectors of the economy (Fortune Business Insights, 2020), thus pointing to their potential in terms of social entrepreneurship.

In terms of environmental sustainability, the mushroom boasts systemic value, because, in addition to the mushroom mycoprotein being able to play an important role in plant-based and organic diets, it also allows for reductions in the consumption of food of animal origin and contributes to the composting process of waste.

This is why mushroom production can constitute a remarkable systemic response to reduce carbon footprints and maintain biodiversity.

The centuries-old tradition of the consumption of edible mushrooms is on the rise due to climate changes and the need to achieve sustainability goals. Unlike some food fads, which are temporary and fade over time, the demand for alternative proteins, namely, those of vegetable origin, emerges as persistent and steady. Overall, the global market sales for plant-based products could reach $160 billion by the end of the decade, which is equivalent to 7.7% of the world protein market. It is foreseeable that, by 2030, vegetable protein options will be worth $74 billion, up from around $4.2 billion currently, led by multinational companies (Wood, 2021). The high demand for mushrooms is reflected in the anticipated $4.653 billion in 2022 and an estimated increase to $6.379 billion by 2028, with an estimated sales rise of 5.4% in the Compound Annual Growth Rate (CAGR) expected for the 2022–2028 period, due to their nutritional composition, bioactive compounds and their properties, which benefit health, directly and indirectly, as natural plant-based food, or as food supplements and medicinal substances (Industry Research, 2022, Sganzerla et al., 2022). Edible mushrooms are rich in carbohydrates (sucrose, xylose, rhamnose, mannose and fructose), protein, amino acids (glutamic acid, aspartic acid, glutamate, methionine and cysteine), fatty acids (linoleic acid, stearic acid, palmitic acid, adrenic acid and nervous acid) and phenolic compounds (gallic acid, caffeic acid, protocatecnic acid, p-coaric acid, p-hydroxybenzoic acid and pyrogallol), vitamins (folic acid, riboflavin, niacin, thiamine, ergocalciferol and cyanocobalamin) and salt minerals (zinc, cadmium and molybdenum), with functions that support the proper functioning of bodies, the maintenance of health and the prevention and treatment of diseases. They are functional food and nutraceutical, that is, food that, given its nutritional composition, contributes to the prevention and treatment of diseases, without side effects. Among others, they stimulate immunomodulators, which act against degenerative diseases; and they also have a range of antibacterial, antifungal, antimicrobial, antioxidant, anti-inflammatory, anticancer, hepatoprotection and prebiotic properties (Das & Prakash, 2022; Kahleova et al., 2020). Mushrooms are fungi, appreciated for their flavour in cooking and gastronomy, and they are exploited for their biotherapeutic and nutraceutical properties. As such, they are used in the development of drugs (Vamanu et al., 2021). In cancer treatment, they can be utilized as a complementary means to counteract the side effects of chemotherapy and radiotherapy treatments, namely, nausea, bone marrow suppression, anaemia and reduced resistance (Prasad et al., 2015; Zhang et al., 2016).

For these reasons, mushrooms can be understood as a nature-based solution that promotes health, immaterial culture, social inclusion and entrepreneurship based on the circular economy. They also foster dialogue between universities, social organizations and civil society through projects. As a systemic asset, mushrooms can boost multifaceted social entrepreneurship processes that embrace all the dimensions shown in Fig. 1 below.

The *Mushroom House* is, thus, an incubator, a cultural setting to strengthen the relationship between several actors and the natural world, celebrating the role of

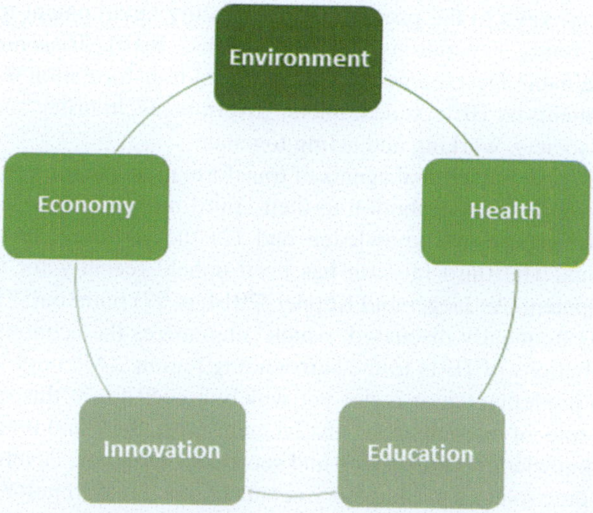

Fig. 1 Mushroom life cycle. (Source: Elsa Alves, author elaboration)

fungi in nature and in our daily lives and using mycological literacy to provide health and environmental education, thus contributing to community awareness of the importance of preserving ecosystems and biodiversity. Fungi, like animals and plants, are threatened with extinction, with this entailing serious consequences for our planet and well-being. In the twenty-first century, in the face of hunger, human and soil malnutrition, water scarcity and the growing increase in the global population (which is estimated to double by 2050), it is imperative to reverse this scenario through the involvement of people with nature, to preserve biodiversity and produce food that benefits people and the planet's health (Balkrishna et al., 2022). It is important to emphasize that mushrooms serve humanity as both food and as medicine, so it is vital to raise awareness and optimize human health through healthy and sustainable diets that protect our planet. All this means that we must promote mushroom consumption, for its benefits for health, both human and environmental (Taculao, 2020).

Higher Education Institutions and Sustainability Practices

In a global commitment and engagement to mitigate current and emerging social, economic and environmental challenges, the UN has defined an action plan, the 2030 Agenda, to curb and regenerate the climate, which is at risk of degradation, threatening human well-being. The primary objective is global transformation for the achievement of sustainable development in all the dimensions that constitute it: environmental, social and economic. Thus, 17 objectives are set forth that are intended to be achieved through the pursuit of 169 tasks, with people, partnerships and prosperity as keys to the promotion of a healthy environment today, without compromising future generations (United Nations, 2016). To achieve the 2030 Agenda's Sustainable Development Goals, a global transformation is needed, with strong commitments by HEIs, stakeholders, governments, industry, social organizations and civil society, working and acting together.

HEIs are potential and crucial agents of transformation and innovation in society and the economy. Their contribution to their "third mission" involves (1) the production and aggregation of knowledge and (2) the transference of knowledge through students. The third mission has been upheld recently for its active role in local development. As Jaeger and Kopper (2014, p. 97) point out: "The third mission of HEIs is frequently discussed, which summarizes the economic, social and cultural contributions of HEIs to the surrounding region". Although this concept's definition and implementation is not yet widely agreed to, in this case study, we focus on the role of participatory involvement with the community in the co-production of knowledge, cooperation and social and economic development. Thus, HEIs must assume roles as participatory agents in this transformation by producing and transferring knowledge and cocreating solutions that adapt to real societal problems and promote local development. Student internships are an important moment for the exchange of empirical and scientific knowledge, as well as to enhance and adapt students' skills to the needs of organizations and society. Dealing with concrete problems, students become more confident and present as entrepreneurs, decision-makers and challenge solvers. Students' academic and pedagogical supervision is an important time for them to reflect and to develop critical thinking skills as they consolidate their knowledge and improve their capacities to do their jobs as social workers more effectively. The presence of students in social organizations as interns empowers them to use theoretical academic knowledge and action/research methodologies, which allow for reflective praxis, follow-up on interventions and evaluations of the impact of their efficiency and effectiveness to make sustainability happen. In this way, through the involvement of HEIs and different actors in communities, we support the third mission (Nölting et al., 2020; Stolze & Sailer, 2022).

Methodological Framework of the *Mushroom House* Project

As mentioned before, the *Mushroom House* project was the result of a curricular internship in Social Work at the University of the Azores, in collaboration with the Kairós Cooperative, a non-governmental organization (NGO) that aimed to carry out ecosocial work practices through an active project-based learning (PBL) process, in which a trial-and-error approach was employed, understood as a problem-solving method in which several attempts are made to arrive at a solution. The University of Azores' Sociology/Social Work Department received an Erasmus student from Thomas More University, Belgium, who travelled to Azores to complete an internship in ecosocial work. This took place at Kairós, which hosted him for 4 months and created the conditions for the project's success.

Methodologically, we base our ideas on those of Edward Lee Thorndike (1874–1949), concerning learning by trial and error, understood as a method of observation to evaluate the success or failure of procedures, which can be repeated until success is achieved. This methodological approach was aligned with project-based learning (PBL), which is based on "collaborative work, interdisciplinarity and the ability to solve open problems" (Mota & Rosa, 2018, p. 270). Given that the project was carried out in a context of professional practice in ecosocial work, in which learning is as important as professional practice, we opted for PBL "as a teaching and learning strategy that involves problem identification in complex situations, based on real life, and the search for possible solutions" (Lopes et al., 2019, p. 50). In the context of the institutional internship at Kairós, there was an interest in testing new solutions to the problem of unemployed persons in situations of social vulnerability through new social businesses, operationalized in a context of promoting the circular economy and aligned with the SDGs and the European ESCUTA[2] project.

The *Mushroom House* project was an initiative co-developed by the professionals at the Kairós Cooperative and the student's pedagogical supervisor at the University of the Azores. The basic idea of the project, as well as diverse documentation about environmental social work/eco social work, social entrepreneurship, the circular economy, mushroom production and stakeholders in the field of mushroom production, was shared with the Social Work student. After a tripartite exploratory meeting with the student, he was asked for a problem diagnosis, built from the following starting questions: "How do we produce biological mushrooms from coffee grounds in the Azores?" and "How do we use the project in the context of entrepreneurship and environmental education?"

The project unfolded in the following stages:

Step 1: Establishing relationships with the problem—This step aimed to motivate the student regarding the topic by watching the movie *Fantastic Fungi*,[3] directed by Louie Schwartzberg (2019), which describes the magical, mysterious and medicinal world of fungi and their power to heal, sustain and contribute to the regeneration of life on Earth. After viewing the film, a discussion was held with the project team about the main ideas and problems addressed in the film and its relationship with ecosocial work. After this, the student was asked to read Paul Stamets' *Fantastic Fungi, How Mushrooms Can Heal, Shift Consciousness, and Save the Planet* (2019), and *Mycelium Running: How Mushrooms Can Help Save the World* (2005), as well as a more technical book entitled *How to Grow Mushrooms from Scratch: A Practical Guide to Cultivating Portobellos, Shiitakes, Truffles* (2018), by Magdalena Wurth and Herbert Wurth. In the context of an internship supervision meeting, ideas, challenges and problems related to the subject were discussed with the student, and good practices in mainland

[2] https://escuta.com.pt

[3] https://fantasticfungi.com

Portugal regarding the cultivation of mushrooms using coffee grounds, entitled NÃM,[4] were shared and discussed.

Step 2: Design an action plan to solve the problem—In this step the student was asked to create a chart as an organizing tool for the work to be done. The chart was composed of four columns and with the following fields: Ideas, Facts, Learning Issues and Action Plan. From the discussions held, with the help of the academic and Kairós Cooperative supervisors, the student drew up his action plan, systematizing ideas and scheduling actions.

Step 3: Tackling the problem—The operationalization of the solution to the problem was implemented in this phase, in which several activities of a collaborative nature were carried out. Among the main ones, we can mention the following: (1) in a first phase, a market study was carried out, with visits to various commercial establishments that sell mushrooms. A network was designed to collect coffee grounds, and several interviews were carried out with local, national and international producers to reinforce knowledge acquisition and get recommendations from those with experience in mushroom production, to avoid risks and mistakes. (2) In a second phase, the identification of sellers for the acquisition of mycelium was carried out to commence the inoculation of the mushroom in the coffee grounds. Then we washed and prepared the containers in which to grow the mushrooms. In this phase it is crucial to ensure the appropriate environment and space, which means that we had to ensure a dark room for production, with the right conditions as regards light, heat, humidity and asepsis. In a third phase, the mycelium was inoculated into the coffee grounds by the project team. However, children attending the institution's Extra Time Activities were also involved in the inoculation process as part of the environmental education dimension integrated into the project.

The *Mushroom House*, hence, featured two dimensions: social entrepreneurship and environmental education. It was in this latter one that the children were involved, and this included the collection of coffee grounds and containers for cultivation with their families, and, later, participation in the awareness and environmental education sessions specifically prepared for the purpose, which included a practical part on inoculation in the coffee grounds substrate by the children. The aim of this activity was to reach out to families through their children, thus raising awareness among young and old alike of the environmental issues that our society faces. It was also to convey an idea of hope through the promotion of small initiatives that we can undertake to further the sustainability of our ecosystems, while also improving our diets as a source of well-being. This dimension of the project is very important, as it generates knowledge and experience as regards the role of social workers as environmental educators. The strategies included in the Action Plan proved to be successful and allowed the project to flourish as a collaborative learning practice.

[4] https://nammushroom.com/

Step 4: Re-evaluating the problem—We identified mistakes and reinforced learning at this stage, given the relative failure in the production of the first mushrooms. A report was produced with all the evidence and the procedures and resources used, environmental conditions, schedule of actions and the size and amount of the mushroom production. Through this analysis, we concluded that we were not using the proper bottles (plastic) and humidity and light conditions to allow the mushrooms to grow. The evidence and information collected were an action/learning activity. The evidence was reviewed and compared with the action plan developed, which triggered and motivated the team to design solutions, in situ and on time. This made it possible to analyse and evaluate the process, to share and discuss evidence and knowledge and to follow up, correct mistakes and design adapted activities and strategies. A new action plan was then drawn up, with the above considerations in mind, which made possible strong mushroom production in the coffee grounds substrate in terms of their size and quality, which allowed us to move on to the next phase.

Step 5: Developing and presenting the products—This is the final phase of a PBL that allowed us to evaluate the success of what had been done beyond a quantitative dimension, refocusing the evaluation on the learning process, based on a self-assessment by the team and the intern. In this stage the work included picking the mushrooms and organizing a culinary event with local chefs, embodied in a dinner featuring dishes made with the mushrooms already produced. Several local stakeholders were invited to the event, as well as the extended project team. The general opinion of the participants in the event was very positive, and the entrepreneurial, inclusion dimension and the process of food and social education, showed its potential, while the value and importance of a project like this was demonstrated.

Figures 2, 3, 4, 5, 6, 7, 8, 9, 10, 11, and 12 below present a photographic review of the project, with photos taken by the project team.

Results

The *Mushroom House* produced evidence and results indicating that ecosocial work is possible when multidisciplinary work is carried out in accordance with the *People's Charter for an Eco-Social World* (2022):

(a) The project was overseen by a Social Work trainee student who was integrated into a local NGO. He set up an experimental project within the field of ecosocial work. This internship was supervised by Professor Eduardo Marques, from Azores University, co-author of this chapter, and Pedro Gouveia, the placement supervisor from Kairós.

(b) None of those involved in this project had previous experience in mushroom production, so it was necessary to acquire knowledge about the entire produc-

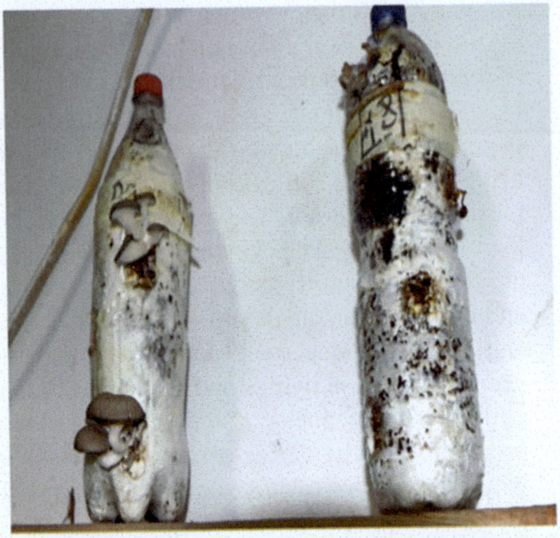

Fig. 2 The first mushrooms. (Source: Authors)

Fig. 3 Testing different containers. (Source: Authors)

tion process and to design environmental and food education strategies based on it.

(c) It was demonstrated that it is possible to produce mushrooms with a minor investment (less than €100) and, given the results obtained, the support organization (Kairós) will scale up the project in order to produce organic mushrooms for the local market.

(d) The project focused not only on production but also on marketing and environmental education, thus upholding the role of social work and social entrepreneurship in responding to social problems through job creation and the promotion of the circular economy and nature-based solutions.

(e) The connection between gastronomy and nutrition, as well as its potential in terms of a tourism product, is an important asset. These links can be used to boost revenues and healthy diets in low-income families. It is important to train

Fig. 4 Beholding the beauty of nature. (Source: Authors)

Fig. 5 Quality evaluation. (Source: Authors)

Nova Experiência Lúdico-Pedagógica na Quinta Kairós

Casa do Cogumelo - produção em baixa tecnologia de cogumelos em borras de café

Fig. 6: Information disseminated in the press. (Source: Authors)

Fig. 7 Mushroom night with local chefs. (Source: Authors)

Fig. 8 Tasting dishes with mushrooms. (Source: Authors)

Fig. 9 Environmental education for children. (Source: Authors)

Fig. 10 Environmental education for children. (Source: Authors)

Fig. 11 Preparation for mycelium inoculation. (Source: Authors)

Fig. 12 Training children as family educators. (Source: Authors)

and empower families in situations of economic lack and food shortages to embrace self-production. This project undoubtedly revealed the potential to contribute to healthy, affordable and delicious alternative plant-based diets (Harvard Health, 2021), in addition to social inclusion, environmental awareness and education.

(f) The project was carried out as a real collaborative laboratory for ecosocial learning, which allowed for more thorough collaboration between the University of the Azores and the Kairós NGO, transferring knowledge and learning in both directions, to contribute to an agroecological transition that is increasingly necessary and urgent.

Discussion

The *Mushroom House* was a pilot initiative that demonstrated that it is possible to carry out ecosocial work, thereby inspiring Kairós to adapt and replicate learning outcomes using other social work interns as human resources to adopt new ecosocial approaches. The outstanding results achieved by this project made it possible to create a sustainable link between Azores University and Kairós, which inspired several more students to choose this NGO to do internships. Ultimately, they were recruited as ecosocial workers. This project also made possible a new ecosocial intervention known as the "EKO Space",[5] an incubation centre for seven houses, namely, Ugly Fruit House, Beer House, Compost House, Recycling and Reuse House, Plastic Reuse House, Soap House and Weaving House. These outcomes highlight the need for practical ties between "ecology" (the science of the relationship between living things and their habitats) and "economy" (domestic management). This project made it possible to demonstrate the relationships between social work and the implementation of some SDGs and how entrepreneurship, jobs and social inclusion can be promoted through green initiatives centred on food and nutrition, food sovereignty, health, well-being and environmental education. The "Mushroom House" project also spawned knowledge transfer between HEIs (University of Azores) and an NGO (Kairós) in a collaborative process focused on Social Work students, helping them to "Be the change they wish to see in the world" (Gandhi). The social work students used their time and knowledge, acquired during the academic internship, to influence cultural and organizational leaders and to promote community, green care and ecosocial work. It offered a framework for social workers to engage in meaningful and effective action addressing social and environmental problems, recognizing that these issues are interconnected and cannot be fully addressed in isolation. We believe that this project will be used in a transversal way in other settings, such as community mushroom houses in poor neighbourhoods, using existing expertise and experiences to tackle new challenges effectively, whether in the establishment of social enterprises or in the field of environmental education. The "Mushroom House" is an active teaching and experiential learning strategy project showing us, through its positive results, that it is possible to put ecosocial work into practice.

This project allowed us to integrate ecological perspectives into practice, inspire new ecosocial policy development, engage in advocacy efforts to adopt NBS and emphasize the importance of interdisciplinary and institutional cooperation, thereby building bridges with communities.

Conclusions

As a holistic framework, the ecosocial transition calls for strong commitments by social work practitioners. To achieve this result, Social Work education should be able to include ecosocial work principles into its curricula, ascertaining the best pedagogical methodology and strategies and undertaking action research to understand what is and what is not working. According to the People's Charter for an Eco-Social World (2022, p. 1), "Education to rebuild the relationship between nature and the planet is key to achieving an ecosocial world". To achieve this, we can use fieldwork training (curriculum internships and project-based learning) to

[5] https://633ac0b42d348.site123.me

spur organizations to cocreate a different way to do social work, integrating an environmental perspective into the work. This is what we have done with the "Mushroom House", which constitutes a living space to teach and learn about ecosocial work, guided by *buen vivir* principles, understood as a "way of life and a form of development that sees social, cultural, environmental, and economic issues working together and in balance" (People's Charter for an Eco-Social World, 2022, p. 5). The principles of sustainability were understood as a set of concepts that promote the long-term well-being of individuals, society and the environment, while ensuring the efficient and responsible use of resources, while interdisciplinarity guided us to deliver a project that can be used to promote environmental education in social work.

Sustainability is vital to guide us to choose the best strategies to foster economic growth and address a range of societal needs based on open dialogue between NGOs and universities. Interdisciplinarity is needed to pool knowledge and methods from multiple disciplines in order to address complex problems and questions that cannot be adequately answered by any single discipline alone and to bridge gaps between different fields, cultures and communities.

It is a challenge to help and empower people to integrate an ecological perspective into the way they live, consume and eat. According to Levin and Nevo (2022, p. 7) "Ecological social work and approaches associated with it provide institutions seeking to shape social work toward a more ecocentric, sustainable direction—what is an already quite large scope of theories, practice wisdom, evidence-based insights, and redesigned ethics".

There is a need to find new paths, dialogue and tools for ecosocial workers and educators, professionals and students to collaborate on deep reflections about the big challenges that our societies face today, dealing with climate change and environmental problems, and how social work can be part of the solution. Our "Mushroom House" is an example of how students can be bridges that connect HEIs to real-world problems and how science can be part of the solution.

In the contemporary scenario of conflict, humanitarian crises, food insecurity, social and environmental risks and health pandemics, there is a need for robust actions, inter- and transdisciplinary dialogues, and drawing upon science in communities in order to halt and, if possible, reverse the "landscape" of degradation and natural and human decline resulting from the "predatory" global economy and agriculture (Clark et al., 2018).

Mushroom House thus functioned as a practical case study to research how to teach ecosocial work, to assess and validate its benefits and challenges for NGOs and to integrate green entrepreneurship, sustainability and nutrition into a solidarity economy perspective. Cooperation between HEIs and NGOs can unleash social innovation through the transfer of knowledge between supervising professors, students and social organizations. HEIs also have a lot to learn from NGOs, given that proximity to these institutions allows for the deepening and development of new knowledge, thus improving teaching and learning contexts in response to current environmental and societal problems.

This partnership made it possible to demonstrate the importance of an ecosocial work approach to promote sustainable and equitable practices, addressing the root causes of social and environmental problems. Through this project we confirmed the importance of ecosocial work as an approach that recognizes the interconnectedness between social, environmental and economic factors in shaping the well-being of individuals and communities.

Bibliography

Balkrishna, A., Sharma, G., Sharma, N., Kumar, P., Mittal, R., & Parveen, R. (2022). Global perspective of agricultural systems: From ancient times to the modern era. In A. Balkrishna (Ed.), *Sustainable agriculture for food security. A global perspective* (1st ed., pp. 3–46). Apple Academic Press.

Besthorn, F. (2012). Vertical farming: Social work and sustainable urban agriculture in an age of global food crises. *Australian Social Work, 66*(2), 187–203. https://doi.org/10.1080/0312407X.2012.716448

Butler-Warke, A., & Bolger, J. (2021). Fifty years of social work education: Analysis of motivations and outcomes. *Journal of Social Work, 21*(5), 1019–1040. https://doi.org/10.1177/1468017320911603

Cañeque, F. (2006). *La RSE ante el espejo : carencias, complejos y expectativas de la empresa responsable en el siglo XXI*. Prensas Universitarias de Zaragoza.

Chankseliani, M., & McCowan, T. (2021). Higher education and the sustainable development goals. In *Higher education* (Vol. 81, Issue 1). Springer Science and Business Media B.V. https://doi.org/10.1007/s10734-020-00652-w

Chigangaidze, R. K. (2022). Environmental social work through the African philosophy of Ubuntu: A conceptual analysis. *International Social Work, 0*(0). https://doi.org/10.1177/00208728211073382

Clark, M., Hill, J., & Tilman, D. (2018). The diet, health, and environment trilemma. *Annual Review of Environment and Resources, 43*(1), 109–134. https://doi.org/10.1146/annurevenviron-102017-025957

Das, S., & Prakash, B. (2022). Chapter 11 – Edible mushrooms: Nutritional composition and medicinal benefits for improvement in quality life. Em B. Prakash, Research and Technological Advances in Food Science (pp. 269–300). Academic Press. https://doi.org/10.1016/B978-0-12-824369-5.00013-0

Dubey, P. K., Singh, A., Chaurasia, R., & Abhilash, P. C. (2020). Climate adaptive agricultural interventions for food, nutritional, health and livelihood security. In: S. Dhyani, A. Gupta, M. Karki (Eds.), Nature-based solutions for resilient ecosystems and societies. Disaster resilience and green growth. Springer, Singapore. https://doi.org/10.1007/978-981-15-4712-6_16

EAT-Lancet. (2019). Retrieved March 15, 2023, from https://eatforum.org/content/uploads/2019/07/EAT-Lancet_Commission_Summary_Report_Portugese.pdf

Fortune Business Insights. (2020). *Mushroom market size, growth, share* | Global Analysis (2028). Retrieved March 16, 2023, from https://www.fortunebusinessinsights.com/industry-reports/mushroom-ma

Frantzeskaki, N., Mahmoud, I. H., & Morello, E. (2022). Nature-based solutions for resilient and thriving cities: Opportunities and challenges for planning future cities. In I. H. Mahmoud, E. Morello, F. Lemes de Oliveira, & D. Geneletti (Eds.), *Nature-based solutions for sustainable urban planning. Contemporary urban design thinking*. Springer. https://doi.org/10.1007/978-3-030-89525-9_1

Gardiner, P. (2020). Learning to think together: Creativity, interdisciplinary collaboration and epistemic control. *Thinking Skills and Creativity, 38*. https://doi.org/10.1016/j.tsc.2020.100749

Godden, N. J., Farrant, B. M., Yallup Farrant, J., Heyink, E., Carot Collins, E., Burgemeister, B., Tabeshfar, M., Barrow, J., West, M., Kieft, J., Rothwell, M., Leviston, Z., Bailey, S., Blaise, M., & Cooper, T. (2021). Climate change, activism, and supporting the mental health of children and young people: Perspectives from Western Australia. *Journal of Paediatrics and Child Health, 57, 1759–1764*. https://doi.org/10.1111/jpc.15649

González-Domínguez, J., Sánchez-Barroso, G., Zamora-Polo, F., & García-Sanz-Calcedo, J. (2020). Application of circular economy techniques for design and development of products through collaborative project-based learning for industrial engineer teaching. *Sustainability (Switzerland), 12*(11). https://doi.org/10.3390/su12114368

Hart, C. S. (2019). Education, inequality and social justice: A critical analysis applying the Sen-Bourdieu Analytical Framework. *Policy Futures in Education, 17*(5), 582–598. https://doi.org/10.1177/1478210318809758

Harvard Health. (2021, March 30). *The right plant-based diet for you*. Retrieved March 14, 2023, from https://www.health.harvard.edu/staying-healthy/the-right-plant-based-diet-for-you

Industry Research. (2022, September 28). *Global mushroom industry research report, growth trends and competitive analysis 2022–2028*. Retrieved May30, 2023, from https://www.industryresearch.biz/global-mushroom-industry-21797061

Jaeger, A., & Kopper, J. (2014). Third mission potential in higher education: Measuring the regional focus of different types of HEIs. *Review of Regional Research, 34*, 95–118. https://doi.org/10.1007/s10037-014-0091-3

Kahleova, H., Becerra-Tomas, N., Sonia, Mejia B., Glenn, AJ, de Vriese, S., Jenkins, David JA, Kendall, Cyril WC, Salas-Salvadó, J., Sievenpiper, JL, & Arnoldi, A. (2020). Scientific update on plant-based eating and cardiometabolic health.

Latouche, S., & Hapagès, D. (2011). *La Hora del decrecimiento*. Octaedro.

Levin, L., & Nevo, M. (2022). The nature of social work: An ecocentered analysis of key social work statements. *Social Work, 67*(4), 371–380. Retrieved March 30, 2023, from https://doi.org/10.1093/sw/swac027

Lino, C., & Carvalho, P. (2018). Património micológico e micoturismo. Revista Turydes: *Turismo y Desarrollo, 24*. Retrieved March 17, 2023, from http://www.noticiasdonordeste.pt/2006/07/jornadas-de

Lopes, R., Alves, N., Pierini, F., & Filho, M. (2019). Características gerais da aprendizagem baseada em problemas. In R. Lopes, N. Alves, F. Pierini, & M. Filho (org.). *Aprendizagem baseada em problemas: fundamentos para a aplicação no ensino médio e na formação de professores* (pp. 47–74). Publiki.

López Peláez, A., Erro-Garcés, A., & Pérez-García, R. M. (2022). Food security and social protection in times of COVID-19. *International Social Work, 65*(3), 421–433. https://doi.org/10.1177/00208728221086388

Löwy, M. (2015). *Ecosocialism: A radical alternative to capitalist catastrophe*. Haymarket Books.

Luke & THL. (2017). *Green care quality manual*. The Natural Resources Institute Finland and the National Institute on Health and Welfare. Retrieved May 4, 2023, from https://www.gcfinland.fi/tiedostopankki/343/Green-Care-Quality-Manual.pdf

McBride, E., Arden, M. A., Chater, A., & Chilcot, J. (2021). The impact of COVID-19 on health behaviour, well-being, and long-term physical health. *British Journal of Health Psychology, 26*(2), 259–270). John Wiley and Sons Ltd.. https://doi.org/10.1111/bjhp.12520

McGeeney, A. (2016). *With nature in mind: The ecotherapy manual for mental health professionals*. Jessica Kingsley Publishers.

Mehta, A., Mittal, R., & Verma, R. (2022). An insight of biological and ecological effects of agrochemicals: A global scenario. In A. Balkrishna (Ed.), *Sustainable agriculture for food security. A global perspective* (1st ed., pp. 122–149). Apple Academic Press.

Mota, A. & Werner da Rosa, C. (2018). *Ensaio sobre metodologias ativas: reflexões e propostas.* Espaço Pedagógico, v. 25, n. 2, Passo Fundo, pp. 261–276, maio/ago. 2018. Retrieved March 4, 2023, from www.upf.br/seer/index.php/rep

Närhi, K. (2002). *The eco-social approach in social work and the challenges to the expertis of social work.* University of Jyväskylä, Publishing Unit.

Närhi, K., & Matthies, A. L. (2018). The ecosocial approach in social work as a framework for structural social work. *International Social Work, 61*(4), 490–502. Retrieved April 14, 2023, from https://doi.org/10.1177/0020872816644663

Nölting, B., Molitor, H., Reimann, J., Skroblin, J. H., & Dembski, N. (2020). Transfer for sustainable development at higher education institutions-untapped potential for education for sustainable development and for societal transformation. *Sustainability (Switzerland), 12*(7). https://doi.org/10.3390/su12072925

ONU. (2018). *Harmony with nature.* In ONU (Vol. 12223, Issue July). https://doi.org/10.1163/9789462095847_074

Pereira, R., Ribeiro, M. C., Rua, O. M., & Martins, D. (2019). Is there entrepreneurship within the public sector?: A literature review. In O. Rua (Ed.), *Entrepreneurial orientation and opportunities for global economic growth* (pp. 92–118). IGI Global. https://doi.org/10.4018/978-1-5225-6942-8.ch005

Prasad, S., Rathore, H., Sharma, S., Yadav, A., & Yadav, A. S. (2015). Medicinal mushrooms as a source of novel functional food. *International Journal of Food Science, Nutrition and Dietetics, 04*(5), 221–225. Retrieved April 14, 2023, from https://doi.org/10.19070/2326-3350

Rambaree, B. (2020). Ecosocial work through corporate social responsibility: the case of company engagement with civil society in coastal communities in Mauritius. *Journal of the Indian Ocean Region, 16*(1), 47–62. https://doi.org/10.1080/19480881.2020.1709375

Renn, O. (2018). Real-world laboratories – The road to transdisciplinary research? *Gaia-Ecological Perspectives for Science and Society, 27*(1), 1. https://doi.org/10.14512/gaia.27.S1.1

Sganzerla, W. G., Todorov, S. D., & Da Silva, A. P. G. (2022). *Research trends in the study of edible mushrooms: Nutritional properties and health benefits.* www.begellhouse.com

Slattery, M., Ramsay, S., Pryor, A., Gallagher, H., Norton, C. L., Nikkel, L., Smith, A., Knowles, B., & McAuliffe, D. (2022). Nature-based interventions in social work practice and education: Insights from six nations. *International Social Work.* https://doi.org/10.1177/00208728221123160

Stamets, P., & Schwartzberg, L. (2019). *Fantastic fungi: how mushrooms can heal, shift consciousness & save the planet. Earth Aware.*

Stephens, J. C., Hernandez, M. E., Román, M., Graham, A. C., & Scholz, R. W. (2008). Higher education as a change agent for sustainability in different cultures and contexts. *International Journal of Sustainability in Higher Education, 9*(3), 317–338. https://doi.org/10.1108/14676370810885916

Stolze, A., & Sailer, K. (2022). Advancing HEIs' third mission through dynamic capabilities: The role of leadership and agreement on vision and goals. *Journal of Technology Transfer, 47*(2), 580–604. Retrieved April 24, 2023, from https://doi.org/10.1007/s10961-021-09850-9

The People's Global Summit. (2022, 29 June–2 July 2022). *The people's charter for an eco-social world.* Retrieved March 9, 2023, from https://newEcosocialworld.com

Taculao, P. B. (2020). *Community-based mushroom production and processing, part 2: Growing mushroom through the tissue culture method.* Retrieved April 1, 2023, from https://mb.com.ph/2020/10/28/community-based-mushroom-production-and-processing-part-2-growing-mushroom-through-the-tissue-culture-method/

United Nations. (2016). Transforming our World: the 2030 agenda for sustainable development. *Agenda.* Retrieved April 24, 2023, from https://doi.org/10.1201/b20466-7

Vamanu, E., Dinu, L. D., Pelinescu, D. R., & Gatea, F. (2021). Therapeutic properties of edible mushrooms and herbal teas in gut microbiota modulation. *Microorganisms,* 9(6). MDPI AG. Retrieved April 18, 2023, from https://doi.org/10.3390/microorganisms9061262

Wanner, M., Hilger, A., Westerkowski, J., Rose, M., Stelzer, F., & Schäpke, N. (2018). Towards a cyclical concept of real-world laboratories: A transdisciplinary research practice for sustainability transitions. *disP – The Planning Review, 54(2), 94–114.* https://doi.org/10.1080/02513625.2018.1487651

Willett, W., et al. (2019). Food in the anthropocene: The EAT–Lancet Commission on healthy diets from sustainable food systems. *The Lancet, 393*(10170), 447–492. https://doi.org/10.1016/S0140-6736(18)31788-4

Wood, J. (2021, August 26). *Why fungi could become the go-to meat-free alternative*. Retrieved January 7, 2023, from https://www.weforum.org/agenda/2021/08/mushrooms-plant-based-meat-free-proteins/

Zhang, J. J., Li, Y., Zhou, T., Xu, D. P., Zhang, P., Li, S., & Li, H.-B.. (2016). Bioactivities and health benefits of mushrooms mainly from China. *Molecules, 21*(7). MDPI AG. Retrieved April 18, 2023, from https://doi.org/10.3390/molecules21070938

Case Study 5: The Inner Anti-hero and Eco-anxiety: A Student Movie from Sweden

Olivia Ahlin

Background of the Project and the Event

The World Social Work Day celebration is annually held at the University of Gävle. This event for the Year 2022 was held remotely due to the COVID-19 pandemic, and my contribution to the celebration was through a short movie regarding eco-anxiety. Prior to the event, our bachelor programme had a course in ecological social work, where eco-anxiety was frequently discussed. In an engaging dialogue with a teacher from this course, we formulated the notion of my contribution to a creative project. The project was foremost aimed at social workers, providing an example of how expressive art can be a teaching tool.

Why an Anti-hero? A Student's Reflexivity

I applied for the bachelor programme in international social work with some kind of hero instinct. This aspiration has followed me throughout my studies. However, various life events during the pandemic drained my energy, which decreased my ability to reach my goal. It became challenging to prioritise my mental health and ecological social work added to this negative spiral. I found it hard to maintain a balance between being kind to myself while still being environmentally friendly. I realised that I had to have empathy for my own inner anti-hero to get myself back on track. My creative projects, aside from my studies, became my comfort zone, and I became more interested in working in an interdisciplinary manner. My motivation increased significantly when my teachers expressed a great interest in my passion for the expressive arts. It was liberating to conduct social work in a way that made

O. Ahlin (✉)
Gävle University, Gävle, Sweden

C. Forde et al. (eds.), *Teaching and Learning in Ecosocial Work*,
https://doi.org/10.1007/978-3-031-58708-5_19

very much sense to me as a person. I became more aware of my personal relationship with climate change. A song lyric, 'Ben Kött och Känsler' (bones, flesh, and feelings) by Markus Krunegård, inspired me to create a character who suffers from eco-anxiety. This character was supposed to embody human imperfection in environmentally related actions. This was a peculiar process since I had to get to know an exaggerated fictional version of myself. To add additional depth to the character, I included a spoken poem based on the original song lyrics by Krunegård. This project led me to reflect on how unforgiving eco-anxiety is. Unlike other forms of anxiety, it cannot be repaired with an apology or somebody's forgiveness.

Movie Summary and Plot

The inner anti-hero concerns a Swedish girl in her mid-20s. She is the main character who suffers from eco-anxiety, and her struggles to navigate everyday life are exacerbated by constant reminders of the climate crisis. To get rid of these feelings, she decides to take up the role of an environmental anti-hero for her mental health purpose. The pandemic of COVID-19 worsens her mental health since the media is a frequent reminder of how the pandemic and the climate crisis are interrelated. This short movie of 3 min investigates the complexity of eco-anxiety and seeks to evoke empathy for individuals who are going through this type of self-sabotage. The following section will describe the movie in detail. The main messages conveyed through a poem at different stages of the film are presented below in each scene description.

Scene 1

The film begins in outer space. The main character arrogantly blows cigarette smoke towards the Earth, which serves as a symbol of air pollution. The intention is to draw parallels between environmental degradation and abuse:

> I am an abuser. I torture the things that I love the most, and I'm not even noticing. I am the worst kind of person. A person who is aware of her habits even though she knows how toxic they are. I am a toxic person. I am polluting mother, earth and soon I'm about to be an orphan, and this is my own fault.

Scene 2

This part integrates archival footage from LDL Media AB, which is a media production company that focuses on creative ingenuity. A friend of mine is employed there, and he was inspired by the narrative I had created for this film. In exchange for knowledge, I was provided a film camera, a drone, and archival footage of the climate crisis. The material incorporates polluted oceans, wildfires, and traffic-filled roadways. The purpose of this scene was to create an uncomfortable setting while portraying what the world currently looks like due to the consequences of environmental neglect:

> A uterus shouldn't be filled with plastic. The oceans shouldn't die because of us. People's conveniences shouldn't be the end.

Scene 3

This scene explores the challenge of being exposed to conflicting information regarding climate change. Fictional tweets pop up in the sky to highlight the confusion that could occur on social media:

> The keyboard is the new silent freedom of speech. Fake news, the hate, the threat. I don't even know what to think anymore.

Scene 4

This scene reflects on gender inequality. The character claims that there is no point in making herself heard as a woman. Regardless of political agenda, she will most likely be silenced by men anyway. A face mask is burned in this scene:

> But it doesn't matter what I say though, because I am a girl and a girl's voice is always choked.

Scene 5

The COVID-19 pandemic is peaking during this scene. It explores the balance of following restrictions given by the government while simultaneously coping with loneliness and anxiety. Street signs that encourage social distancing are shown, along with face masks:

> And I know that people are suffering. I should stay inside, keep my distance from everything beyond my power to solve. And it's not that I don't care it's just that I can't relate, and someone is just going to build higher chimneys anyway to hide from the smoke.

Scene 6

This scene starts with a man who consumes a hamburger filled with plastic. The purpose of this clip is to emphasise blindness to climate-related issues. The scene shifts to a panic attack in a bathtub, emphasising how eco-anxiety can literally drain you:

> But you know it's all smoke and mirrors and I'm denying everything I see. Disgusted by the reflection who is staring back at someone who used to be me.

Scene 7

The main character cuts out a small slice of a cake to another person. This is first interpreted as a nice gesture, but then she eats the rest of the cake with a spoon. The cake is consumed in an increasing speed, and the scene ends with the main character digging into the cake with her bare hands. The scene also goes back and forth to the previous panic attack in the bathtub. The metaphor with the cake refers to social and environmental injustices where the character is taking advantage of her privilege:

> I am sick, but I need dessert. Scoop on more fat, more, more, more, just so I can show that I stand above you.

Scene 8

Similar to scene two, archival content is shown to emphasise the result of indifference to other people's suffering. There are bomb sounds in the background that causes a baby to cry. Some people pray, while others can buy their happiness at the cost of social and environmental justice:

> If there's a hole in the boat, then give me a scoop because I don't belong in your group. I only pray when I need something beyond hope. Money has become my God and i'd do anything to get some applause. I don't want to believe in anything I can't see with my own eyes. I cover my ears when the babies on the TV cry.

Scene 9

This scene starts with a series of selfies. The pictures are meant to emphasise the right to social and environmental justice. The setting switches to the character lying surrounded by rubbish in the snow. The film closes with Earth suffering from a heart attack in outer space:

> Bones, flesh and feelings we are all human beings.

Evaluation of Movie

The audience consisted of international professors and social work students. I was provided with various feedback that sparked interesting discussions. Overall, the audience expressed inspiration and emotional connection, which was what I had hoped for. Some participants disagreed on whether expressive art would increase the emotional impact and message delivery. The seminar was fruitful for me as a social worker and artist. I increased my awareness of content adaption and learned more about ecological social work. The movie attempted to illustrate occurring social issues from the perspective of an everyday person. The plot and content were selected to portray an example of how the environmental discourse can be perceived. The archival content focused on eco-anxiety and covered topics such as pollution, flooding, and inequality. Vulnerable groups, such as women in low-income countries, were incorporated into the story to highlight the devastating domino effect of climate-related issues. The global influence of the COVID-19 pandemic was explored in this movie. The world experienced challenges maintaining social and environmental justice during this time, and inequality increased. The character's decision-making was an important part of the narrative since it emphasised the challenges a person can face when they experience eco-anxiety. Additionally, the film investigated social media as a knowledge field. Cook et al. (2017) emphasise the risks associated with media portrayal of different discourses on climate change. This could potentially lead to scepticism and neglect of climate science, whereby it becomes the individual's responsibility to access accurate information. In my movie, the main character does not have the energy to live up to that responsibility.

Teaching and Learning Ecosocial Work

Creative learning involving art often focuses on feelings and evoking empathy (Leonard et al., 2018). This project made me look at ecosocial work from a perspective I did not initially expect. My instant reaction to the subject was that I had to become a climate hero, which did not align with my mental health at the time. As a learner in ecosocial work, I became more self-aware as I formed the narrative of this character. The movie sparked discussions between teachers and students who could relate to the inner anti-hero I created and portrayed in the movie. I want to argue that an openness for art-based creative, reflective, and transformative learning is essential in ecosocial work. A practitioner within ecosocial work will meet individuals who may have a different narrative on their situation than the social worker does. Applying art-based teaching and learning methods and approaches to social work can convey different perspectives on topics such as climate change, pollution, and eco-anxiety. A teacher cannot take for granted that their lectures are perceived the way they are intended by the students. Going back to myself, I was initially afraid of getting judged for my opinions about eco-anxiety, but the film project gave me control to critically reflect and talk openly about it. Professors may also encounter students with conflicting opinions about ecosocial work. Promoting self-reflective activities where student get opportunities to express themselves is an essential component in transformative teaching and learning (Jones, 2011; Mezirow, 2009; Rambaree, 2020). It is not always a guarantee that a social work student necessarily thinks ecosocial work is important or a priority area within social work education. After all, the climate crisis is a product of human actions. Creative methods in teaching and learning of ecosocial work could lead to efficient communication that could spark new learning as well as intervention methods. Culturally speaking, expressive art has connected people throughout history. Combating ecosocial work problems and issues can only be successfully executed if the teacher is empathetic to new teaching methods, approaches, and tools.

Acknowledgement Thanks to everyone involved in this project.

Bibliography

Cook, J., Lewandowsky, S., & Ecker, U. K. H. (2017). Neutralizing misinformation through inoculation: Exposing misleading argumentation techniques reduces their influence. *PloS One, 12*(5), e0175799–e0175799. https://doi.org/10.1371/journal.pone.0175799

Jones, P. (2011, January). Transformative learning, ecopedagogy and learning ecology: Pathways for social work education. *Eco-social justice: Issues, challenges and ways*. https://www.researchgate.net/profile/Peter_Jones64/publication/283008475_Transformative_learning_ecopedagogy_and_learning_ecology_Pathways_for_social_work_education/links/5625fc3a08aeedae57db3478.pdf

Leonard, K., Hafford-Letchfield, T., & Couchman, W. (2018). The impact of the arts in social work education: A systematic review. *Qualitative Social Work, 17*(2), 286–304. https://doi.org/10.1177/1473325016662905

Mezirow, J. (2009). An overview of transformative leaning. In *Contemporary theories of learning: Learning theorists in their own words* (pp. 90–105).

Rambaree, K. (2020). Environmental social work: Implications for accelerating the implementation of sustainable development in social work curricula. *International Journal of Sustainability in Higher Education, 21*(3), 557–574. https://doi.org/10.1108/IJSHE-09-2019-0270

Index

C. Forde et al. (eds.), *Teaching and Learning in Ecosocial Work*,
https://doi.org/10.1007/978-3-031-58708-5

SPRINGER NATURE

GPSR Compliance

The European Union's (EU) General Product Safety Regulation (GPSR) is a set of rules that requires consumer products to be safe and our obligations to ensure this.

If you have any concerns about our products, you can contact us on ProductSafety@springernature.com

In case Publisher is established outside the EU, the EU authorized representative is:

Springer Nature Customer Service Center GmbH
Europaplatz 3
69115 Heidelberg, Germany

The manufacturer's authorised representative in the EU is Springer Nature Customer Service Centre GmbH, Europaplatz 3, 69115 Heidelberg, Germany. If you have any concerns regarding our products, please contact ProductSafety@springernature.com

Printed and bound by CPI Group (UK) Ltd, Croydon, CR0 4YY
23/09/2026
02221541-0004